TIME AND BEHAVIOUR
Psychological and Neurobehavioural Analyses

ADVANCES IN PSYCHOLOGY

120

Editors:

G. E. STELMACH

P. A. VROON

ELSEVIER

Amsterdam – Lausanne – New York – Oxford – Shannon – Tokyo

TIME AND BEHAVIOUR
Psychological and Neurobehavioural Analyses

Edited by

C.M. BRADSHAW

and

E. SZABADI

Department of Psychiatry
University of Nottingham
Nottingham, UK

1997

ELSEVIER
Amsterdam – Lausanne – New York – Oxford – Shannon – Tokyo

NORTH-HOLLAND
ELSEVIER SCIENCE B.V.
Sara Burgerhartstraat 25
P.O. Box 211, 1000 AE Amsterdam, The Netherlands

ISBN: 0 444 82449 9

Contents

Preface

That time is both a dimension of behaviour and a ubiquitous controlling variable in the lives of all living things has been well recognized for many years. Several excellent books have been published on the topic of time and behaviour during the last 20 years, including the influential monograph by Richelle and Lejeune (1980), edited volumes in the New York Academy of Sciences series (Gibbon & Allan, 1984) and the Psychology of Learning and Motivation series (Bower, 1991), and the conference proceedings volume *Time, action and cognition* (Macar et al., 1992). Nevertheless, there are several reasons for believing that the time is ripe for a new volume on the same topic.

The last decade has seen a burgeoning of interest in the quantitative analysis of timing behaviour, and progress during the last five or six years has been particularly impressive, with the publication of several major new theoretical contributions. A conspicuous feature of these modern theoretical developments is the integration of models of timing with insights derived from other areas of behaviour analysis, notably memory and motivation. Moreover, the field has benefited greatly from the erosion of the traditional boundaries separating behaviour analysis and cognitive psychology.

There has also been considerable progress in behavioural methodology during the past decade. In the area of reinforcement schedules, for example, the venerable interresponse–time schedule, fixed–interval peak procedure and interval bisection task have been complimented by a 'second generation' of incisive instruments for analyzing timing behaviour. Furthermore, experimental ingenuity and mathematical sophistication have been brought to bear to extract new information on the fine structure of timing behaviour from the older generation of schedules.

Another area of recent development is the analysis of the neurobiological substrate of timing behaviour. Several research groups are currently studying the involvement of various central neurotransmitter systems in the timing behaviour, and the ability of centrally acting drugs and discrete brain lesions to alter timing processes.

Yet another recent development in timing research is the growing dialogue between two fields that have grown up separately, although, superficially at least, they seem to have much in common: the experimental analysis of 'interval timing', traditionally the province of experimental psychology, and behavioural chronobiology. The last few years have seen a growing interest in the comparative properties of the internal 'clocks' that regulate biobehavioural rhythms with time bases in the circadian range or longer, and those that are entailed in timing of intervals in the range of seconds or minutes.

All these areas of research, and others, are represented in the chapters that make up this volume. We hope that, in collecting these essays within the covers of a single volume, this book will help to promote further interactions among researchers who hail from disparate disciplines, but who share a common interest in the temporal properties of behaviour.

It has been a pleasure as well as a privilege to edit this volume, and we wish to express our gratitude to the authors, all active researchers in the field, for finding time to provide a wealth of exciting material on their chosen topics, and for their patience with the editorial process.

Anyone with a passing familiarity with the central themes of this book will need no reminding of the unique and invaluable contributions made to the psychology of timing behaviour by Professor Marc Richelle, and to the study of circadian behaviour by Professor Jürgen Aschoff. The influence of their pioneering work pervades many of the chapters in this book. On behalf of the authors and ouselves, we take this opportunity to express our appreciation of their work, and to extend to both our best wishes for their retirements.

Bower, G.H, ed. (1991). *The psychology of learning and motivation: advances in research and theory, vol. 27.* San Diego, Academic Press.

Gibbon, J. & Allan, L., eds. (1984). *Timing and time perception. Annals of the New York Academy of Sciences, vol. 423.*

Macar, F., Pouthas, V. & Friedman, W.F., ed. (1992). *Time, action and cognition: towards bridging the gap.* Dordrecht, Kluwer.

Richelle, M. & Lejeune, H. (1980). *Time in animal behaviour.* Oxford, Pergamon.

C.M. Bradshaw
E. Szabadi

December, 1996

Contributors

Al–Ruwaitea, A.S.A.
Department of Psychiatry, University of Nottingham, Queen's Medical Centre, Nottingham, NG7 2UH, U.K.

Al–Zahrani, S.S.A.
Department of Psychiatry, University of Nottingham, Queen's Medical Centre, Nottingham, NG7 2UH, U.K.

Bizo, L.A.
Department of Psychology, Arizona State University, Box 8711104, Tempe, AZ 85287–1104, U.S.A.

Bradshaw, C.M.
Department of Psychiatry, University of Nottingham, Queen's Medical Centre, Nottingham, NG7 2UH, U.K.
[e–mail: c.m.bradshaw@nottingham.ac.uk]

Campbell, S.S.
Laboratory of Human Chronobiology, New York Hospital – Cornell Medical Center, 21 Bloomingdale Road, White Plains, NY 10605, U.S.A.
[e–mail: sscampb@med.cornell.edu]

Carr, J.A.R.
Department of Psychology, University of British Columbia, 2136 West Mall, Kenny Building, Vancouver, BC, V6T 1Z4, Canada
[e–mail: carrjaso@unixg.ubc.ca]

Church, R.M.
Department of Psychology, Box 1853, Brown University, Providence, RI 02912, U.S.A.
[e–mail: Russell_Church@brown.edu]

Fairhurst, S.
New York State Psychiatric Institute and Columbia University, Biopsychology, Unit 50, 722 West 168th Street, New York, NY 10032, U.S.A.

Fetterman, J.G
Department of Psychology, LD 124, Indiana University / Purdue University at Indianapolis, 402 N. Blackford Street, Indianapolis, IN 46202, U.S.A.
[e–mail: gfetter@indyvax.iupi.edu]

Gibbon, J.
New York State Psychiatric Institute and Columbia University, Biopsychology, Unit 50, 722 West 168th Street, New York, NY 10032, U.S.A.
[e–mail: jg34@columbia.edu]

Goldberg, B.
New York State Psychiatric Institute and Columbia University, Biopsychology, Unit 50, 722 West 168th Street, New York, NY 10032, U.S.A.

Grant, D.S.
Department of Psychology, University of Alberta, Edmonton, Alberta, T6G 2E9, Canada
[e–mail: dgrant@psych.ualberta.ca]

Higa, J.J.
Department of Psychology: Experimental, Duke University, Box 90086, Durham, NC 27708–0086, U.S.A.
[e–mail: jennifer@psych.duke.edu]

Hinton, S.C.
Department of Psychology: Experimental, Duke University, Box 90086, Durham, NC 27708–0086, U.S.A.
[e–mail: sean@psych.duke.edu]

Ho, M.Y.
Department of Psychiatry, University of Nottingham, Queen's Medical Centre, Nottingham, NG7 2UH, U.K.
[e–mail: mcxmyh@unix.ccc.nottingham.ac.uk]

Kelly, R.
Department of Psychology, University of Alberta, Edmonton, Alberta, T6G 2E9, Canada
[e–mail: dgrant@psych.ualberta.ca]

Killeen, P.R.
Department of Psychology, Arizona State University, Box 8711104, Tempe, AZ 85287–1104, U.S.A.
[e–mail: killeen@asu.edu]

Meck, W.H.
Department of Psychology: Experimental, Duke University, 9 Flowers Drive, Durham, NC 27708–0086, U.S.A.
[e–mail: meck@psych.duke.edu]

O'Boyle, D.J.

Department oof Psychology, University of Manchester, Coupland Street,Manchester, M13 9PL, U.K.

[e–mail: oboyle@psy.man.ac.uk]

Roberts, W.A

Department of Psychology, Faculty of Social Science, University of Western Ontario, London, Ontario, N6A 5C2, Canada

Spetch, M.L.

Department of Psychology, University of Alberta, Edmonton, alberta, T6G 2E9, Canada

[e–mail: mspetch@psych.ualberta.ca]

Staddon, J.E.R.

Department of Psychology: Experimental, Duke University, Box 90086, Durham, NC 27708–0086, U.S.A.

[e–mail: staddon@psych.duke.edu]

Szabadi, E.

Department of Psychiatry, University of Nottingham, Queen's Medical Centre, Nottingham, NG7 2UH, U.K.

[e–mail: elemer.szabadi@nottingham.ac.uk]

Wilkie, D.M.

Department of Psychology, University of British Columbia, 2136 West Mall, Kenny Building, Vancouver, BC, V6T 1Z4, Canada

[e–mail: dwilkie@cortex.psych.ubc.ca]

Time and Behaviour: Psychological and Neurobehavioural Analyses
C.M. Bradshaw and E. Szabadi (Editors)
© 1997 Elsevier Science B.V. All rights reserved.

1

CHAPTER 1

Dynamic Models of Rapid Temporal Control in Animals

J.J. Higa & J.E.R. Staddon

Timing refers to a wide range of behaviors that reflect a learned sensitivity to the duration of a stimulus such as a tone or light, or the time between successive events. Our paper focuses on one of the earliest examples of timing called *temporal control* (see Killeen & Fetterman, 1988, and Richelle & Lejeune, 1980, for classifications and descriptions of other forms of timing). An example of temporal control is responding during fixed interval (FI) reinforcement schedules. Here, a reinforcer (usually food for a hungry pigeon) is given for the first response that occurs after a fixed amount of time has elapsed since the delivery of the previous reinforcer. Animals readily detect the regularity of the interval of time between two reinforcers (called an *interfood interval*, IFI), so that they usually wait after a reinforcer for about two–thirds of the IFI duration before responding (e.g., Richelle & Lejeune, 1980; Schneider, 1969).

A striking feature of temporal control is its ubiquity. A variety of animals ranging from rats and pigeons (e.g., Harzem, 1969; Innis, 1981) to captive starlings (e.g., Brunner, Kacelnik, & Gibbon, 1992) to some fish and turtles (Lejeune & Wearden, 1991) show behavioral sensitivity to events separated in time by intervals ranging from seconds to a few minutes. Its prevalence in so many species suggests that the ability to detect, learn, and use temporal information is an essential and basic process of animal behavior and learning. Indeed, the time between events – stimuli, responses, and rewards – exerts an overwhelming influence on what associations are made, and how learning progresses. Understanding temporal control may, therefore, also contribute to and clarify the operation of other learning and memory processes.

Traditional Approaches: Steady–State Timing

Research on animal timing has a long and venerable history. The majority of studies and most approaches have examined timing from a psychophysical point of view (e.g., Gibbon & Allan, 1984) with a concentration on conditions producing steady–state properties of timing. For example, consider the *peak procedure* (e.g., Catania, 1970; Roberts, 1981). This is a trial–by–trial version of the standard FI reinforcement schedule. On "food" trials, after a fixed period (an IFI), response–contingent food reinforcement is delivered. "Probe" trials last for a period between two to four times the value of the training IFI, and food is not delivered. Food and probe trials are mixed within a session, separated by intertrial intervals of random length. Exposure to a target interval usually lasts for 20 to 30 sessions until steady–state responding is achieved. Response rate within a probe trial is characterized as a roughly normal distribution with maximal responding (peak rate) around the time of reinforcement (peak time). A characteristic result from the peak procedure is *scalar timing*: 1) superposition of response rate distributions when normalized along both axes (time within an interval and response rate); and 2) a constant coefficient of variation (CoV; ratio between the standard deviation and mean) across a range of interval values (e.g., Gibbon, 1977, 1991). Despite recent questions about the generality of scalar timing (e.g., see Zeiler & Powell, 1994 for a notable exception), it is held to be a defining characteristic of all forms of timing and is also observed in regular FI procedures (e.g., Dews, 1970; Richelle & Lejeune, 1980; Schneider, 1969) and differential–reinforcement of low rates schedules (e.g., Staddon, 1965).

   Using methods like the peak procedure, questions about timing have focused on understanding *what* an animal has learned about the interval(s) presented during training. Unfortunately, the *dynamics* – how timing is accomplished –has been largely overlooked. There are at least three reasons for the current state of research: 1) historical assumptions about the speed with which timing develops; 2) concentration on procedures that generate steady–state, static, results; and 3) a focus on quasi–dynamic models.

Time discrimination develops slowly. Conventional wisdom, with few exceptions (e.g., Gibbon & Balsam, 1981), holds that time discrimination develops slowly. Exposing animals to 30 or more sessions per condition, for example, is not uncommon (e.g., Cheng & Westwood, 1993; Innis, 1981; Meck & Church, 1984; Schneider, 1969). This practice may be due to early observations of the kind reported by Skinner (1938). He stated that while post–reinforcement pauses (depressions in a cumulative record) during short IFIs developed within a few days, the process was substantially retarded during longer IFIs (p. 125). The apparent slowness in the adaptation of post–reinforcement pauses (and other measures of time discrimination) to a given interval duration, may explain why most studies use extensive training procedures where effects are measured after numerous trials. We show, later in this paper, that the assumption that time discrimination always develops slowly is misguided.

Concentration on procedures that generate steady–state, static, results. Under the usual conditions, response profiles reflect a mixture of effects from training and test intervals and intertrial intervals (ITIs). For example, in the peak procedure an animal receives training trials in which reinforcement is given for the first response after, say 50 s, has elapsed. To test what it has learned, training trials are interspersed with "empty" probe trials that last for 130 s and do not end with reinforcement. Within a session, up to 50% of the trials may consist of the longer probe–trials. Training and test trials are randomized in a session and each are separated by long ITIs (e.g., Meck & Church, 1984). In the bisection procedure animals are trained to discriminate between a short and long stimulus, for example, between a 2–s and 8–s tone. Concurrent with presentation of training stimuli, animals are also presented test stimuli of intermediate duration each separated by an ITI (e.g., Church & Deluty, 1977). For example, 50% of the stimuli presented in a session can consist of intermediate (test) stimuli. Of interest are the psychometric functions that describe the probability of responding "long" following these test stimuli.

It is important to note that although the results from these procedures tell us what an animal learns after extensive training, less understood is how timing develops. Response profiles typically represent the average effect (end–result) of training– and test– intervals. Therefore, by obscuring the effects from individual intervals, standard methods do not show how timing progresses or the path by which an end–state is reached.

Quasi–dynamic models. The leading models of animal timing are Scalar Expectancy Theory (SET; e.g., Church & Broadbent, 1991; Gibbon, 1977; Gibbon & Church 1984, 1990) and a Behavioral Theory of timing (BeT, Fetterman & Killeen, 1991; Killeen & Fetterman, 1988). Briefly, SET is based on three cognitive processes – a clock, memory, and comparison process. On each trial (of the peak procedure for example) an accumulator receives a periodic signal (pulses) from a pacemaker. The content of the accumulator is compared with a sample from a memory distribution of each reinforced (target) time. The comparison process determines whether the elapsed time has reached a criterion duration; specifically, when the difference between elapsed time and a value sampled from memory (or its relative difference) exceeds a threshold the animal switches from a "no–responding" state to a "responding" state. Variance from any of the processes may determine an animal's estimate of time (e.g., Gibbon, Church, & Meck, 1984).

A competing, alternative, theory to SET is BeT. According to BeT, behavior itself serves as a cue for the passage of time and mediates time discrimination (e.g., Killeen & Fetterman, 1988). Reinforcement generates adjunctive behaviors (i.e., collateral behavior, broadly defined to include both elicited and emitted responses) whose frequency rises and falls at different rates within an interfood interval.

Each behavior is assumed to be associated with an underlying state (n), and transitions between states are produced by pulses from a Poisson (random) pacemaker. The pacemaker speed depends on the rate of reinforcement in a given context so that increases in reinforcement rate increase the speed of the pacemaker. The amount and probability of reinforcement may also change the pacemaker speed (e.g., Fetterman & Killeen, 1991).

SET and BeT have been tested under a range of timing procedures (e.g., Bizo & White, 1994; Killeen & Fetterman, 1993; Leak & Gibbon, 1995), and the results have advanced our knowledge about what is learned during time discrimination. However, despite the considerable body of experimental and theoretical work, relatively little is known about the dynamics of timing. The dynamics may have received less attention perhaps because SET and BeT are based on molar properties of the experimental situation and theoretical processes that are insensitive to moment–to–moment changes in the interval duration. SET's assumptions about memory for time intervals, memory sampling rate, and thresholds of time estimates are based on statistical distributions derived from molar features of the pacemaker system or reinforcement schedule (e.g., Church & Broadbent, 1991; Gibbon, 1995; Gibbon & Church, 1984; Gibbon, Church, & Meck, 1984). BeT, too, is based on molar properties of the experimental situation. The speed of BeT's pacemaker depends on the *average* time between reinforcers in a given context and the overall amount and probability of reinforcement (e.g., Fetterman & Killeen, 1991). Both are molar models and do not specify the time–window in which intervals are supposed to have their effect.

As far as we can determine, SET and BeT are at best *quasi– dynamic* models of timing (Staddon, 1988). They allow us to conclude something about "equilibrium behavior – whether there are equilibria or not, and whether or not they are stable – but lack features necessary for a full dynamic analysis ... The distinctive feature of such models is that they are not specified well enough to permit predictions about trajectories or periodic behavior, that is, about learning curves, sequential statistics, or responses to rapidly changing conditions (p. 305)." In other words, SET and BeT tell us about the end result of some training procedure, but they have less to say about the path by which animals reach that end–state – that is, about the dynamics and processes involved in the development of time discrimination.

It is perhaps noteworthy that newer models (e.g., Church & Broadbent, 1990, 1991; Machado, in press) are dynamic and real–time in contrast to trial–level models (see Sutton & Barto, 1987, for a discussion about advantages and limitations of different kinds of models). However, they have not yet been seriously applied to dynamic data or predictions about how learning should progress. For example, Church and Broadbent (1990) summarize their research agenda as: "The first fact that any theory of temporal generalization must account for is the smooth peak function in which the probability of a response gradually increases to a maximum near the time of reinforcement then decreases in slightly asymmetrical fashion ... The second fact ... is that performance on individual trials ... is characterized by an abrupt change from a state of low responding to a state of high responding ... break–and–run ... The third fact ... is that the mean functions are very similar with time shown as a proportion of the time of reinforcement (p. 58)". All three properties are steady–state results and are based on the overall effects of intervals an animal receives during training.

Dynamics: An Example of Rapid Timing

In contrast to traditional approaches that focus on steady–state and molar features of timing, there is a set of studies that concentrate on understanding the dynamics of timing. These studies typically use IFIs programmed according to response–initiated delay (RID) reinforcement schedules and wait–time duration as a measure of temporal control by the IFI duration. RID intervals are a modified version of conjunctive fixed ratio (FR) 1 fixed time (FT) reinforcement schedules (see Figure 1). The first response following a reinforcer is accompanied by a change in the color of the response key from red to green (in the case of pigeons pecking a key for food reinforcers). The time between the reinforcer and the first response marks the wait time duration (*t*). Subsequent responses are recorded but have no programmed effect on when reinforcement occurs. While wait time is under the control of the subject, the experimenter controls the delay to reinforcement (*T*). For example, *T* can be fixed or vary according to *t* (e.g., Wynne & Staddon, 1988).

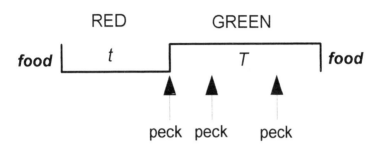

Figure 1. Illustration of a single response–initiated delay (RID) interfood interval.

Higa, Wynne, and Staddon (1991) presented pigeons RID IFIs that changed in duration, within a session, according to a sinusoidal pattern. To eliminate control of responding by cues other than the input sequence, they varied the number of cycles and the starting point within a cycle across sessions. A "free–running" period of constant IFIs occurred at the end of each day to test for any residual periodicity induced by the sinusoidal schedule. Figure 2 presents the result from all subjects and the group average. The upper graph shows the results from a condition in which the IFIs ranged from 5 to 15 s (Short condition). The results from a Long condition, where the IFIs changed from 30 to 90 s, appear in the lower graph. The data are mean wait times from the last cycle of a session and the first 15 IFIs from the free–running period. Mean wait times were normalized so that $t' = (t-t_{min})/(t_{max}-t_{min})$. The values for t_{min} and t_{max} were calculated separately for each bird according its shortest and longest wait time in each condition, respectively.

Overall, pigeons *tracked* the changes in IFI duration in both conditions. They tracked in the sense that the output pattern of wait times matched the input series of IFI duration. During a cycle, wait time decreased as the IFI duration decreased, and increased as the IFIs increased. During the free–running period, wait times were more variable, but there was no clear evidence of continued periodicity in the pattern of wait times: The birds did not appear to "learn" or memorize the input cycle. Instead, their wait time appeared to be controlled by the prevailing IFI duration. Two additional features of the results indicate temporal control was rapid. First, a correlation between the input–series and output of wait times was sometimes highest between wait time in IFI_{n+1} and the duration of IFI_n, which suggests that wait time for some birds was strongly determined by the most recent (just–preceding) IFI duration. Second, temporal control by the sinusoidal input occurred during the first few training sessions (sometimes by the second and third sessions) and did not improve with training.

The speed of the temporal control of wait times was an unusual finding. Numerous studies have established that animals (given extensive training on periodic reinforcement) will wait approximately a constant proportion of the IFI duration under a range of conditions (e.g., Harzem, 1969; Innis, 1981; Richelle & Lejeune, 1980; Staddon, 1964). However, typically 30 sessions or more were presented and believed to be required for stable responding during a single interval within the range of intervals used in the sinusoidal sequence (e.g., Schneider, 1969). The discovery that temporal control developed rapidly – even when the input sequence varied in length and phase across sessions – provided evidence of a timing mechanism that is sensitive to rapid changes in IFI duration.

Dynamic Models: Linear Waiting

What kind of process could account for the sinusoidal results? The simplest possible mechanism consistent with these results is that wait time is determined by the preceding interval (IFI_n) –– *linear waiting* (Wynne & Staddon, 1988):

$$t_{n+1} = \alpha IFI_n + \beta, \tag{1}$$

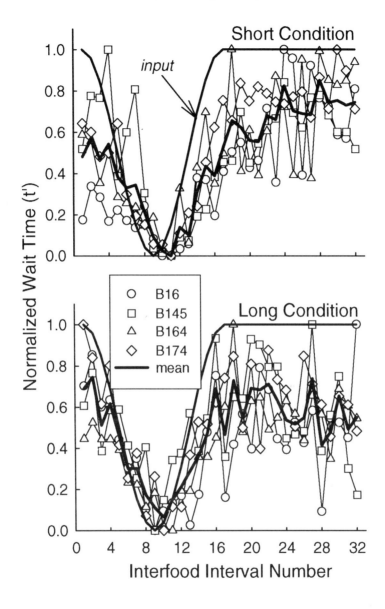

Figure 2. Results from Higa, Wynne, and Staddon (1991), Experiment 1 (sinusoidal input series of interfood intervals). Pigeons' wait times from the last cycle of a session and the first 15 constant IFIs after the cycle are given. The data are re-plotted and normalized so that $t' = (t - t_{min})/(t_{max} - t_{min})$. The values for t_{min} and t_{max} were calculated separately for each bird according its shortest and longest wait time in each condition, respectively. The heavy solid line shows the group mean.

where α and β are constants. In practice, α is about 0.25 and β is very small. Wynne and Staddon tested a surprising implication of Equation 1. Their logic was as follows. Suppose that Equation 1 holds and we program IFIs according to the following relation:

$$\text{IFI}_n = kt_{n,} \tag{2}$$

where k is a constant. Combining Equations 1 and 2 yields

$$t_{n+1} = \alpha kt_n + \beta. \tag{3}$$

If β is negligible, Equation 3 predicts that t will increase or decrease, depending on whether the quantity $k\alpha$ is greater or less than 1.0. Wynne and Staddon showed both effects: progressive increases or decreases in wait time duration as a function of k, in individual animals within a single session.

Higa et al. (1991) provided additional evidence for linear waiting, using a *single-impulse* procedure and pigeons as subjects. The procedure involved presenting subjects an occasional short IFI (5 s in duration called an *impulse*) intercalated into a longer series of 15–s IFIs (called non–impulse intervals). The location of an impulse varied across sessions and so its occurrence was relatively unpredictable. They looked to see whether an impulse decreased wait times in the *next* IFI as implied by a linear waiting process. Figure 3 presents the results from a baseline condition in which all IFIs were 15 s and the

---

Figure 3. Results from the first baseline and impulse condition from Experiment 3 of Higa, Wynne and Staddon's (1991) study. For the impulse condition (lower graph) pigeons' wait times during an impulse (5–s IFI, location shown as dashed vertical line) and the preceding and following 15 non–impulse (15–s) IFIs are shown. An impulse occurred once per session and its location was randomized across sessions. For the baseline condition (upper graph) an arbitrary IFI was selected and wait time during it and the preceding and following 15 non–impulse IFIs were used in the analysis. The data are re–plotted and normalized so that $t' = (t-t_{min})/(t_{max}-t_{min})$. The values for t_{min} and t_{max} were calculated separately for each bird according its shortest and longest wait time in each condition, respectively. The heavy solid line shows the group mean.

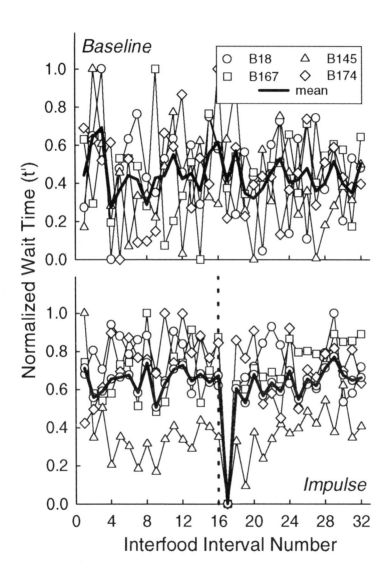

single–impulse condition. Mean wait time during the impulse (its occurrence is marked by a dashed vertical line) and the 15 preceding and following IFIs are given for each subject and the group. Wait times were normalized as in Figure 2. According to linear waiting, an impulse decreased wait time in the next IFI only: For most subjects, wait time in subsequent IFIs returned to levels observed in the IFIs preceding an impulse. The effect of an impulse did not change systematically across sessions.

It is worth noting that one–back IFI tracking by wait times resembles a recent choice result. Mark and Gallistel (1994) reported that rats rapidly track wide and unsignaled fluctuations in relative rates of (brain stimulation) reinforcement. Rats lever pressed on concurrent variable–interval (VI) schedules. One schedule provided a higher reinforcement rate at the start of the session. After a period of time the levers retracted for 10 s, and the programmed VI was reversed so that the previously "lean" lever now provided the "richer" density of reinforcement and *vice versa*. Mark and Gallistel found that adjustment of time spent responding on two levers is complete within one or two interreinforcer intervals on the lever associated with the leaner schedule. It is too early to determine the relation between these kinds of results and results from the present study. Nonetheless, there are a few interesting differences. For one, the rate of adjustment to a change in relative reinforcement rate depends on the amount of exposure to a prolonged period of stability before the change (Mark & Gallistel, 1993). Rats appear to adjust slowly after a long stability period, and adjust rapidly if they experience several changes in their recent past. The data from experiments on temporal control indicate a different effect of training. For example, in one experiment, when IFIs changed frequently, pigeons initially tracked the sequence by basing wait time on the preceding IFI duration. However, tracking eventually disappeared and was replaced by a variable–interval like performance; that is, a postfood wait time appropriate to the shortest IFI duration (Higa, Thaw, & Staddon, 1993).

Nonetheless, despite its simplicity, linear waiting fits a considerable range of results and seemed a reasonable approximation to the dynamic process involved in rapid temporal control. However, as we pointed out in previous papers (Higa, 1996; Staddon & Higa, 1991) linear waiting fails to predict other findings from interval schedules of reinforcement. Briefly, linear waiting does not explain performance on VI schedules with successive IFIs that vary in duration. It predicts that wait times should track this random series, but they do not: Wait time during VI schedules is typically constant, almost the same after every food delivery with no relation to the preceding IFI (e.g. see Ferster & Skinner, 1957, for examples). Results showing that pigeons cannot learn some simple, predictable sequence of IFI durations pose another problem for a strict linear waiting process. For example, pigeons cannot track a repeating sequence of 12, 1–min intervals followed by four, 3–min intervals (Staddon, 1967). Linear waiting predicts short wait times during the 1–min intervals, longer wait times during the 3–min intervals, and a delayed change in wait times from short to long and back down to short by one IFI (as the square–wave sequence cycled through a session). Instead, Staddon's pigeons waited approximately the same amount of time in all IFIs, 30 s on average, even after many training sessions. Yet, when successive IFIs vary smoothly and cyclically, wait time does track the changes in the IFI duration in a way reasonably consistent with linear waiting, and tracking is immediate (Higa, et al., 1991; Innis, Cooper & Mitchell, 1993; see also Innis, Mitchell & Staddon, 1993). Third, an intermediate case of linear waiting has been recently reported, where pigeons were trained during a repeating cycle of four IFIs – a 15, 5, 15, 45 s series. Pigeons initially tracked the sequence according to a linear waiting process. However, tracking eventually disappeared and was replaced by a VI–like performance – that is, a postfood wait time appropriate to the shortest IFI duration (Higa, Thaw & Staddon, 1993).

What is Missing from a Linear Waiting Account?

In sum, the results indicate that under certain conditions wait time is controlled by the prior IFI duration (e.g., single impulse effect), but under other conditions wait time is determined by more than the just–preceding interval (e.g., intermediate forms of linear waiting). Linear waiting appears to be an approximation to a more general process. What is the more general process? Some studies, both old and recent, suggest time discrimination depends on: 1) the frequency of individual IFIs; 2) the recency of IFIs or how long ago an IFI occurred (either a few interval or several sessions ago); and 3) the direction in which IFIs change – either increasing or decreasing. These properties are not modeled by a strict linear waiting process.

Frequency of intervals. Indicators about the role of IFI–frequency on timing behavior come from early studies on interval reinforcement schedules. For example, Catania and Reynolds (1968) discovered that increasing the probability of shorter intervals disrupted the response rate profile during longer intervals. They presented pigeons with an FI 240–s reinforcement schedule and varied the probability (0, .05 or .5) that some of the intervals would be 30 s. When all the intervals were 240 s, pigeons produced a standard FI scallop with peak responding near 240 s. When the probability of short intervals increased, responding started sooner and response rates peaked at 30 and 240 s during a 240–s interval. Similar effects are found in Pavlovian conditioning paradigms (e.g., with interstimulus intervals, Millenson, Kehoe & Gormezano, 1977). Recently, Wynne, Delius, and Staddon (1996) showed that wait times (in pigeons) can be disproportionately affected by the number of short intervals in the recent past: Median wait times were shorter when a set of intervals consisted of more 2–s intervals than 10–s intervals. Together, these results show that changes in the frequency with which intervals occur can alter when an animal starts to respond.

Recency of intervals. The results from the single–impulse experiment (Higa et al., 1991) suggest that how long ago a particular interval occurred can have differential effects on temporal control. In that study, the effect of an impulse in the single–impulse experiment did not systematically change across sessions. Each impulse shortened wait time in the next IFI by approximately the same amount, and there was little evidence of interactions among impulses across sessions. Interactions could be one of two types: An impulse in one session might attenuate or enhance the effect of an impulse in the next session. There was no evidence of either effect. In comparison, when short intervals occur frequently, as in a cyclic triangular series (15, 5, 15, 45) pigeons initially track the series according to linear waiting but eventually wait times reduce to a duration that is shorter overall than that observed during baseline when all intervals were 15 s in duration (Higa et al., 1993). This pattern of temporal control suggests effects from individual IFIs combine within and across sessions.

Direction of Transition. Experimental results show that the development of temporal control also depends on the direction in which IFIs change in duration. Specifically, wait times adjust almost immediately during a decrease in IFI duration and more slowly in response to an increase in the IFI under conditions in which IFIs change across session (Wynne & Staddon, 1992). Higa et al. (1993) also reported similar effects when changes occur within sessions. They gave pigeons short-term training with an IFI series containing a single unpredictable downward (Step–Down) or upward (Step–Up) transition in IFI duration. The occurrence of a transition varied across sessions and was unsignaled. In a No–Step (baseline) condition IFIs were of the same duration. All conditions were in effect for 10 sessions.

The results are given in Figure 4. During a Step–Down transition, wait time significantly decreased after the first transition-interval. Furthermore, adaptation of wait times was largely complete by the second transition–IFI. In contrast, during Step–Up, wait time increased after the first transition–IFI but continued to increase (gradually) across several IFIs. Hence, while initial changes in wait time can be described by a linear waiting process (a decrease or increase in IFI duration caused wait time to decrease or increase in the next IFI), it does not explain the development of different patterns of wait times during subsequent IFIs of a Step–Down and Step–Up transition.

We have since replicated these results with rats (Higa, in press). Despite differences in the programming of the IFIs (RID for pigeons and standard FI for rats) and the overall duration of the IFIs used, Step–Down and Step–Up transitions produced similar dynamic effects. Figure 5 presents the results. It shows data for all subjects from each condition (a baseline, No–Step, condition occurred before each experimental condition). As with pigeons, rats' wait times continue to change across more transition–IFIs during Step–Up than Step–Down.

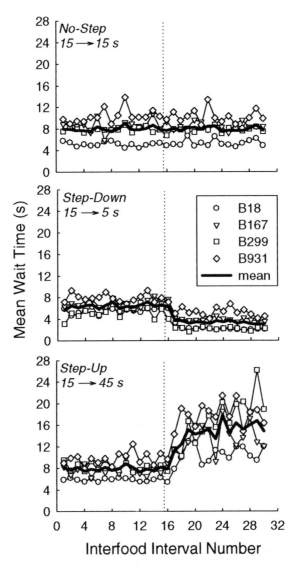

Figure 4. Results from Higa, Thaw, and Staddon (1993), Experiment 1 (Steps study). Pigeons' wait times during the 15 IFIs preceding and following a transition in IFI duration are shown. During Step–Down, the IFIs changed from 15 to 5; during Step–Up the IFIs changed from 15 to 45 s; and in a baseline condition (No–Step) all IFIs were 15-s in duration. A transition occurred once per session and its location was randomized across sessions. For the baseline condition an arbitrary transition point was selected and wait time from the preceding and following 15 IFIs were used in the analysis. The heavy solid line shows the group mean.

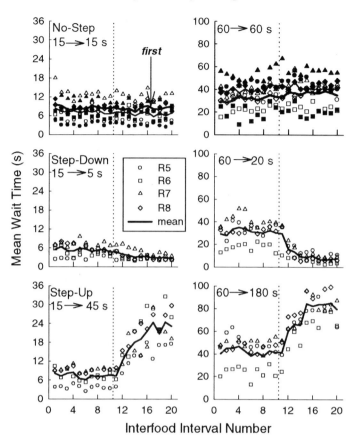

Figure 5. Results from a study with rats in which IFIs either decreased (Step–Down) or increased (Step–Up) once per session, at an unpredictable point. During a baseline condition (No–Step) IFIs were of the same duration throughout a session. The left column shows conditions in which the IFIs ranged from 5 to 45 s, the right column presents results from conditions with IFIs ranging from 20 to 180 s. All figures show rats' wait times during the 10 IFIs preceding and following a transition in IFI duration. For the No–Step conditions an arbitrary transition point was selected and wait time from the preceding and following 10 IFIs were used in the analysis (for No–Step conditions, the heavier solid line shows the group wait time from the first No–Step condition). The heavy solid line shows the group mean.

A Model of Temporal Dynamics

The results from several studies point to a timing process that is sensitive to both recent and remote changes in individual IFIs an animal experiences, including the frequency, age, and transition among IFIs. Wynne et al. (1996) tested four expanded versions of linear waiting for how prior IFIs could determine wait time. Of the models considered, an exponentially weighted moving–average of prior intervals that gave greater weight to shorter intervals best accounted for the data from experiments in which the pattern of IFIs varied across cycles (e.g., LLLL, LLSS, SSSS, and so forth, where S represents a 2–s interval and L a 10–s interval). By setting the parameter controlling the effect of the just–preceding IFI, the model also predicted the general features of the single–impulse data: a decrease in wait time following a 5–s interval intercalated in a series of 15–s intervals. However, the model also predicted a shortening of wait time that was not confined to the next interval. There is less evidence for this effect in the data. For most pigeons an impulse consistently shortened wait in the next IFI only. Nevertheless, the results raise an important theoretical issue: What kind of process can produce the single–impulse (one–back) data and the results from experiments that show wait time is also sensitive to events in an animal's remote past?

Our initial attempt at a model for how the dynamics of timing might occur is the diffusion–generalization model (e.g., Staddon & Higa, 1991). We proposed that time discrimination is linked to an animal's memory for reinforced postfood times (i.e., IFIs). Postfood times are represented in memory by a discrete linear continuum of "units" (i), each with an activation strength (X_i). X_i changes with time according to the equation given below written in discrete time (see Staddon & Higa, 1991; Staddon & Reid, 1990):

$$X_i(t) = \alpha X_i(t-1) + [(1-\alpha)/2][X_{i-1}(t-1) + X_{i+1}(t-1)] + S_i(t) \qquad (4)$$

where $X_i(t)$ is the activation strength of a unit (i) at real time t; $X_{i-1}(t-1)$ and $X_{i+1}(t-1)$ represent activation strengths of two adjacent units at time t–1; and α is a diffusion rate parameter. $S_i(t)$ is set to either 1 or 0: If reinforcement occurs at postfood time t, then unit i corresponding to t is incremented. In other words, reinforcement at a particular postfood time increases the activation strength of the unit that represents this time. Activation strength of that unit generalizes and diffuses to other neighboring units according to the equation given above.

Figure 6 illustrates the mechanics of the model. It shows the effect of a single reinforcement at postfood time 5 s. Each panel shows the state of the model (i.e., activation strengths of units) at different times since reinforcement. Immediately after reinforcement, specifically one iteration (a time step, equivalent to a second of real time in this example) after reinforcement, unit 5 increases in activation strength. In the next panel, 2 iterations after reinforcement, the activation strength of unit 5 decreases, but activation has spread to surrounding units 4 and 6. The remaining panels show the how the gradient centered on unit 5 spreads with time. If no other reinforcement occurs, the gradient will eventually flatten and the strength of all units will approach 0.

Response strength [output of the system, $V_i(t)$] depends on a clock–like process. The units are tied to real postfood time and they are potentially active one at a time: At real time t, the corresponding unit i is active if it exceeds a threshold, θ. The dashed horizontal lines in Figure 6 mark a threshold that remains constant in time. Vertical arrows mark which unit is active at each iteration. This arrow moves toward the right with each iteration or clock tick. According to this example, then: at real time 1, the activation strength of unit 1 is below the threshold and is not active so the output will be 0; at real time 2 unit 2 is inactive; at real time 3 unit 3 is active (this translates to a measure of wait time); at real time 4 unit 4 is active and it has a higher activation strength than unit 3 did when it was active; finally at real time 5 unit 5 is active.

Unlike current timing models (i.e., SET and BeT) the DG model is not based solely on the molar properties of the experimental condition. It is sensitive to properties of individual IFIs and can account for several timing effects from interval reinforcement schedules (e.g., see Staddon & Higa, 1991). The model also suggests how the dynamics depend on the frequency and recency of individual IFIs. First, because the activation strength (height) of the gradient increases after each reinforcer, the overall form of the gradient (and

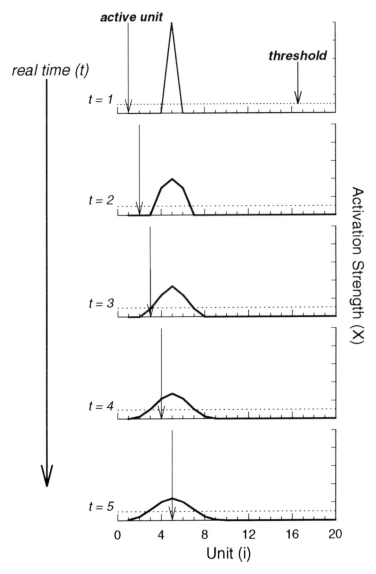

Figure 6. Illustration of mechanics of DG−model. Each panel shows the state of the model after different time−steps following a single reinforcement at postfood time 5 (α = .4, θ = .1). The dashed horizontal line marks the threshold for responding.

wait time) depends on the frequency of reinforcement at a particular postfood time. For example, the gradient of activation after four 5−s intervals will be higher than after a single 5–s interval. While, the two gradients diffuse at the same rate, the effect of many 5–s intervals will be evident in more subsequent IFIs, because of the higher level of activation. Second, because the DG–process is dynamic –– gradients rise and decay between each reinforcement according to α –– the overall form of the gradient also depends on how long ago reinforcement at a postfood time occurred. For example, separating several 5–s intervals by four 15–s intervals prevents the accumulation of the tendency to respond short from each 5–s interval, by allowing the gradient of activation (surrounding units for postfood time 5) to diffuse and dissipate (for a given, fixed, set of values for α and θ).

Test of the DG Model: Frequency and Recency of IFIs. We tested the DG model's predictions about the frequency and age of IFIs, by evaluating the model under four conditions that were extensions of the single–impulse procedure (see Higa, 1996 for more details). The frequency of IFIs was tested by varying the number of consecutive impulse IFIs intercalated in a series of longer non–impulse IFIs. To test for recency effects, the spacing among a constant number of impulses varied. Hence during two–close and eight–close conditions, pigeons were given either two or eight consecutive impulses intercalated in a series of non–impulse intervals, respectively. During a two–far and eight–far condition, each impulse was separated by four non–impulse intervals. In one set of conditions the impulses were 5 s and non–impulse IFIs 15 s; in another set the impulses were 15–s and non–impulse 45 s.

Figure 7 presents the simulation results (see Higa, 1996 for detailed explanation on how simulations were conducted). For illustrative purposes, only the results for the first set of conditions are given (with 5–s impulses and 15–s non–impulse IFIs). Simulations of the second set of conditions were qualitatively similar. Furthermore, because the simulations did not reveal differences during pre–impulse IFIs, wait time only during and after a set of impulses are shown. Finally, wait times were normalized (t) so that: $t' = t/t_{mean}$, where t_{mean} is the mean wait time in the intervals preceding a set of impulses (calculated separately for each condition). Hence, wait time is shown as changes relative to pre–impulse levels.

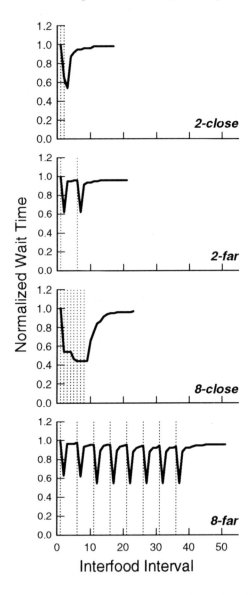

Figure 7. Simulation results for a condition in which: 1) two consecutive 5–s impulses were intercalated into a long series of 15–s IFIs (2–close condition); eight consecutive impulses (8–close); two impulses separated by four 15–s IFIs (2–far); and eight impulses separated by four 15–s IFIs (8–far). Only wait times (normalized, see text for details) from the set of impulses and the next 15 IFIs are shown. The

The simulations reveal four main effects. First, each impulse shortens wait time in the *next* IFI, according to linear waiting. Second, unlike linear waiting, the model predicts that the recovery of wait times (to pre–impulse levels) is slower after eight than two impulses (compare two–close and eight–close). Third, the slow recovery after eight close impulses is attenuated when each impulse is separated by four (longer) non–impulse IFIs (compare eight–close and eight–far). Finally, in all but the two–far condition, successive impulses should produce a successively shorter wait time in the next IFI (compare wait time after the first and last impulse of these conditions).

The results from pigeons trained on the same conditions appear in Figures 8 (two–close and eight–close) and 9 (two–far and eight–far). The data have been normalized in line with the simulation results, on an individual basis. The data from actual subjects are consistent with several of the model's predictions. For one, the majority of impulses caused a significant decrease in wait time in the next IFI. Also, wait times were slower to recover after eight than two consecutive impulses. Separating impulses by longer IFIs appeared to attenuate interactions between each impulse: In comparison to wait times after eight consecutive impulses, recovery after two or eight spaced impulses is relatively quick. However, unlike the simulations, the results from birds do not show successively shorter wait times after each impulse, in any condition. Wait times appear shorter after the second impulse of the two–close condition, but this is not a significant effect.

It is not yet clear why pigeons do not show successively shorter wait times during a set of eight–close impulses. Not finding a decreasing pattern may be due to a floor effect, however, increasing

---

location of an impulse within a series is shown as dashed vertical lines. For each set of simulations: a) an iteration (time–step) was equivalent to a second of real time; b) $\alpha = .18$ and $\theta = 2.1$; c) 500 "units" were used; d) each simulation lasted for 10 "sessions" per condition (100 intervals per session) and the state of the model after each session was carried over to the next session; e) reinforcers were presented for two iterations; and f) the location of first impulse (of a set of impulses) varied across simulated sessions.

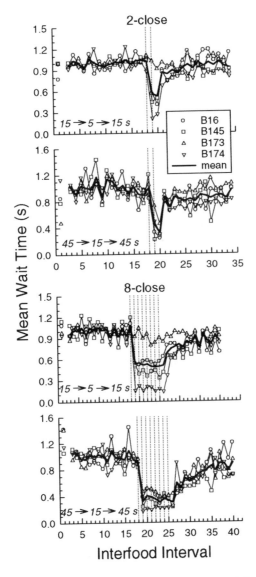

Figure 8. Results from pigeons exposed to 2–close and 8–close conditions in which the number of impulse IFIs varied but their spacing was held constant. In one set of conditions (15→15→15 s) non–impulse and impulse IFIs were 15 and 5 s, respectively. In another set (45→15→45 s) non–impulse IFIs were 45 s and impulses 15 s. Data shown are normalized mean wait time from a set of impulses and the preceding and following 15 non–impulse IFIs . The results for each subject are shown as well as the group mean (heavy solid line). Dashed vertical lines mark the occurrence of an

the overall duration of all IFIs by a factor of three did not change the pattern of wait times. It is probably too early to conclude that the model is wrong in its predictions because rats do show this pattern under comparable conditions. Figure 10 presents the results from a preliminary study in which rats were given a series of IFIs that changed from 120 to 30 back to 120 s (short square wave, SSW, condition) or from 120 to 480 back to 120 s (long square wave, LSW, condition). Given are mean wait time from transition IFIs and the preceding and following 10 IFIs. The figure also shows the actual (mean) IFI duration each animal experienced. Rats tracked the SSW condition that is similar to the eight-close condition from the pigeon study. However, unlike the pigeon data, there is evidence for decreasing wait times during the "impulse" IFIs. Procedural differences may explain the difference between these studies. For example, the IFIs in the rat study were between two to three times longer, and were programmed according to a standard FI reinforcement schedule. Future studies are needed to determine the importance of the differences between pigeons and rats.

Figure 10 also shows the results from another kind of eight-close condition, where the IFIs increase for a brief period (LSW). The same condition has not yet been performed with pigeons, but the basic result can be explained by a DG model. Its explanation for these data rests in is implication about the asymmetry between short and longer IFIs -- that the effect of a longer IFI may be overwhelmed by shorter IFIs. Consider the following example. Suppose an animal is given several 120-s IFIs. Its tendency to respond will be a fraction of that IFI, say 60 s. If the next IFI is 480 s, the model predicts a wait time based on the preceding IFI duration so it predicts a wait time 60 s. But what will the animal do if presented with another 480-s IFI? It will probably respond "short" at 60 s. Why? Although it has a tendency to

---

impulse, and symbols near the y-axis indicate mean wait time during the previous baseline condition in which all IFIs were either 15 or 45 s. The data were normalized according to the same method used to generate Figure 7.

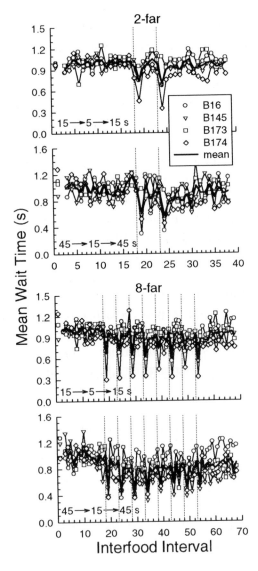

Figure 9. Results from pigeons exposed to 2–far and 8–far conditions in which the spacing of impulse IFIs varied but their number varied. In one set of conditions (top row) non–impulse and impulse IFIs were 15 and 5 s, respectively. In another set (bottom row) non–impulse IFIs were 45 s and impulses 15 s. Data shown are normalized mean wait time from a set of impulses and the preceding and following 15 non–impulse IFIs . The results for each subject are shown as well as the group mean (heavy solid line). Dashed vertical lines mark the occurrence of an impulse,

respond "long" (based on the preceding 120-s IFI), it still has a strong, though weakening, tendency to respond short (based on all the preceding 120-s IFIs it has recently experienced). Thus, according to the DG model, an animal fails to respond "long" not necessarily because it fails to learn about the 480-s IFI, but because its weakened tendency to respond short preempts responding at longer wait times. Eventually, the tendency to respond short will dissipate altogether, and we may eventually see longer wait times.

Therefore, finding no increase in wait times during 480-s IFIs does not necessarily mean the rats failed to learn about the 480-s interval. Instead, it suggests that the smaller tendency to respond short preempts responding at longer wait times. In a sense, then, discrimination may be affected by a kind of proactive interference from recent experience with short IFIs. The DG model is a suggestion about how that process may work. For similar reasons given above, it predicts that intercalating a single long IFI into a series of short IFIs (single-impulse experiment with a long impulse) will not be detected, that is, wait time following a long IFI will not necessarily increase because of preemption from shorter wait times. Experiments addressing this issue have not yet been conducted, but it would be an interesting test of the model.

Test of the DG Model: Step-Down and Step-Up transitions. Does the model predict the different transitional effects during an abrupt decrease or increase in IFI duration? We simulated the Step-Down (IFIs changed from 15 to 5s) and Step-Up (IFIs changed from 15 to 45 s) conditions from Higa's (in press) study with rats and present the results in Figure 11 ($\alpha = .12$ and $\theta = 1.6$) along with the results from subjects in the study (group means). Our simulations closely

---

and symbols near the y-axis indicate mean wait time during the previous baseline condition in which all IFIs were either 15 or 45 s. The data were normalized according to the same method used to generate Figure 7.

matched the conditions of the experiment (an iteration represented a second of real time), and the state of the model was "frozen" between sessions.

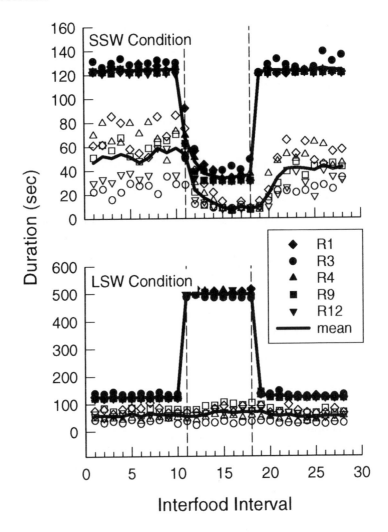

Figure 10. Rats' mean wait time during a square wave (SSW) in which IFIs changed once per session from 120 to 30 back to 120 s and long square wave (LSW) condition in which the IFIs changed from 120 to 480 back to 120 s. Unfilled symbols show wait time for individual subjects whereas filled symbols present the actual (experienced) IFI duration. Heavy solid lines represent the mean of all animals, and dashed vertical lines mark the start and end of a transition.

The simulations are qualitatively consistent with the pattern of wait times in Figures 4 (pigeons) and 5 (rats), but there are two noticeable differences. First, simulated wait times in pre–step IFIs are longer during Step–Up than Step–Down. The data from rats we chose to simulate do not show as large a shift in the overall level of pre–step wait times. However, the results from rats when IFIs change from 60 to 20 s and 60 to 180 s (Figure 5, bottom panel of graphs) do show an increase in wait times during 60–s IFIs when they are presented in the context of a shift to longer (180–s) intervals. Data from pigeons also show this tendency (Figure 4), although the difference is less remarkable.

Second, there is a decreasing trend in wait times during pre–transition IFIs of the Step–Up condition that is not evident in either sets of data from rats and pigeons. We did not fully explore the entire parameter space nor did we attempt to find the best–fitting parameter values. Nonetheless, other values for α and θ that we did test also produced a similar decreasing trend. Additional simulations showed that because we "froze" the state of each model between sessions, the tendency to respond "long" at the end of a session (during 45–s IFIs) interacted with the IFIs at the start of the next session that always began with 15–s IFIs. Wait times during the 15–s (pre–transition) IFIs decreased as the session elapsed and the 45–s IFIs became more remote. Eventually, a transition from 15 to 45 s would occur, and post–step wait times would begin to increase again.

This discrepancy between animals and simulations reveals a key theoretical element: A model of temporal control should simulate multiple time scales ranging from a few seconds and minutes to hours. Our current simulations do not incorporate 24 hours between sessions. For a more precise account of the Steps data, one possibility is to assume different diffusion rates during a session and for the time that elapses between sessions. Otherwise, the activation gradients either decay between sessions and have no carry–over effects, or do not decay enough as in the Step–Up simulations. In other words, the model is not able to handle context effects, and requires an additional rule for specifying when diffusion should cease or slow down.

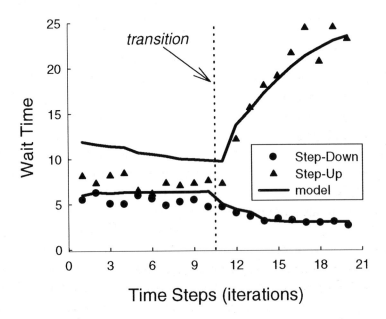

Figure 11. Simulation of Step–Down and Step–Up conditions (DG model). For comparison, data from rats are also given (symbols).

Future Directions and Conclusions

Few models address the available dynamic data. One possible model for temporal dynamics is the DG model. While it predicts several properties of temporal dynamics, it fails in a potentially important way: Without additional assumptions, It does not represent a wide enough range of time scales, and therefore cannot handle effects that occur within and between training sessions. Interestingly, a recently proposed model for the temporal properties of *habituation* (Staddon & Higa, in press) may provide a basis for temporal control during interval reinforcement schedules.

Habituation refers to a decrease in responding to repeated

presentation of a stimulus (e.g., Davis 1970). The original model for habituation consists of a series (cascade) of *integrators* with different time scales. It successfully accounts for several features of short–term habituation (within session effects of the interstimulus interval, ISI) and long–term habituation (effects seen across sessions). That the model is able to handle effects across multiple time scales suggests it is a good candidate for behavioral sensitivity to IFIs of different durations – i.e., temporal control. How might the model be adapted for timing? At this point we can only demonstrate the feasibility of such a model and a sketch of the main ideas. The version we present here does not depend on a pacemaker system and represents a departure from most approaches – modeling temporal regularities in behavior without using a clock.

The idea of a clock or pacemaker for timing intervals on the order of seconds and minutes used to be a theoretical tool for explaining how animals estimate the passage of time (e.g.,Roberts, 1981). However, recent neurophysiological work points to a biological mechanism for the clock: Research suggests there is a cortico–striato circuit mediating timing whose effects seem to depend on the neurotransmitter dopamine (e.g., Church, 1984). The clock appears to function like a kind of stopwatch: Pulses are emitted from the substantia nigra and dopamine is accumulated in the caudate putamen, and the frontal cortex "reads" and integrates the information in the accumulator and decisions about whether or not to respond (see Meck, 1996 for a detailed description of the circuits and system). However, given the prevalence of temporal control, and sensitivity to time–dependent properties of learning, across a wide range of species with different brain structures (i.e., invertebrates, mammals, and birds all show behavioral sensitivity to temporal events), it seems equally reasonable to contemplate processes that show temporal patterning without a pacemaker.

An integrator model. A diagram of a single integrator appears in Figure 12 and its equation (in discrete time) is:

$$V_i(t) = aV_i(t-1) + bX(t) \tag{5}$$

where V (response strength) is the integrated effect of past stimuli (X) at a moment in time (t). X represents the input at time t. when a stimulus is presented, X is set to 1, otherwise it is 0. The parameter *a* is a time constant representing the period over which the effect from a stimulus is integrated, and *b* is the weighted effect of a stimulus. The figure shows, for a given set of values for *a* and *b*, the strength of responding over time to a single stimulus. Changes in *a* and *b* determine the height and rate at which Vi changes. Staddon (1993) showed that a modified version of the integrator in Figure 12 accounted for basic properties of habituation such as faster habituation with shorter interstimulus intervals (ISIs). However, a single integrator could not simulate the *rate–sensitive* property of habituation. Rate–sensitivity refers not only to the finding that habituation is more rapid and complete when the ISI is short than long (which is easily described by a single integrator), but that recovery from habituation is also faster after short than after long ISIs (e.g., Davis, 1970; Staddon, 1993). In a recent paper we showed that a series of two to three linked integrators reproduces several features of habituation and most notably its rate–sensitive property (Staddon & Higa, in press). Hence, for the present demonstration we begin with several integrators.

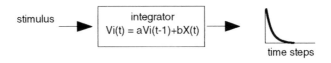

Figure 12. Illustration of a single integrator and its output.

Architecture. In our initial attempts to develop an integrator model for timing, we assume a cascaded series of integrators (i=6, see

Figure 13). We began with six integrators described by Equation 5. Except for the first integrator, the input to each integrator (X_i) is the output of the prior, adjacent, integrator. For the first integrator, its input is either 1 (with reinforcement) or 0 (no reinforcement). Each integrator is connected to an output node and the connections have different weights (w_i). The net response of the model (V_o) is based on the weighted sum of the output of each integrator and a threshold (θ, $\theta = 0$ in our simulations):

$$V_o = \Sigma V_i w_i, \qquad \text{if } \Sigma V_i w_i > \theta, \tag{6}$$
$$= 0 \text{ otherwise.}$$

The model is similar in many respects to a one proposed by Machado (in press). However, (1) we do not assume that the state of the system resets with each reinforcement, and 2) we do not assume a clock process (Machado assumes a Poisson pacemaker).

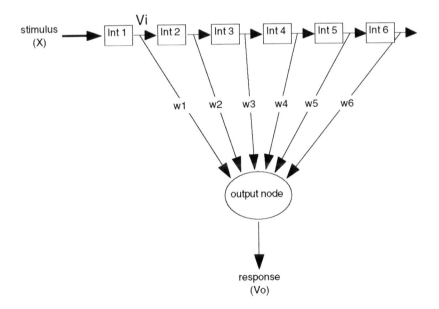

Figure 13. Architecture of an integrator model for timing, with six integrators each with a weighted connected to an output node.

Weights. We have not yet determined a learning rule for the weights, and for now assume fixed weights for each integrator. Guided by evidence that reinforcement has an initial inhibitory effect on responding(e.g., Kello, 1972; Staddon & Innis, 1966) we selected the

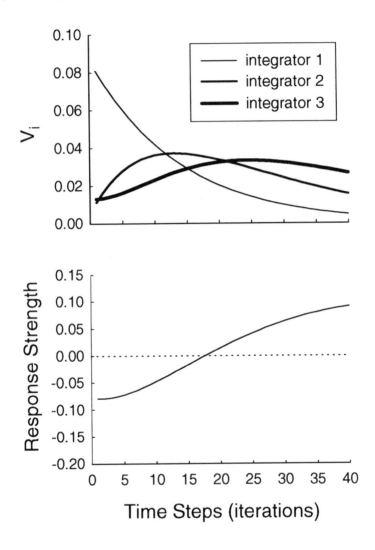

Figure 14. Top figure presents the state (V_i) of the first three integrators after a single reinforcer. Bottom figure shows output of system before a threshold: sum of the V–value of all integrators multiplied by its weight.

weights so that the strength of responding to a single reinforcer over time looked like the function in the bottom of Figure 14. The top graph shows the Vi values for three (of the six) integrators after a single reinforcer and illustrates how the output (bottom graph) is generated. Immediately after a reinforcer, the first integrator has the highest response strength. But, because its weight is −2, it contributes a large negative value to the output node after reinforcement (see lower graph). Other integrators such as Integrator 3 peaks later in the interval between reinforcers. Because its weight is 1.0, it contributes a positive value to the output node. Together, the cascade of integrators produces the function in the bottom graph of Figure 14.

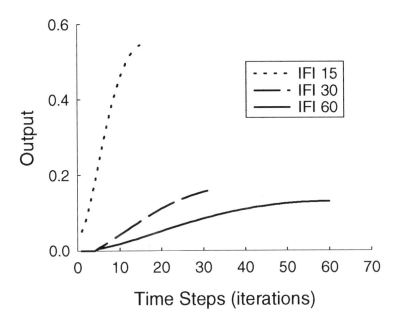

Figure 15. Simulation of different IFIs. Each time step is equivalent to a second of real time. For the IFI 15 condition all the parameter values for all integrators were: $a=.83$, $b=.21$; for IFI 30 $a=.93$, $b=.08$; and for IFI60 $a=.95$ and $b=.06$. The weights remained the same across conditions: −2, −2, 1, 1, 1, 1, for Integrators 1 though 6, respectively.

Once the response to a single reinforcer was determined, we simulated different IFIs – specifically, an FI 15, 30, and 60 s (Figure 15). The results look like standard FI patterns of responding and show that wait times increase with increases in the IFI duration, for a given threshold value. Our demonstration suggests it is possible to generate temporal "behavior" without reliance on a clock. We do not mean to suggest that organisms have structures similar to integrators: Only that the process by which they time events – and the ability to represent multiple time scales – has the properties of a cascaded integrator system, which may be accomplished differently across species and nervous systems.

In summary, standard research on time discrimination has mainly focused on steady–state responding that reflects what an animal has learned about intervals of time after a period of training. Recent studies indicate timing is rapid under certain conditions, and show it is possible to study how animals learn about intervals of time. A substantial amount of dynamic data is now available that show time discrimination can be understood from interactions among IFIs. We know that the process is more complex than linear waiting. We presented a DG model for these effects. Although it does not capture all the dynamic effects of time discrimination, its failure brought to light the importance of modeling a process with multiple time scales. A model based on a small number of simple integrators provides a starting point for understanding and explaining the processes involved in the temporal organization of behavior.

Author Notes

The authors thank Erika Holz and Lynn Talton for their help in the preparation of the manuscript, Nancy Innis and George King for comments on an earlier draft of the paper. The research was supported by a Duke University Arts and Sciences Research Grant to J. Higa and grants from NIMH, NSF to J. Staddon. Address correspondence to either author at the Department of Psychology: Experimental, Duke University, Box 90086, Durham, NC 27708–0086.

References

Bizo, L.A., & White, K.G. (1994). Pacemaker rate in the behavioral theory of timing. *Journal of Experimental Psychology: Animal Behavior Processes, 20*, 308–321.

Brunner, D., Kacelnik, A., & Gibbon, J. (1992). Optimal foraging and timing processes in the starling, *Sturnus vulgaris*: Effect of inter–capture interval. *Animal Behavior, 44*, 597–613.

Catania, A. C., & Reynolds, G. S. (1968). A quantitative analysis of the responding maintained by interval schedules of reinforcement. *Journal of the Experimental Analysis of Behavior, 11*, 327–383.

Catania, A.C. (1970). Reinforcement schedules and psychophysical judgments: A study of some temporal properties of behavior. In W.N. Schoenfeld (Ed.), *The theory of reinforcement schedules* (pp. 1–42). New York: Appleton–Century–Crofts.

Cheng, K., & Westwood, R. (1993). Analysis of single trials in pigeons' timing performance. *Journal of Experimental Psychology: Animal Behavior Processes, 19*, 56–67.

Church, R.M. (1984). Properties of the internal clock. In J. Gibbon & L. Allan (Eds.), *Timing and Time Perception* (pp. 52–77). New York: New York Academy of Sciences.

Church, R. M. Broadbent, H.A. (1990). Alternative representations of time, number, and rate. *Cognition, 37*, 55–81.

Church, R.M., & Broadbent, H.A. (1991). A connectionist model of timing. In M.L. Commons, S. Grossberg, & J.E.R. Staddon (Eds). Hillsdale, NJ: Lawrence Erlbaum Associates.

Davis, M. (1970). Effects of interstimulus interval length and variability on startle–response habituation in the rat. *Journal of Comparative and Physiological Psychology, 72*, 177–192.

Dews, P. (1970). The theory of fixed–interval responding. In W.N. Schoenfeld (Ed.), *The Theory of Reinforcement Schedules* (pp. 43–61). New York: Appleton–Century–Crofts.

Ferster, C.B., & Skinner, B.F. (1957). *Schedules of reinforcement*. New York: Appleton–Century– Crofts.

Fetterman, J.G., & Killeen, P.R. (1991). Adjusting the pacemaker. *Learning and Motivation, 22*, 226–252.

Gibbon, J. (1977). Scalar expectancy and Weber's law in animal timing. *Psychological Review, 84*, 279–325.

Gibbon, J. (1991). Origins of scalar timing. *Learning and Motivation, 22*, 3–38.

Gibbon, J. (1995). Dynamics of time matching: Arousal makes better seem worse. *Psychonomic Bulletin & Review, 2*, 208–215.

Gibbon, J., & Allan, L. (Eds.). (1984). *Annals of the New York Academy of*

Sciences: Vol. 423. Timing and Time Perception. New York: New York Academy of Sciences.

Gibbon, J., & Balsam, P. (1981). Spreading association in time. In C. M. Locurto, H.S. Terrace, & J. Gibbon (Eds.). *Autoshaping and Conditioning Theory* (pp. 219–251). San Diego, CA: Academic Press.

Gibbon, J., & Church, R.M. (1984). Sources of variance in an information processing theory of timing. In H.L. Roitblat, T.G. Bever, & H.S. Terrace (Eds.), *Animal Condition* (pp. 465–488). Hillsdale, NJ: Erlbaum.

Gibbon, J., & Church, R.M. (1990). Representation of time. *Cognition, 37*, 23–54.

Gibbon, J., Church, R.M., & Meck, W.H. (1984). Scalar timing in memory. In J. Gibbon & L. Allan (Eds.), *Timing and Time Perception* (pp. 52–77). New York: New York Academy of Sciences.

Harzem, P. (1969). Temporal discrimination. In R. M. Gilbert & N. S. Sutherland (Eds.), *Animal Discrimination Learning* (pp. 299–334). New York, Academic Press.

Higa, J.J (in press). Rapid timing of a single transition in interfood interval duration by rats. *Animal Learning & Behavior*.

Higa, J.J., (1996). Dynamics of time discrimination: II. The effects of multiple impulses. *Journal of the Experimental Analysis of Behavior, 66*, 117–134.

Higa, J.J., Thaw, J.M., & Staddon, J.E.R. (1993). Pigeons' wait–time responses to transitions in interfood interval duration: Another look at cyclic schedule performance. *Journal of the Experimental Analysis of Behavior, 59*, 529–541.

Higa, J.J., Wynne, C.D.L., & Staddon, J.E.R. (1991). Dynamics of time discrimination. *Journal of Experimental Psychology: Animal Behavior Processes, 17*, 281–291.

Innis, N.K. (1981). Reinforcement as input: Temporal tracking on cyclic interval schedules. In M.L. Commons & J.A. Nevin (Eds.), *Quantitative analysis of behavior: Discriminative properties of reinforcement schedules* (pp. 257–286). New York: Pergamon Press.

Innis, N.K., Cooper, S.J., & Mitchell, S.K. (1993). The determinants of postreinforcement pausing. *Behavioural Processes, 29*, 229–238.

Innis, N.K., Mitchell, S.K., & Staddon, J.E.R. (1993). Temporal control on interval schedules: what determined the postreinforcement pause? *Journal of the Experimental Analysis of Behavior, 60*, 293–311.

Kello, J.E. (1972). The reinforcement–omission effect on fixed–interval schedules: Frustration or inhibition. *Learning and Motivation, 3*, 138–147.

Killeen, P.R. & Fetterman, J.G. (1988). A behavioral theory of timing. *Psychological Review, 95*, 274–295.

Killeen, P., & Fetterman, J.G. (1993). The behavioral theory of timing: Transition analyses. *Journal of the Experimental Analysis of Behavior, 59*, 411–422.

Leak, T. M., & Gibbon, J. (1995). Simultaneous timing of multiple intervals: Implications of the scalar property. *Journal of Experimental Psychology: Animal Behavior Processes, 21*, 3–19.

Lejeune, H., & Wearden, J.H. (1991). The comparative psychology of fixed–interval responding: Some quantitative analyses. *Learning and Motivation, 22*, 84–111.

Machado, A. (in press). Learning the temporal dynamics of behavior. *Psychological Review*.

Mark, T.A., & Gallistel, C.R. (1993, November). *The microstructure of matching*. Paper presented at the 34th Annual Meeting of the Psychonomics Society, Washington, DC.

Mark, T.A., & Gallistel, C.R. (1994). Kinetics of matching. *Journal of Experimental Psychology: Animal Behavior Processes, 20*, 79–95.

Meck, W.H. (1996). Neuropharmacology of timing and time perception. *Cognitive Brain Research, 3*, 227–242.

Meck, W.H., & Church, R.M. (1984). Simultaneous temporal processing. *Journal of Experimental Psychology: Animal Behavior Processes, 10*, 1–29.

Richelle, M. & Lejeune, H. (1980). *Time in animal behavior*. Oxford: Pergamon Press.

Roberts, S. (1981). Isolation of an internal clock. *Journal of Experimental Psychology: Animal Behavior Processes, 7*, 242–268.

Schneider, B.A. (1969). A two–state analysis of fixed–interval responding in the pigeon. *Journal of the Experimental Analysis of Behavior, 12*, 677–687.

Skinner, B.F. (1938). *The Behavior of Organisms*. New York: Appleton–Century–Crofts.

Staddon, J. E. R. (1964). Reinforcement as input: Cyclic variable–interval schedule. *Science, 145*, 410–412.

Staddon, J.E.R. (1965). Some properties of spaced responding in pigeons. *Journal of the Experimental Analysis of Behavior, 8*,19–27.

Staddon, J.E.R. (1967). Attention and temporal discrimination: Factors controlling responding under a cyclic–interval schedules. *Journal of the Experimental Analysis of Behavior, 10*, 349–359.

Staddon, J.E.R. (1983). *Adaptive Behavior and Learning*. Cambridge: Cambridge University Press.

Staddon, J.E.R. (1988). Quasi–dynamic choice models: Melioration and ratio invariance. *Journal of the Experimental Analysis of Behavior, 49*, 303–319.

Staddon, J.E.R. (1993). On rate–sensitive habituation. *Adaptive Behavior, 1*, 421–436.

Staddon, J.E.R., & Higa, J.J. (in press). Multiple time scales in simple habituation. *Psychological Review*.

Staddon, J.E.R., & Higa, J.J. (1991). Temporal learning. In G.H. Bower (Ed.), *The psychology of learning and motivation* (Vol. 27, pp. 265–294). Academic Press, Inc.: San Diego.

Staddon, J.E.R., & Innis, N.K. (1966). An effect analogous to "frustration" on interval reinforcement schedules. *Psychonomic Science, 4*, 287–288.

Staddon, J.E.R., & Reid, A.K. (1990). On the dynamics of generalization. *Psychological Review, 97*, 576–578.

Sutton, R.S., & Barto, A.G. (1987). A temporal–difference model of classical conditioning. Proceedings of the Ninth Conference of the Cognitive Science Society, pp. 355–378.

Wynne, C.D.L. & Staddon, J.E.R. (1988). Typical delay determines waiting time on periodic–food schedules: Static and dynamic tests. *Journal of the Experimental Analysis of Behavior, 50*, 197– 210.

Wynne, C.D.L., & Staddon, J.E.R. (1992). Waiting in pigeons: The effects of daily intercalation on temporal discrimination. *Journal of Experimental Analysis of Behavior, 58*, 47–66.

Wynne, C.D.L., Staddon, J.E.R., & Delius, J.D. (1996). Dynamics of waiting in pigeons. *Journal of the Experimental Analysis of Behavior, 65*, 603–618.

Zeiler, M.D., & Powell, D.G. (1994). Temporal control in fixed–interval schedules. *Journal of the Experimental Analysis of Behavior, 61*, 1–9.

Time and Behaviour: Psychological and Neurobehavioural Analyses
C.M. Bradshaw and E. Szabadi (Editors)
41

CHAPTER 2

Timing and Temporal Search

Russell M. Church

The study of animal timing represents a successful combination of experimental investigation and quantitative modeling. The purposes of this chapter are:

i.   To describe the goals for a theory of timing and temporal search,

ii.   To describe the basic concepts involved in three timing theories (Scalar Timing Theory, the Behavioral Theory of Timing, and a Multiple Oscillator Model of Timing),

iii.   To describe what timing theories are currently able to do, and what they fail to do, and

iv.   To speculate on the nature of the next generation of timing theories.

Goals for a Theory of Timing and Temporal Search

A long–term goal is to attempt to develop a single theory of timing (1) for humans and all other animals, (2) for perception and performance, (3) for all scales of time, and (4) for fixed and random times, and all other interevent distributions. Of course, such a general theory may be unobtainable. It is possible that humans use timing mechanisms unavailable to other animals, or that different animals have evolved different mechanisms for timing; it is possible that the principles of perception of time are different from the principles of timed performance; it is possible that principles of timing in the second–to–minute range are different from those in the range of milliseconds or hours, or in the circadian range; and it is possible that animals time relatively fixed interevent intervals but ignore relatively random interevent intervals.

Species generality

The extensive research on time perception and timed performance of rats and pigeons has revealed many fundamental similarities, and the few differences that have been reported have not been attributed to differences in the timing mechanisms. Comparative studies have demonstrated the common timing abilities of different species, and revealed few, if any, systematic species differences that would restrict the applicability of a general theory of timing and temporal search (Lejeune & Wearden, 1991). The possibility that the same theory of timing may be applied to humans and other animals now seems much more likely than it did prior to the mid–1980's. Although the study of human and animal timing have long histories, they were conducted quite independently. Different investigators studied different problems using different methods. They used different theories to explain the data; articles about human and animal timing had very few cross references. Fraisse (1963) provided a good treatment of the knowledge of human timing; Richelle and Lejuene (1980) provided a good treatment of the knowledge of animal timing. Probably the symposium organized by Gibbon and Allan (1984) was the first opportunity for many investigators to consider the possibility of an integration of the study of human and animal timing. One of the co–organizers of that symposium was an expert in animal timing; the other was an expert in human timing. They invited an approximately equal number of participants from the two fields and then organized sessions around topics in which there was significant participation from both investigators of animal and human timing. Following this symposium there have been collaborative efforts by experts in animal and human timing and, more importantly, there has been the attempt to apply methods and theories primarily associated with one field to the other (Allan & Gibbon, 1991; Wearden, 1991; Wearden, 1992; Wearden & Towse, 1994). This integration is particularly important for the development of understanding of the neural basis of timing which can be studied most effectively in animal subjects, but which is relevant to human neuropsychology.

Perception and performance

In a time perception task the experimenter controls the duration of an event or interevent interval and the animal classifies its perceived duration by making different responses following the different durations; in a timed performance task the experimenter controls the time–of–occurrence of events and the animal controls the time–of–occurrence of responses. The timed performance task may be considered to be a temporal search task, analogous to a spatial search task. In a spatial search task an animal responds in different locations; in a temporal search task an animal responds at different times. Conceptualization of spatial and temporal search as similar tasks may encourage the identification of common principles. The problem of timed performance is to identify the determinants of when a response occurs; the problem of time perception is to determine why, after a particular time interval, the animal chooses to make one response rather than another. A complete list of the times–of–occurrence of inputs (e.g., lever presses and other recorded responses) and outputs (e.g., stimuli, reinforcers, and other changes in the animal's environment) provide the data for either type of study. For the study of timed performance, such as a fixed–interval schedule of reinforcement in which food is delivered following the first lever press that is at least 30 seconds after the last reinforcement, it is necessary to relate the time–of–occurrence of a response to the occurrence of the last reinforcement; for the study of time perception, such as a discrimination task in which food is delivered following a left lever response if a noise stimulus lasted 2 seconds but it is delivered following a right lever response if a noise stimulus lasted 8 seconds, it is necessary to relate the response that is made (left or right) to the duration of the stimulus. Although some attempts have been made to develop different theories for timed performance than for time perception, the two phenomena are so closely related that theories of timing typically attempt to account for both.

Scales of time

Most studies of human timing deal primarily with durations in the range of 100 to 1000 ms, while most studies of animal timing deal primarily with durations in the range of seconds to minutes. This difference is due both to the different historical traditions, and to some practical considerations. The use of short durations in human timing minimizes counting artifacts and reduces the boredom that might interfere with optimal performance if long durations were used. But the theories that are developed for different ranges are not generally thought to be restricted to these ranges. The major exception is that many investigators of circadian rhythms believe that timing near 24 hours is qualitatively different from other timed behavior. Whether or not a single theory of timing may be developed that applies both to interval and periodic timing is a major question considered later in this chapter.

Fixed and random interevent intervals

Most studies of timing deal with fixed interevent intervals which are, in principle, perfectly predictable. But should a timing theory also attempt to explain performance with random interevent intervals which are locally, in principle, totally unpredictable? Probably they must do so because there are an infinite number of possible distributions between the completely predictable and the completely unpredictable. Thus, theories of timing should be evaluated not only on their ability to account for performance with fixed interevent intervals, but with their ability to account for performance on random times, and all other interevent intervals.

Measurement of the Time-of-occurrence of a Response

The emphasis of this chapter will be on timed performance, especially the time-of-occurrence of responses of rats in the second-to-minute range. One of the basic questions in the study of timing is to determine why a response occurs at the time that it does, rather than earlier or later. Two requirements for such a study are (a) to define the response to be timed and (b) to record the times-of-occurrence. In laboratory situations, neither of these requirements pose any difficulties for investigators.

The response may be defined as the closure of a switch. For example, a rat may close a switch by mechanically pressing a lever, by completing a circuit by contacting a drinking tube, or by breaking a photobeam. The time-of-occurrence of a response can be measured with virtually any accuracy; it is now common to record times-of-occurrence to the nearest millisecond. Of course, a time-stamped record of every response may generate an enormous amount of data, but standard laboratory computers now have hard disks with gigabyte capacity, and various inexpensive ways are available, such as tape, removable hard disks, optical disks, and floppy disks, to store the original data for later analysis.

Prior to the availability of laboratory computers, times could be measured accurately but there was no convenient way to deal with large amounts of times-of-occurrence data. Thus, investigators collected only summary data. At one extreme, only the total number of responses during a session would be recorded with a single counter which, when combined with the total session time, would produce a mean response rate. With a bank of counters, the number of responses during a well–defined interval of time could be recorded. For example, with 10 counters one could record the number of responses (and calculate the local response rate) as a function of successive tenths of a session. Such a bank of counters was also used to record the number of responses as a function of successive intervals of time since a well defined event (such as the presentation of food). Thus, perhaps with the addition of a bank of timers to record the total time in which the animal had the opportunity to respond, the investigator could calculate the local response rate as a function of successive intervals of time since food. Such summary measures of times–of–occurrence of responses provided very limited information and, of course, they could not be used to reproduce the sequence of times–of–response that actually occurred.

One of the virtues of the cumulative record (a record of the cumulative number of responses as a function of the time since the session began) was that it preserved all of the time–of–occurrence information. But, because it preserved this record only on paper, it could not be conveniently used for quantitative analysis. When computers became available, many investigators recorded the times–of–occurrence more precisely. For example, investigators of circadian rhythms often counted the number of wheel–turns in 10–min intervals (over successive 24–h periods), and investigators of animal timing in the second–to–minute range often counted the number of lever–presses in 1–s intervals, or in time bins that were some fixed proportion of the interval to be timed. The use of time bins may have been due to conventions established when the apparatus restricted the precision to which times–of–occurrence could be recorded. With such summary measures one can calculate interresponse times only with assumptions that are known to be inexact (such as the assumption of a uniform distribution of times–of–occurrence within a time bin). Some investigators prefer to have a known number of intervals in each session for analysis, but they can be created from the original time–of–occurrence data. With standard modern equipment, it is always desirable to record precise times–of–occurrence of each response and all other events. These may be used for the calculation of summary measures.

Scalar Timing Theory

Scalar Timing Theory is the most completely developed general quantitative model of animal timing. It is a contender as a model that will achieve the four goals for a theory of timing and temporal search described at the beginning of this chapter. (1) Although it was developed to account for the data from animal timing experiments (e.g., Gibbon, 1977; Gibbon, Church, & Meck, 1984), it has been applied to similar data from human timing experiments (Allan & Gibbon, 1991; Wearden, 1991; Wearden, 1992; Wearden & Towse, 1994). (2) It has been used extensively to account for the data from perceptual experiments in which an animal chooses a response based upon some time interval; it has also been used extensively to account for the data from timed behavior experiments in which the time–of–occurrence of a response is related to the time of availability of a reinforcement (Gibbon, 1991). (3) Although Scalar Timing Theory has been primarily used to account for timing in the range of seconds to minutes, no changes in the theory are necessary to apply it to much shorter or much longer durations. (4) Although it has been used primarily to account for performance in conditions in which there is a fixed time from some event until reinforcement is available, no changes in the theory are necessary to apply it to other interevent distributions, and it has been applied to exponential distributions of interevent intervals (Gibbon, Church, Fairhurst, & Kacelnik, 1988).

The next section describes the essential features of Scalar Timing Theory, based primarily on the version described by Gibbon, Church, & Meck (1984). In that article the theory was applied to three experimental procedures: temporal generalization (Church & Gibbon, 1982), the peak procedure (Roberts, 1981), and the time–left procedure (Gibbon & Church, 1981). For the expository purposes of this section, the description of the theory will be in the context of the peak procedure. In this procedure, as originally described by Catania (1970) there are two types of trials: food and nonfood. Following an intertrial interval, a stimulus (such as white noise) begins. On food trials the first response after a fixed interval is followed by food and the termination of the noise; on nonfood trials, no responses are reinforced and the noise remains on much longer. On nonfood trials, the mean response rate increases gradually to a maximum near the time when food is sometimes received, and then it decreases in a slightly asymmetrical manner (Meck & Church, 1984; Roberts, 1981). Such data have been well fit by Scalar Timing Theory.

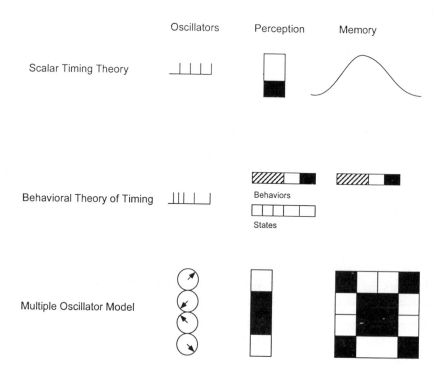

Fig. 1. Three timing theories. In Scalar Timing Theory (top panel) pulses from an oscillator are summed in an accumulator and stored in a distribution. In the Behavioral Theory of Timing (middle panel) pulses from an oscillator advance behavioral states, each of which has some strength. In the multiple oscillator model, half–phases from multiple oscillators are stored in an autoassociation matrix.

Pacemaker and accumulator

Scalar Timing Theory uses a pacemaker–accumulator system for the perceptual representation of time. The pacemaker is an oscillator that emits pulses in time, and the accumulator is a counter that sums the pulses. The animal is assumed not to have any direct information from the oscillator; its representation of time is solely based on the value in the accumulator. The number of counts in the accumulator after a given time interval depends on the characteristics of the oscillator. Three alternatives were considered: (1) the oscillator may emit pulses at some fixed interpulse interval, (2) the oscillator may emit pulses at some random interpulse interval, an exponential waiting time process in which the expected time to the next pulse is a constant regardless of the time since the last pulse, or (3) the oscillator may emit pulses at some fixed interpulse interval on any particular trial, but there is a normal distribution of rates across successive trials. If the mean interpulse interval is very short relative to the durations being measured, the form of the interpulse distribution will have only minimal effect on the accumulator count. On the basis of data currently available, it has not been possible to determine the distribution of interpulse intervals on individual trials. It might be fixed or random, or any other distribution. Thus, it is simplest to assume that the pacemaker puts out a fixed interpulse interval but to recognize that the theory would not be invalidated if it were determined that the neural pulses serving as the output of the oscillator were distributed as a Poisson, or other, process. The third alternative – that there is a normal distribution of mean oscillator rates between trials – is a useful assumption for fitting data. It is, however, difficult to distinguish from memory storage variability so, in practice, only between–trial clock variability or between–trial memory variability is assumed. There are some estimates that the mean pacemaker rate is about 5 pulses/second (Meck, Church, & Gibbon, 1985). The accumulator has two characteristics: It can be incremented and reset. Thus, it can be used repeatedly to make a temporary record of a duration. Animals may be trained to reset the accumulator under particular conditions; it is also possible that such reset may occur automatically after some brain lesions, such as fimbria fornix lesion or other interference with hippocampal processing (Meck, Church, & Olton, 1984), and after some pharmacological manipulations, such as systemic atropine, or other cholinergic receptor blockers (Meck & Church, 1987).

Switch

Although the pacemaker may be continuously running, pulses are not always being added to the accumulator. There is a switch between the pacemaker and the accumulator; if the switch is open no pulses enter the accumulator, but if it is closed, pulses are added to the accumulator. Presumably, when the stimulus begins there is a latent period before the animal begins to time the stimulus. In the peak procedure, presumably the switch is closed shortly after the stimulus begins and it is opened shortly after the stimulus ends (or food is delivered). Both the closing and opening of the switch may occur with some latency (t_1 and t_2, respectively). If the switch were closed as soon as the stimulus began and opened as soon as it ended, then the number of pulses in the accumulator would be λt, where λ is the oscillator rate (number of pulses per second) and t is the stimulus duration in seconds. But, if the switch took t_1 seconds to close when the stimulus began, and if it took t_2 seconds to open when the stimulus ended, the effective duration of the stimulus would be $t + t_2 - t_1$. Thus, the number of pulses in the accumulator would be $\lambda(t + t_2 - t_1)$. Typically, the latency to close the switch has been found to be longer than the latency to open it.

Rats can learn to control the operation of the internal clock, presumably by controlling the switch. For example, if rats have learned to time a white noise stimulus, if that stimulus is presented for some interval (a), followed by a gap interval (g), and then a restoration of the stimulus for a duration (b), the rats may sum pulses in the accumulator during the entire period (a + g + b), or during the stimulus (a + b), or after the gap (b). These are known as the "run", "stop", and "reset", modes, respectively (Roberts & Church, 1978).

Memory

Although working memory was shown in the figures describing Scalar Timing Theory, it was not used to transform information in any way. In contrast, reference memory has a critical role in the theory. Reference memory is assumed to be composed of a distribution of values based on counts that had been in the accumulator. Thus, temporal memory is a distribution of elements, and each element represents a time–of–occurrence. When reinforcement is received, it is assumed that the current count in the accumulator (an integer) is multiplied by a real number (k^*), and that this number is put into the distribution of values in reference memory. The concept of the memory storage constant (k^*) was developed by Meck (1983) to account for the fact that animals may appear to remember time intervals as longer and shorter than they actually are, and that these temporal memory distortions are not eliminated by training. For example, physostigmine (a drug that increases the effective level of acetylcholine) made rats remember time intervals as shorter than they were, and atropine (a drug that blocks cholinergic receptors) made rats remember time intervals as longer than they were. The magnitude of the effect was not a constant number of seconds, but a constant proportion of the interval being timed. Thus, k^* was introduced as a multiplicative, rather than an additive, constant. The intuitive idea is that a perceived time interval (the number of counts in the accumulator) may be transformed when it is stored in reference memory, and this, possibly distorted, remembered duration is used by the animal to decide whether or not to respond. The mean value of k^* can be affected by various lesions of the cholinergic system (Meck, Church, Wenk, & Olton, 1987), and it is higher in old rats than young or mature rats (Meck, Church, & Wenk, 1986). Although the mean value of k^* of a group of mature rats is approximately 1.0, there are reliable individual differences in this value.

Decision Rule

When a rat decides whether or not to respond at a particular instant, it compares the perceived current time (the number of counts in the accumulator) with a sample of the remembered time (a sample from reference memory). Two extreme possibilities were considered: The animal might be using some measure of central tendency of the distribution of remembered times, such as the mean, or it might be using a random sample of a single element. Although it seems intuitively plausible that the animal would use some combination of all of the information available (such as a mean), the standard deviation of this representation of remembered time would decrease as a function of the number of elements in the distribution (the central limit theorem). Because there appeared to be substantial and fairly stable variability in the remembered time of reinforcement even after considerable training, the other extreme proposal was made. That is, the assumption is that, at the beginning of each trial the animal takes a random sample of a single element as the remembered time of reinforcement. The standard deviation of this representation of remembered time is unaffected by the number of elements in the distribution, which seems to correspond better with the observation of asymptotic variability in temporal performance. Thus, the proposal is that the remembered time–of–reinforcement is based on a single specific example on each trial.

The elements for a decision regarding whether or not to respond at a particular time is based on two values: the number of pulses in the accumulator (a) and the example from reference memory (m). The next step was to combine these two values into a measure of "subjective discrepancy". Two combination rules were considered. One was a difference (m − a); the other was a ratio ((m−a)/m). The ratio combination rule (a relative subjective discrepancy) was clearly necessary. The major empirical generalization is that performance on temporal tasks is similar when time is scaled in relative units (that is, time relative to the duration being timed) rather than in absolute units (seconds). This is called the superposition rule, and it is an example of Weber's Law in timing (Gibbon, 1977; Gibbon, 1991). There is evidence that superposition is even greater if performance is scaled in psychologically, rather than physically, relative units. Thus, in the peak procedure, greater superposition occurs if time is scaled relative to the observed time of maximum responding rather than relative to the time that reinforcement is primed or delivered.

Now that a measure of relative subjective discrepancy has been defined, a value between 0 and 1, the final problem is how to translate it into action. The proposal is that there is a normally distributed threshold distribution which is set at some mean subjective discrepancy between 0 and 1. A response occurs if the relative subjective discrepancy is less than a sampled threshold value.

Sources of variance

As described above, Scalar Timing Theory includes many factors involved in the perception and memory of temporal intervals, and the decisions based on this information. Any of these factors may be variable. There is potential variability in the pacemaker rate (within and between trials); in the switch latency to start and stop the pacemaker; in the translation of the accumulator count to the value in reference memory (k^*); and in the threshold. For convenience, it is assumed that the distribution of each of these parameters is normal, so that each can be completely described by a mean and standard deviation. Explicit solutions are available for this theory (see Appendix of Gibbon, Church, & Meck, 1984).

A Behavioral Theory of Timing

Scalar Timing Theory was developed in the spirit of quantitative information–processing models in human cognitive psychology. It has modules that correspond to various psychological processes (perception, memory, and decision) and it has explicit transformation rules. An alternative theory of timing, a behavioral theory of timing, was developed in the spirit of quantitative behavioral theories, in which emphasis is placed on observed stimuli and behavior (Killeen & Fetterman, 1988). These comments will be based upon this version of the theory; the most fully developed version has been presented by Machado (in press). The Behavioral Theory of Timing has been applied to both timed performance and time perception. As in the case of the treatment of Scalar Timing Theory, the example will be taken from a timed performance study, the peak procedure. Such data may also be well fit by the Behavioral Theory of Timing. For the purpose of facilitating comparisons, the description of the theory that follows will use categories compatible with an information–processing model.

Pacemaker

The Behavioral Theory of Timing, like Scalar Timing Theory, assumes that a pacemaker emits pulses as a function of time. The theory has been primarily developed with the assumption that the pulses occur according to a random waiting–time distribution with some fixed mean. This type of pacemaker, also postulated by Creelman (1962), has several positive features. It is a precisely specified, mathematically tractable extreme in which the expected time to the next pulse is independent of the time since the last pulse. From the assumption of such a random emitter of pulses, the distribution of times at which the nth pulse occurs may be calculated.

States

The assumption is that each pulse of the pacemaker is associated with a behavioral state. Consider a rat in a lever box with the opportunity to eat, to drink, to move about in the box, to run in a running wheel, or to press a lever. If each of the behavioral states in the theory were directly observable and distinguishable, a rat might do the following on a 30–s fixed interval schedule of reinforcement (in which food is delivered following the first lever press that occurs more than 30 s since the last food delivery). After food is delivered, at the first pulse of the pacemaker the rat might eat the food; on the next pulse of the pacemaker it might drink some water; on the next pulse it might move about the box; on the next pulse it might run in the wheel; and on the next it might press the lever. Because the pulses are assumed to occur at random intervals, each of these behaviors would not occur at exactly the same time after the last food was delivered, but on average the time of lever pressing would be later than the times of the other behaviors. In fact, the Behavioral Theory of Timing does not assume that each state is necessarily directly observable, and it does not assume that the behavior associated with every state is necessarily distinguishable. Thus, although each pulse of the pacemaker advances the state number, and each state number is associated with some behavior (possibly unobservable), the observed behavior is not an infallible indicator of the state. In the example of the rat on the 30–s fixed–interval schedule of reinforcement, the successive states might be eat–silent–drink–drink–activity–wheel–lever.

Motivation

The mean pacemaker rate is assumed to be controlled by the level of arousal. Although many environmental and pharmacological variables can affect arousal, the rate of reinforcement is the one of greatest importance for the theory. If the mean pacemaker rate (τ) was controlled by the reinforcement rate, an animal would progress through the same sequence of states when the interreinforcement interval was short as when it was long, but it would do so more quickly. Under the assumption that the distribution of pacemaker pulses is random, Killeen & Fetterman (1988) have shown that the probability of the rat being in State n at an absolute time t is distributed approximately as a Gamma distribution with a mean of $(n+1)\,\tau$, and a standard deviation of $[(n+1)\,\tau]^{1/2}$. They then demonstrated that, if the pacemaker rate is a constant proportion of the reinforcement rate, then the probability of the rat being in State n at any relative time (absolute time divided by mean interreinforcement time) is the same for all reinforcement rates.

This is consistent with the extensive data demonstrating the approximate superposition of timed behavior when that behavior is plotted as a function of relative, rather than absolute, time. Although arousal is increased by an increase in the reinforcement rate, the magnitude of increase is not great enough to explain superposition, which is presumably due to multiple factors (Gibbon, 1992).

Memory

Reinforcement occurs after the animal has performed some sequence of behavior. In the example above the sequence was eat–silent–drink–activity–wheel–lever. In addition to its effect of increasing pacemaker rate, a second effect of a reinforcement is to strengthen the particular sequence of behaviors that has occurred. Note that the reinforcement strengthens the sequence of behaviors, not of states. Thus, the particular durations of each behavior due to the particular random interpulse interval or to the number of states that led to the same behavior are not stored in memory. What is strengthened is a particular sequence of behaviors. A third effect of the reinforcement is to develop a behavioral discriminative stimulus. In the example above, the rat was running in the wheel just before the reinforced lever press. Thus, the behavior of running in the wheel may serve as a discriminative stimulus for lever pressing. People and other animals can use their own behavior as a way to estimate the time to some event. A fully developed behavioral theory of timing requires a fully developed learning theory that provides a specification of the conditions under which response sequences are strengthened, generalized, and developed as conditioned stimuli. A major point of the Behavioral Theory of Timing is to suggest that the same principles of learning apply to temporal learning as apply to other associative learning.

Decision Rule

Finally, it is necessary to translate the strengths of various behavioral sequences into action. The usual assumption is that the behavior sequence with the greatest strength will occur. Many simple rules might be considered, and the Behavioral Theory of Timing suggests that the same principles of action apply to temporal learning as to other associative learning.

A Multiple–oscillator Model of Timing

In his textbook in psychology, William James (1890) explained behavior in two rather different ways. He used the concept of "soul" in much the way we now use "mind", to refer to the various modular faculties that include sensation, perception, memory, etc. His textbook includes a description of the principles of each of these faculties, as known at that time. The modern reader will find in this textbook many principles that are now well–known, but attributed to more recent investigators. But, in addition to his treatment of faculty psychology, he also had chapters on association. These were applied to habits, ideas and other activities. These two approaches to the understanding of psychological processes were based on different philosophical traditions (Boring, 1950), and they are still present in different approaches that have been used in the study of animal behavior. Most studies of animal cognition use modular explanations; most studies of animal learning use associationist explanations.

A purpose of the development of a multiple–oscillator theory of timing was to determine whether an associationist theory of timing could be developed that would account for the data that was explained by Scalar Timing Theory (Church & Broadbent, 1990).

Multiple Oscillators

Instead of a single pacemaker, with a particular mean rate, the multiple–oscillator theory of timing postulates many different periodic processes. In his integration of periodic and interval timing, Gallistel (1990) proposed a computational model that involved such multiple oscillators. The existence of periodicities at many, very different, time scales is well established; a multiple oscillator theory proposes that they are used as an essential part of the representation of time. To provide a concrete proposal that could be simulated, Church & Broadbent (1990) proposed that the oscillators had mean periods of 200 ms, 400 ms, 800 ms, etc., and there were 11 such oscillators so that the slowest one had an interpulse interval of 204.6 s. If there were 30 such oscillators, the slowest one would not complete one period within the lifetime of a rat, so this system does not require an excessive amount of neural material. When an event occurs, the oscillators are reset and then advance at their mean rates (plus or minus a random normal variable with a mean of zero and a standard deviation of some proportion of a period). Thus, the assumptions are that there is some trial–to–trial variation in oscillator speed, and that all oscillators are completely coupled.

Status Registers

Although the phase of the oscillators conveys more information, this model assumes only that the animal has access to the half–phase of each of the oscillators (–1 or +1). Thus, the perceptual representation of time is a vector of –1s and +1s. The status registers associated with the fastest oscillators typically convey little information. Because of the random variability they are quite likely to be in either half–phase after a short time. The status registers associated with the slowest oscillators also typically convey little information because they are quite likely to be in the same half–phase even after a long time. But the status registers associated with oscillators with intermediate periods usually do contain useful information for discrimination between different intervals.

Working Memory

Working memory is assumed to be an autoassociation matrix of the status register vector. This type of memory representation has been used productively in neural modeling (see, for example, Anderson et al., 1977). In this timing model, the matrix representation of the duration retains information about the relationship between adjacent status registers. With variability, some times are not reliably associated with the particular half–phase of any oscillator, but they are closely related to the agreement of adjacent oscillators. One can think of the status register as a binary counter with variability. No single bit provides a good representation of the number 7 (with variability), but the third and fourth bits are very likely to be different. For example, if the values are likely to be 6, 7 or 8 (0110, 0111, 1000) the third and fourth digits (from the right) are likely to be different. Thus, if there is some variability in the oscillators, the representation of a particular physical time in the status register may be quite different on different occasions, but the representation of this physical time in working memory will be quite similar on different occasions.

Reference Memory

Reference memory is a matrix with the same dimension as working memory. When reinforcement is received, each element of reference memory is updated on the basis of the corresponding element in working memory according to a linear averaging rule. Thus, the value in reference memory might be 99% of the value in reference memory and 1% of the value in working memory. This is a standard stochastic learning rule (Bush & Mosteller, 1955).

Decision Rule

For retrieval, the representation of the current time is compared to the memory of the reinforced time that has been stored in reference memory. The representation of the current time is a vector (based on the half–phase of multiple oscillators); the outer product of this vector and the reference memory matrix is compared with the representation of current time. The comparison is a similarity measure (a cosine or, equivalently, a correlation coefficient). The value of this similarity measure is compared with a threshold parameter; if the similarity is greater than the threshold, a response will be made at this particular time.

The Multiple Oscillator Model, as well as Scalar Timing Theory and the Behavioral Theory of Timing, is able to fit the mean response functions produced by the peak procedure.

Common features of the three theories

Oscillators

Both Scalar Timing Theory and the Behavioral Theory of Timing assume a single pacemaker. Typically, Scalar Timing Theory assumes that the pacemaker emits pulses with a fixed interpulse interval and the Behavioral Theory of Timing assumes that the pacemaker emits pulses at random intervals, with a fixed mean. But this is not a fundamental difference. Scalar Timing Theory readily incorporates different distributions of interpulse intervals, including random interpulse intervals, although it is difficult to distinguish among them empirically (Gibbon, 1991). And the Behavioral Theory of Timing is not restricted to the assumption of random interpulse intervals. The quantitative assumptions about the pacemaker in the two theories, however, do differ. Typically, the interpulse interval in Scalar Timing Theory is assumed to be about 5 pulses per second (Meck, Church, & Gibbon, 1985), but the interpulse interval in the Behavioral Theory of Timing is assumed to be about 3 or 4 pulses per reinforcement. Under most conditions, these are radically different rates. The Multiple Oscillator Model of Timing includes oscillators at many different periods. In the simulations, values of .2, .4, .8, 1.6, 3.2, 6.4, 12.8, 25.6, 51.2, 102.4, 204.6 s have been frequently used.

Perceptual representation of time

In Scalar Timing Theory, the perceptual representation of time is an accumulation of discrete pulses. In the Behavioral Theory of Timing, the perceptual representation of time is a behavior sequence (or, in some cases) a single behavior. In the Multiple Oscillator Model of Timing, the perceptual representation of time is a vector that contains information regarding the phase of each of the oscillators. It has been implemented with minimal information about each phase, i.e., only the half–phase has been used. Although each of the theories has a perceptual representation of time, the representations are quite different. Scalar Timing Theory uses a quantity; the Behavioral Theory of Timing uses a nominal scale; and the Multiple Oscillator Model uses a pattern that has some of the features of each of the other representations.

Temporal Memory

In Scalar Timing Theory, reference memory consists of a distribution of elements. Each element consists of the number of pulses that were in the accumulator, multiplied by a memory storage value. In the Behavioral Theory of Timing, temporal memory is a list of behavior sequences (or, in some cases) single behaviors, each associated with a frequency of reinforcement. In the Multiple Oscillator Model of Memory, temporal memory is a single matrix. Although each of the theories has representations of temporal memory, the representations are quite different. Scalar timing theory preserves the discreteness of the individually remembered elements by using a distribution; the Behavioral Theory of Timing aggregates similar behaviors and provides a weight for each of these aggregates; and the Multiple Oscillator Model uses a single matrix for temporal memory.

Decision Process

For retrieval from memory, Scalar Timing Theory uses a sampling procedure in which a single element from the reference memory distribution is drawn on each trial and compared with the number of elements in the accumulator by a ratio rule. The Behavioral Theory of Timing uses a winner–take–all rule in which the behavior sequence (or, in some cases, behavior) with the greatest frequency of reinforcement is made. The Multiple Oscillator Model uses a comparison procedure in which the representation of current time is compared with the entire matrix of remembered times by a similarity measure. Thus, Scalar Timing Theory is an exemplar theory of retrieval; the Behavioral Theory of Timing is a classified example theory of retrieval; and the Multiple Oscillator Model is a prototype theory of retrieval.

Conclusion

In terms of their psychological modularity, the three timing theories can be considered to be quite similar: They all have information-processing stages of perception, memory, and decision. But their representations of each of these stages are quite different. The unique strength of Scalar Timing Theory is that it has explicit solutions for several experimental procedures, and it has provided precise fits not only to mean functions but also to correlation patterns between indices of behavior, such as the time at which an animal begins to respond and when it stops responding on individual trials of the peak procedure. The unique strength of the Behavioral Theory of Timing is that is provides a parsimonious account of data with emphasis on observed behavior. The unique strength of the multiple oscillator model is that it provides qualitative fits to some features of timing behavior, such as the periodicities and systematic residuals described below.

Strengths of Current Theories

They are precisely defined

Many theories in psychology are not fully specified. They provide basic concepts, and a general approach to an explanation, but they are not specified with sufficient precision that different analysts will obtain the same predictions using the same theory. This lack of clarity is sometimes represented as a scientific virtue of unwillingness to make premature conclusions. At the early stages of the development of a theory this may be justifiable because it provides the flexibility that may encourage others to be creative with these concepts and approaches. This is a view of a theory as a toy with which to interact, rather than a game with rules. Without a precisely defined theory, different analysts may use the same theory to obtain conflicting predictions. In practice, such conflicts are seldom exposed because it is uncertain what the theory actually predicts. In contrast to many theories in psychology, the quantitative theories of timing have been described completely. Such clarity is a major virtue, and it is a prerequisite for evaluation. The predictions of any well–specified timing theory may be done by simulation. Such simulations may be done with a standard programming language (C, Pascal, Basic, etc.), with a program with matrices as its basic data structure (such as Matlab), with a spreadsheet (such as Excel), or with other applications. Explicit solutions of Scalar Timing Theory are available for several timing procedures (Gibbon, Church, & Meck, 1984), and more of them could be developed for other timing procedures or for other measures of performance. The predictions based on an explicit solution have several advantages over those generated by simulation: they are more accurate, they are generated more quickly which is useful for exploration of a multidimensional parameter space, and examination of the explicit solution provides information about the basis for the predictions of the model.

They fit the data

All of the timing models fit some of the data quite well, with only small random errors. One of the major strengths of timing theories is that they have been applied to detailed descriptions of the data, not just to a summary measure. In many studies of animal learning and behavior, the primary descriptive measure is the mean response rate or the mean percentage of conditioned responses, measures which can be summarized in a bar graph or with a few numbers. Studies of timing usually provide much more detailed information regarding the performance of the animals. It is standard to report functions relating mean response rate since some well defined event. In addition, the distribution and intercorrelations of indices of performance on individual trials are being reported. Scalar Timing Theory has been applied to the largest number of procedures and the largest number of dependent measures, and it has accounted for a high proportion of the variance in these cases (Church, Meck, & Gibbon, 1994; Cheng & Westwood, 1993, Gibbon & Church, 1990; Gibbon & Church, 1992;). Of course, it must be recognized that any clearly specified theory may be found to make some incorrect predictions. This has been used as an impetus to improve, rather than to abandon, a theory.

They are analyzable

The theories consist of separate, but interacting, modular parts that are intuitively plausible. The major value of psychological modularity is that it permits the analyst to identify the effect of changes in one of the modules without any change in the others.

They are general

Some theories are explanations of performance under the conditions of a particular experiment. The timing theories are not just explanations of results from a particular procedure, but they are easily extended to new procedures. For example, they apply both to time perception (such as a temporal discrimination procedure) and to timed performance (such as the peak procedure).

Two Challenges for Current Theories

The major challenge for current theories is to make correct predictions of all possible dependent variables under all possible procedures. Current theories of timing have usually been applied to a few dependent variables, such as the mean response rate as a function of time and the relative preference between two measured responses. They have usually been applied to a few standardized procedures, such as the temporal discrimination and peak procedures mentioned above. But, a theory of timing should do more than account for a few dependent variables in a few timing tasks. As noted above, one of the strengths of these theories is that they make explicit predictions about performance under any distribution of interevent interval, and they make explicit predictions about all descriptions of the data. The challenge is to make correct predictions about all descriptions of the data under all sequences of interevent intervals.

To Apply to All Sequences of Interevent Intervals

Acquisition. Although all of the theories make explicit predictions about the learning in a time perception task or a timed performance task, most of the tests of the theory have been on steady-state performance. A timing theory should account for the original acquisition of a temporal discrimination and temporal performance, and for the transitions between different timing requirements.

Extinction. Neither Scalar Timing Theory nor the Multiple Oscillator Model provides a basis for extinction. In both of these theories, reinforcement leads to an updating of temporal memory, but nonreinforcement has no consequences. This missing feature of these timing theories is probably due to a style of theory development that is to add no features to the model that are not required to handle the data under consideration. The Behavioral Theory of Timing deals with extinction qualitatively by assuming that reinforcement strengthens a behavior sequence and nonreinforcement weakens it. An explanation of the effect of nonreinforcement on some trials should account not only for the overall decrease in the probability of a response at a particular temporal interval, but also for various other effects. For example, it should account for an increased resistance to extinction after partial reinforcement, an increased speed of relearning after repeated blocks of reinforced and unreinforced responses, and a difference in the function relating response rate to time until scheduled reinforcement in the peak procedure (in which there is a mixture of nonreinforced trials) and in the fixed interval procedure (in which all trials are reinforced). The assumptions of the theory regarding the effect of nonreinforcement should account for such phenomena, both qualitatively and quantitatively.

Scale. The timing theories have been applied primarily in the range of seconds to minutes. Although there is no problem, in principle, in applying any of them to longer intervals (hours, days, or years), whether or not they do apply is an empirical problem. There may be one or more special durations (such as circadian) for which a special timing theory is required, or it may be that a general theory of timing is applicable over a very broad range of durations. There is a problem, in principle, in applying current timing theories to very short intervals. Animals readily discriminate between pairs of stimulus durations below 100 ms, even when perceived stimulus intensity is controlled (Fetterman & Killeen, 1988). A timing theory without a rapid pacemaker does not permit reliable discriminations of such stimulus durations.

Order of conditions. The representation of temporal memory of Scalar Timing Theory is a distribution of examples, without regard to order. But there is ample evidence that the order of conditions makes a difference. If an animal is given a block of trials with a 20–s peak procedure and then switched to a block of trials with a 10–s peak procedure, it responds at the end of training quite differently than if it is given these two blocks of trials in the other order (Meck, Komeily–Zadeh, & Church, 1984). And, undoubtedly, performance is quite different if food is primed on successive trials of a peak procedure in blocks or in a random order. The Multiple–oscillator Model of timing and the Behavioral Theory of Timing do make different predictions based upon the order of trials, but these predictions have not been tested quantitatively. Although the order of conditions has an effect on behavior even when the transition point between conditions is unpredictable, there is evidence that animals can anticipate some regular changes in the interevent intervals (Higa, Thaw & Staddon, 1993; Higa, Wynne, & Staddon, 1991). The postreinforcement pause of pigeons in a gradually changing set of fixed intervals may follow the most–recent interval, but it has sometimes been reported to anticipate the next interval. This is best seen by a comparison of an ascending and descending series to determine if the behavior is more closely related to the most recent experience or the next interval in the series. The temporal anticipation is presumably similar to the anticipation of food magnitude of rats in a runway given a monotonic series of amounts of food reward (Fountain, 1990; Fountain & Rowan, 1995; Hulse, 1978). Such behavior could be based on a rule, an extrapolation based upon the last two (or more) trials, or upon associations between successive trials. None of the three theories of timing now has a mechanism that permits the use of such sequential evidence.

Distributions of interevent intervals. The theories of timing may be readily applied to any distribution of interevent intervals. The distributions may differ in the mean, in the variance, or in other ways that may be referred to as the shape. Most studies of timing have used fixed interevent intervals; with comparisons made between different fixed interevent intervals. But the timing theories may be applied to distributions that have any particular shape, but differ only in the mean, or differ only in the variance. They may also be applied to distributions that have the same mean and variance, but differ only in shape. The theories may also be applied to procedures that do not explicitly refer to time, such as ratio schedules, although they may not make accurate predictions. Either the theories should specify the domain of interevent interval distributions to which they apply, or they should be tested under a much wider range of conditions than fixed intervals.

To Apply to All Descriptions of the Data

Just as theories of timing should apply to all procedures (all sequences of interevent intervals), they should also apply to all descriptions of the data. A theory that may fit one measure of performance may fail to fit others; a general theory of timing should describe the process that generates the times–of–occurrences of identified responses, and not just one description of that time series.

The time–series of events conveys all of the information, but it contains too much information to be comprehended as a whole. One tradition in psychology and other sciences is to present small samples of the data that are considered to be representative. In some cases, all samples can be ordered on some objective measure (such as mean response rate) and then the presented samples can be chosen based upon some rule (such as the first, second, and third quartile). Although such samples may be useful for illustrative purposes, to reduce the influence of random variation, and to apply the normal standards of statistical evidence, some combination of data is required. The data may be combined in many ways, normally first within subjects and then between subjects.

Performance as a function of time. Often performance is described as a function of time since an event (stimulus or reinforcement). For example, in the peak procedure, the mean response rate may be reported as a function of time since food. Alternatively, performance may be described as a function of time since a well-specified response measure. For example, the mean response rate may be reported as a function of time since the last response, or from some index of responding such as the time at which the animal enters a high response state (Schneider, 1969).

Description of indices of performance. Many different indices of responding may be defined. For example, in the peak procedure it is now standard to define an index of the time that the animal enters the high response state (start) and the time it leaves that state (stop), as well as the derived measures of the center (the mean time between start and stop) and the duration of the high state (the difference in the time between starting and stopping). The distributions of these indices may be examined, and a timing theory should be able to account for various descriptive measure of the centers, spreads, and forms of these distributions. The pattern of correlations between these indices on individual trials has been found to be particularly diagnostic of the sources of variability in a timing theory (Church, Meck, & Gibbon, 1994).

Residuals. An index, such as the time of starting the high state, may be related to an independent variable, such as the duration of the fixed interval. This can account for a high percentage of the variance – often well over 95%. But, the strong control of the behavior by the independent variable may be masking important features of the behavior that may be particularly revealing of the timing mechanism being employed. To observe these effects it is necessary to subtract out the effect of the independent variable, and to analyze the residuals. Theories of timing should be able to account for, not only the main effect, but also for the pattern of residuals. Such residuals may be an independent random variable according to one theory, but have a systematic pattern according to another.

Analysis of periodicities. Although most analyses of behavior have been done in the time domain, some regularities are best observed in the frequency domain. Spectral frequency analysis may be used to identify these periods. For example, if food is available at random time intervals (an exponential random waiting time distribution), rats do not respond at random intervals of time. Instead they have a tendency to respond at periodic intervals, one of which is about 30 s (Broadbent,

1994). This long period between bursts of responses is not evident from the examination of the distribution of interresponse intervals because a large preponderance of these are at a rate of about 2 per second and other processes lead to responses at other intervals. Although bursts of responses may often be spaced at approximately intervals of 30 s, it is extremely rare for two responses to be spaced at this interval. Thus, spectral analysis is a useful way to describe the data from a timed performance experiment, and the predictions of the timing theory should correspond with the observed results. The interpretation of the spectral analysis is simpler if the performance is in a steady state. If this stationarity assumption is not met, analysis should be done with a Short–time Fourier Analysis which provides a spectral analysis as a function of time (as a moving window). Other basis functions, such as those provided by a Wavelit analysis, may be applied to the data for similar purposes.

Pattern analysis. The time–of–occurrence of responses in a timed performance task may have pattern, and there is some evidence for self–similarity at many different time scales. The complexity of the pattern may be captured by the fractal dimension, such as the box dimension, and this measure may be related to experimental treatments. A timing theory should account for such measures of complexity of the pattern of responses at different scales of time under different experimental treatments (Broadbent, Maksik, & Church, 1995).

Speculations about the Next Generation of Timing Theories

The standards for description and quantitative evaluation will be increased

Current standards for the exposition of a timing theory are quite relaxed. Some writers combine formal description of the model with information about the history of the ideas, specific cases, alternative possibilities, etc. Current standards for the evaluation of a timing theory are also quite relaxed. In some cases, analysts are satisfied to capture the general functional relationship between some measure of duration and some measure of performance; in other cases parameters are adjusted to maximize the goodness of fit. The standards for the description and evaluation of timing theories are likely to be increased. The description of a theory should be sufficiently clear that two people, after reading the description and without further guidance, should be able to write computer programs in different languages to implement the theory. The programs should produce identical outputs for all inputs. The input to the theory should be the time of occurrence of each stimulus change and of each reinforcement, and nothing else. The output from the theory should be the identification of the type of response and its time–of–occurrence, and nothing else. The theory may either be based on a simulation or an explicit solution.

To use such a theory, it is necessary to prepare another, quite separate, functional programming unit that is a description of the procedure. The description of a procedure should be sufficiently clear that two people, after reading the description and without further guidance, should be able to write computer programs in different languages to implement the procedure. The programs should produce identical outputs for all inputs. The input to the description of the procedure should be the identification of each response, and its time–of–occurrence (i.e., the output from the theory); the output from the description of the procedure should be the time–of–occurrence of each stimulus change and each reinforcement (i.e., the input from the procedure).

To test the correspondence of the predictions of a theory with data, it is necessary to use the procedure to obtain experimental data from individual animals: the time–of–occurrence of each identified response. Then, for any procedure, there are two streams of times of identified responses, one from the theory and one from the experiment. A comparison of the behavior of the animal with the output of the theory provides the basis for evaluation of the fit of the model. A good theory will apply to all procedures in a well–specified domain, and it should apply to all descriptions of the data. (See the previous section regarding the range of procedures and of descriptions of the data that may be used.) There are no absolute standards for the evaluation of a theory although goodness–of–fit, generality, and few parameters are desirable. For evaluation it is useful to compare two or more theories.

There are three steps in a quantitative analysis of a model: (1) the procedure must be applied to the model to generate predicted output (Procedure → Model → Predicted output), (2) the procedure must be applied to animals to generate observed output (Procedure → Animal → Observed output), and (3) the predicted and observed output should be compared.

Neurobiological evidence will be integrated

Timing theories have been developed primarily to account for behavioral data; they have not been closely related to neurobiological evidence. But the processes postulated by any theory must be biologically realizable. Many experimental studies have identified biological manipulations that affect timing behavior. Although such studies provide information about the relationship between brain and behavior, the results of most of these studies have not been well integrated into timing theory. One exception was Meck's (1983) discovery of pharmacological manipulations that led to durations being systematically overestimated or underestimated, and which led to the introduction of a variable memory storage constant into scalar timing theory (Gibbon, Church, & Meck, 1984). Among behavioral neuroscientists there is enormous interest in the functions of the brain, and timing is one of the important functions of the brain. With the rapid growth of knowledge and techniques in anatomy, electrophysiology, pharmacology, and other fields relevant to behavioral neuroscience, it is likely that quantitative neurobiological facts as well as quantitative behavioral facts will be used to constrain timing theories. With the development of brain scanning techniques with improved spatial and temporal resolution, and improved analysis techniques for the EEG, neurobiological evidence from human subjects may also be incorporated in a timing theory. A formal theory of timing that is constrained by neurobiological and behavioral evidence is much more likely to be correct than one that is constrained by only one set of facts. The parameters of such a theory will be identified with both a concept in a cognitive model of the process (such as clock speed or temporal representation) and with a concept in the neurobiological model of the process (such as activity of some neurotransmitter in some brain structure, or some particular activity of single neurons in some brain structure). This provides a scientific approach to the mind–body problem, the corresponding representations of cognitive and biological processes that can be formally described by the same equations.

Current theories will be modified

New experimental evidence, based on different timing procedures or different measures of performance, will identify additional deficiencies in current timing theories. This will lead to modifications of the theories. Each of them has a few core ideas that are the essence of the theory, and many ideas that are readily adjustable based upon the experimental evidence. For example, Scalar Timing Theory must use a pacemaker–accumulator system for the representation of time, a distribution of examples for memory, a sample from memory for retrieval, and a ratio decision rule. But it could add more pacemakers or change the sampling rule without changing the essence of the theory. The Multiple Oscillator System must use multiple oscillators with a vector of phases for the representation of time, a single matrix for memory, and a similarity decision rule. But it could use more than the half–phase information from an oscillator, or it could use entrainable oscillators without changing the essence of the theory. The Behavioral Theory of Timing must use a pacemaker to drive states that produce observable behaviors for the representation of time, strengths of observable behaviors for memory, and a decision rule based on these strengths. But anything else could be changed without changing the essence of the theory. If a theory needs to be modified often because of new experimental evidence, if each modification adds to the complexity of the model, or if the modified form does not continue to predict data from previous experiments, the theory will be abandoned. A theory that correctly predicts new and, especially, surprising results without further modification is likely to be used extensively. The best way to obtain this ideal probably is not to continue to modify existing theories but to develop a new theory of timing based upon the best features of current models plus some new ideas.

A new theory will be developed

Current timing theories are packages of ideas about the perceptual representation of time, the nature of temporal memory, and decision processes. A new theory undoubtedly will emerge that will not be constrained by the particular combination of ideas found in any of the current theories, or by the type of mathematics that has been applied. New combinations of ideas will be seriously considered, and alternative proposals for perception, memory, and decision will be considered. Any model of timing must have something that changes in a regular way with physical time, but there are countless possibilities: For example, there may be a single pacemaker–accumulator system; there may be a single periodic system with a phase that can be read; there may be multiple pacemaker–accumulator systems or multiple periodic systems; and multiple systems may be independent or coupled. Memory for time may be composed of examples organized as a distribution of limited or unlimited capacity or as an ordered list of limited or unlimited capacity; or the examples could update a memory that is represented as a scalar, a vector, or a matrix. The distinction between temporal memory as a distribution of examples or a matrix (a prototypical pattern) is fundamental, and it is not necessarily related to the perceptual representation of time.

Although one cannot anticipate what the new theory will be, one can anticipate what it will accomplish: It will account for the results of all (or most) new procedures, and it will account for all (or most) new descriptive measures of time perception and timed performance.

Preparation of this manuscript was supported from a grant from the National Institute of Mental Health (RO1–MH44234). Correspondence concerning this article should be addressed to Russell M. Church, Department of Psychology, Box 1853, Brown University, Providence, Rhode Island 02912.

References

Allan, L. G., & Gibbon, J. (1991). Human bisection at the geometric mean. *Learning and Motivation, 22,* 39–58.

Anderson, J. A., Silverstein, J. W., Ritz, S. A., & Jones, R. S. (1977). Distinctive features, categorical perception, and probability learning: Some applications of a neural model. *Psychological Review, 84,* 413–451.

Boring, E. G. (1950). *A history of experimental psychology.* New York: Appleton–Century–Crofts.

Broadbent, H. A. (1994). Periodic behavior in a random environment. *Journal of Experimental Psychology: Animal Behavior Processes, 20,* 156–175.

Broadbent, H. A., Maksik, Y. A., & Church, R. M. (1995). A fractal analysis of random interval data. Paper presented at the Meeting of the Society for the Quantitative Analysis of Behavior, Washington, D. C., May 27, 1995.

Bush, R. R., & Mosteller, F. (1955). *Stochastic models for learning.* New York: Wiley.

Catania, A. C. (1970). Reinforcement schedules and psychophysical judgments: A study of some temporal properties of behavior. In W. N. Schoenfeld (Ed.). *The theory of reinforcement schedules* (pp. 1–42). New York: Appleton–Century–Crofts.

Cheng, K., & Westwood, R. (1993). Analysis of single trials in pigeons' timing performance. *Journal of Experimental Psychology: Animal Behavior Processes, 19,* 56–67.

Church, R. M., & Broadbent, H. A. (1990). Alternative representations of time, number, and rate. *Cognition, 37,* 55–81.

Church, R. M., & Gibbon, J. (1982). Temporal generalization. *Journal of Experimental Psychology: Animal Behavior Processes, 8*, 165–186.
Church, R. M., Meck, W. H., & Gibbon, J. (1994). Application of scalar timing theory to individual trials. *Journal of Experimental Psychology: Animal Behavior Processes, 20*, 135–155.
Creelman, C. D. (1962). Human discrimination of auditory duration. *Journal of the Acoustical Society of America, 34*, 582–593.
Fetterman, J. G., & Killeen, P. R. (1992). Time discrimination in Columba livia and Homo sapiens. *Journal of Experimental Psychology: Animal Behavior Processes, 18*, 80–94.
Fountain, S. B. (1990). Rule abstraction, item memory, and chunking in rat serial–pattern tracking. *Journal of Experimental Psychology: Animal Behavior Processes, 16*, 96–105.
Fountain, S. B. (1995). Sensitivity to violations of "run" and "trill" structures in rat serial–pattern learning. *Journal of Experimental Psychology: Animal Behavior Processes, 21*, 78–81.
Fraisse, P. (1963). *The Psychology of time.* New York: Harper and Row.
Gallistel, C. R. (1990). *The organization of learning.* Cambridge, MA: MIT Press.
Gibbon, J. (1991). Origins of scalar timing theory. *Learning and Motivation, 22*, 3–38.
Gibbon, J. (1992). Ubiquity of scalar timing with a Poisson clock. *Journal of Mathematical Psychology, 36*, 283–293.
Gibbon, J. (1977). Scalar expectancy theory and Weber's law in animal timing. *Psychological Review, 84*, 279–325.
Gibbon, J., & Allan, L., eds. (1984). *Timing and time perception.* New York: New York Academy of Sciences.
Gibbon, J., & Church, R. M. (1992). Comparison of variance and covariance patterns in parallel and serial theories of timing. *Journal of the Experimental Analysis of Behavior, 57*, 393–406.
Gibbon, J., & Church, R. M. (1990). Representation of time. Cognition, 37, 23–54.
Gibbon, J., & Church, R. M. (1981). Time left: Linear versus logarithmic subjective time. *Journal of Experimental Psychology: Animal Behavior Processes, 7*, 87–108.
Gibbon, J., Church, R. M., Fairhurst, S., & Kacelnik, A. (1988). Scalar expectancy theory and choice between delayed rewards. *Psychological Review, 95*, 102–114.
Gibbon, J., Church, R. M., & Meck, W. H. (1984). Scalar timing in memory. In J. Gibbon & L. Allan (Eds.), *Timing and time perception* (423, pp. 52–77). New York Academy of Sciences.
Higa, J. J., Thaw, J. M., & Staddon, J. E. R. (1993). Pigeons' wait–time responses to transitions in interfood–interval duration: Another look at cyclic schedule performance. *Journal of the Experimental Analysis of Behavior, 59*, 529–541.
Higa, J. J., Wynne, C. D. L.,, & Staddon, J. E. R. (1991). Dynamics of time

discrimination. *Journal of Experimental Psychology: Animal Behavior Processes, 17,* 281–291.

Hulse, S. H. (1978). Cognitive structure and serial pattern learning by animals. In S. W. Hulse, H. Fowler, & W. K. Honig (Eds.). *Cognitive processes in animal behavior.* Hillsdale, NJ: Erlbaum. pp. 311–340.

James, W. (1890). *The principles of psychology. Vol. 1.* London: Macmillan.

Killeen, P. R., & Fetterman, J. G. (1988). A behavioral theory of timing. *Psychological Review, 95,* 274–295.

Lejeune, H., & Wearden, J. H. (1991). The comparative psychology of fixed–interval responding: Some quantitative analyses. *Learning and Motivation, 22,* 84–111.

Machado, A. (in press) Learning the temporal dynamics of behavior. *Psychological Review.*

Meck, W. H. (1983). Selective adjustment of the speed of internal clock and memory processes. *Journal of Experimental Psychology: Animal Behavior Processes, 9,* 171–201.

Meck, W. H., & Church, R. M. (1987). Cholinergic modulation of the content of temporal memory. *Behavioral Neuroscience, 101,* 457–464.

Meck, W. H., & Church, R. M. (1984). Simultaneous temporal processing. *Journal of Experimental Psychology: Animal Behavior Processes, 10,* 1–29.

Meck, W. H., Church, R. M., & Gibbon, J. (1985). Temporal integration in duration and number discrimination. *Journal of Experimental Psychology: Animal Behavior Processes, 11,* 591–597.

Meck, W. H., Church, R. M., & Olton, D. (1984). Hippocampus, time, and memory. *Behavioral Neuroscience, 98,* 3–22.

Meck, W. H., Church, R. M., & Wenk, G. L. (1986). Arginine vasopressin inoculates against age–related increases in sodium–dependent high affinity choline uptake and discrepancies in the content of temporal memory. *European Journal of Pharmacology, 130,* 327–331.

Meck, W. H., Church, R. M., Wenk, G. L., & Olton, D. S. (1987). Nucleus basalis magnocellularis and medial septal area lesions differentially impair temporal memory. *Journal of Neuroscience, 7,* 3505–3511.

Meck, W. H., Komeily–Zadeh, F., & Church, R. M. (1984). Two–step acquisition: Modification of an internal clock's criterion. *Journal of Experimental Psychology: Animal Behavior Processes, 10,* 297–306.

Richelle, M., & Lejeune, H. (1980). *Time in animal behaviour.* Oxford, UK: Pergamon.

Roberts, S. (1981). Isolation of an internal clock. *Journal of Experimental Psychology: Animal Behavior Processes, 7,* 242–268.

Roberts, S., & Church, R. M. (1978). Control of an internal clock. *Journal of Experimental Psychology: Animal Behavior Processes, 4,* 318–337.

Schneider, B. A. (1969). A two–state analysis of fixed–interval responding in the pigeon. *Journal of the Experimental Analysis of Behavior, 12,* 677–687.

Wearden, J. H. (1991). Human performance on an analogue of an interval bisection task. *The Quarterly Journal of Experimental Psychology, 43B*, 59–81.

Wearden, J. H. (1992). Temporal generalization in humans. *Journal of Experimental Psychology: Animal Behavior Processes, 18*, 134–144.

Wearden, J. H., & Towse, J. N. (1994). Temporal generalization in humans: Three further studies. *Behavioral Processes, 32*, 247–264.

Time and Behaviour: Psychological and Neurobehavioural Analyses
C.M. Bradshaw and E. Szabadi (Editors)

CHAPTER 3

Time's Causes

Peter R. Killeen, J. Gregor Fetterman

& Lewis A. Bizo

What is time? St. Augustine knew: "I know what time is", he said, "but if someone asks me, I cannot tell him" (Landes, 1983, p. 1). Not much help. It is the business of scientists to tell, and another ancient philosopher tells us how to tell: Aristotle sought to understand phenomena – and communicate that knowledge – by identifying their four "[be]causes", which he called *material, final, efficient*, and *formal*. We have interpreted these as questions about *what* (description/ definition and substrate), *why* (function), *how* (mechanism), and *like* (analogs and models). These four causes organize our analysis of time and timing.

What

Time is not a fourth dimension – although it may be measured on one. It is not itself a cause – even though effects unfold in time. Time was not discovered; it was invented. The conceptions of time that have been invented by physicists have been adopted by society at large, so that there is a good correlation between the physicist's time – *SI time* – and that on our clocks – *Big Ben Time*. Both times are measured on cyclic interval scales: Each time they pass 24:00 they increment a count on a calendar – another cyclic interval scale – and reset to 00:00. The cyclicity of temporal experience is studied in detail elsewhere in this volume (see chapters by Gibbon et al., Carr & Wilkie, and Campbell). In this chapter, it is elapsed time – time accumulated from a marker – that we study. By specifying an origin scientists create a ratio scale of elapsed time. What is the relation between this scale and the organism's scale of time? Do animals have a scale of time? Do animals *time* in any sense similar to the activities we represent with that verb? If so, what kind of clock do they use? What would we take as evidence for timing in animals? What, for that matter, do we take as evidence of any activity in another organism? Whether Fido can catch a Frisbee the public can readily decide; what he thinks about a Frisbee is a more contentious issue. In both cases, an operational definition facilitates productive discussions. Consider this operational definition for timing: *An entity is "timing" if our use of a clock helps us to predict its behavior, or if its behavior predicts the reading on our clock.*

This definition specifies a necessary condition for timing, but you may conclude that it is insufficient because you can imagine many scenarios that satisfy the definition but might not qualify. Some activities such as Fido's arising and eating breakfast are highly regular in time, but would not be considered acts of timing: Although their occurrence is predictable using our clock, it is even more predictable from the sunrise, or from his master's alarm clock or coffee machine. We are wont to invoke timing only when we cannot identify other proximal cues (e.g., external clocks) that can better predict the behavior. We may say that grunions time their egg–laying to coincide with a spring tide, but it is a combination of sun and moon that do the timing for them. But this puts too fine a point on the matter. We can give you a stopwatch and ask you to time the next lap of a race without contradiction or miscommunication; it is just that psychologists do not include that kind of timing in the domain of their study.

Thus our definition of timing is qualified as the subset of periodic activities not obviously under the control of other stimuli. We say that *an organism is timing if our clock is a better predictor of its behavior than any other stimulus we can identify.* Our extended operational definition is a kind of stepwise regression analysis: If we find better predictors, we note their control – then remove them. Humans have invented many accessories that tell them the time accurately, but in an experiment they will be asked to put them away. They are also told to refrain from counting. Sometimes they are caught unawares and asked how much time has elapsed since, say, they sat down. These practices are aimed at forcing subjects to time in the "purest" sense (i.e., most naively: without prostheses). Timing, then, is what we call periodic behavior when all of the external stimuli that might improve it have been removed, and, often, when covert behaviors that might control it have been discouraged.

Often an animal is only said to be timing while it is involved in activities that in the past have been reinforced for improving temporal judgments. If we ask you how long you have been reading this chapter, you can give us an estimate, but we hope you weren't timing as you read (why would you have been timing?). You were "timing" only after we asked the question, and that scramble to estimate elapsed time is different than timing how long it takes to finish the rest of this chapter, which you might now initiate if you suspect another psychologist's trick. These different kinds of timing have important similarities, in particular the effects of task demands on accuracy and the effects of differential payoffs on bias. Whether you engage them depends on what reinforces your behavior – depends on what your goals are, and what you expect your environment to demand of you. Answers to *why* questions – why animals time or why they give a particular temporal estimate in a particular situation – constitute our second Aristotelian question.

Why

"By looking at how a clock is built, we can explain [how] it keeps good time, but not why keeping time is important" (Skinner, 1989, p. 18). *Why* questions typically address functional issues, that is, purposes. Timing clearly serves one very basic function long recognized as a fundamental aspect of the learning process – anticipation/prediction. Sometimes important events are associated with exteroceptive cues –

a rustle in the grass, the cast of a shadow, a whiff of smoke, all pluck ancient chords of alertness. In these cases the cues may function as Pavlovian CSs, triggering sympathetic reactions (e.g., Hollis, 1983).

> Without a signal the animal would still be forced to wait in every case for the stimulus to arrive before beginning to meet it. The veil of the future would hang just before its eyes. Nature began long ago to push back the veil. Foresight proved to possess high–survival value, and conditioning is the means by which foresight is achieved. (Culler, 1938, p 136)

In many other cases the only signal is the temporal regularity of the event itself, or its delayed relation to a predictor, as found in temporal conditioning (Pavlov, 1927). When significant unsignalled events occur at regular intervals (e.g., the availability of food; the appearance of predators or potential mates), it is to an organism's advantage to learn the periodicity. Such learning allows organisms to prepare for events by premonitory behaviors. Pigeons and other birds flock to park benches in advance of the lunch hour – anticipating the arrival of food–bearing *Homo sapiens* (Saksida & Wilkie, 1994; Wilkie, 1995), who flock to the park benches at the lunch hour. Nectar–feeding birds anticipate the time taken for replenishment of the nectar source and revisit flower patches when sufficient time has passed to make it worthwhile (see, e.g., Kamil, Krebs, & Pulliam, 1987). Parasitoid wasps use the time it takes them to circumambulate potential hosts as a measure of its volume (Schmidt & Smith, 1987).

Doing the right thing at the right time is an essential aspect of a variety of skilled performances, exemplified in athletics and the performing arts. Coordinated timing also underlies more mundane but important activities such as walking and driving a car. Temporal coordination also may involve the activities of other animals as, for instance, with mating rituals displayed by many birds (e.g., Huxley, 1914), the timing of vocalizations by frogs (Grafe, 1996), and cooperative hunting where the timed coordination of the behaviors of different individuals is critical to the success of the foraging enterprise.

Keeping together in time has served as a mechanism for enhancing group solidarity and military efficacy over the ages (McNeill, 1995). Whereas the "origin" for time may be in the future (anticipatory timing), or may be set by a pattern of responses (coordinated timing), animals also may be under the control of the time elapsed from some prior marker (retrospective timing). For instance, a forager may decide to leave one patch and travel to another when its patch residence time exceeds some criterion value, or when the time since the last prey encounter exceeds a criterion (e.g., Brunner, Kacelnik, & Gibbon, 1992; Jones & Davison, 1995; Shettleworth & Plowright, 1992). Whereas functional questions address the goals around which behavior is organised – both the ultimate causes of evolutionary shaping, and the proximal causes of reinforcement – *how* questions address the machinery that accomplishes the process.

How

This Aristotelian question addresses the *mechanisms* of time perception. Several of the papers in this collection (chapters by Hinton & Meck, O'Boyle, and Al–Ruwaitea et al.) are focused on mechanisms conceived as neural substrates. We may also posit cognitive or behavioral treatments, such as information processing models of an internal clock, or the mediational role of adjunctive behaviors. These are not descriptions of alternative mechanisms, however; they are alternative treatments of mechanism couched at different levels of description. Animals may time in different ways – they may time by keeping track of external events, or of their ongoing behavior; by estimating the amount of information maintained in a memory store; by reading the hands on an internal clock or set of clocks; by deconvoluting a matrix; or by tapping off information off of a neuronal oscillator (Selverston & Moulins, 1985). But except for the last item these are not so much treatments of mechanism as they are alternative models of the same (unknown) mechanism. When the mechanisms are covert, *How* questions involving machinery blend imperceptibly into *Like* questions involving hypothetical analogs – while the models employed blend imperceptibly from alternative representations of observed processes to hypotheses about inferred structures and processes. Whereas the Central Nervous System comprises the mechanism, the Conceptual Nervous System constitutes an analogy, a likeness.

Like

Many models of timing posit a special–purpose pacemaker–counter mechanism (e.g., Treisman, 1963; Church, Broadbent, & Gibbon, 1992). In these systems one component, the pacemaker, emits pulses and another, the counter, accumulates the pulses. Elapsed time is represented as a tally of pulses in the count register, and temporal judgments are based on a reading of this tally (see chapter by Roberts). Other models invoke mechanisms that are less specialized and serve a variety of functions, including timing. For instance, memory models (e.g., Ornstein, 1969) stipulate that judged duration depends on the amount of memory occupied by events that occur during an interval. The more information contained in the memory store, the longer the estimated duration. Information processing models (e.g., Hicks, Miller & Kinsbourne, 1976; Sawyer, Meyers, & Huser, 1994) study how perceived duration depends on the amount of information processed ("cognitive effort") over some amount of time (Zakay, 1993). Tasks that elicit intensive cognitive activity (e.g., solving difficult anagrams) are judged of longer duration than those that evoke low levels of activity (e.g., proofreading).

In the remainder of this chapter we shall focus on the proximate *whys* and *likes* of timing. In addressing the question of why – contingencies – we utilize the Theory of Signal Detectability (*TSD*), a model of decision–making that explicitly incorporates the costs and benefits attendant upon perceptual responses. In our discussion of likes – models – we survey develop a general model for psychometric functions and then consider in some detail a favorite theory. We begin with a brief overview of timing methodology.

Methodologies

Different situations call for different strategies for estimating elapsed time, and researchers have tried to capture these differences through various taxonomies (e.g., Killeen & Fetterman, 1988; Stubbs, 1979; Woodrow, 1951). One of our goals in this chapter is to move toward a standard terminology based on decision theory. As Zeiler (1991) has noted, to the subject, procedures we call "timing experiments" are merely another kind of puzzle box to find a way out of as soon as possible. If we have been successful in arranging the contingencies, the best way for the subject to achieve its goals is to estimate SI time as accurately as it can. The different methodologies are differentially successful in accomplishing this goal.

Scheduling stimuli

In the human timing literature, when participants are asked when some historical event occurred it is called retrospective timing. We may designate this as *MIMQR*, signifying an Interval, Marked at its start and at its end, followed by a Query about the elapsed time, and finally by a Response from the subject. When humans are told they will be asked to estimate the duration of an interval, it is called prospective timing in that literature (Brown, 1985). We may designate it as *QMIMR*. It is difficult to conduct such experiments with nonverbal subjects, as the conditioning necessary to get them to understand the "instructions" and attend to the marker prepares them to engage in ongoing timing behaviors – that is, it gives the game away so they are always "timing" through the interval.   Because this chapter emphasizes research with non–verbal subjects, we shall not bother to write the *Q*, which belongs at the start of all our string representations of timing methodology for animals.

Other taxonomies stress the method by which subjects communicate their estimate. For example, participants may be asked to assign a numerical value to the duration of some event (*estimation: MIMR*), to engage in some activity (e.g., depress a key) for a verbally–specified duration (*production: RIR*), or to engage in an activity for a time equal to the duration of a prior event, (*reproduction: MIM RIR*). Killeen and Fetterman (1988) described a taxonomy of timing based on the subject's location in time with respect to the signaled duration. When a subject judges the duration of an elapsed interval the timing is retrospective (*MIMR*); when it responds during an elapsing duration the timing is immediate (*MIR*), and when it chooses between forthcoming delayed outcomes the choice is prospective (*RIM*; cf. the use of that name in the human timing literatures). Figure 1 provides a picture of Killeen and Fetterman's taxonomy in which the subject is located at the origin of the graph. In the top, the subject must choose between forthcoming intervals as a function of the distance of their endpoints, and is always instructed – either by the experimenter or by appetite – to choose the closest.

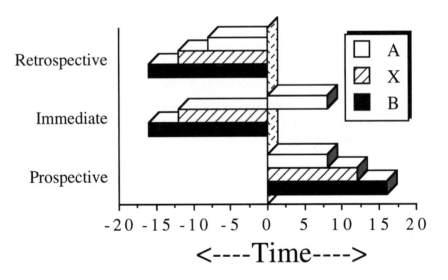

Figure 1. The three types of timing noted by Killeen & Fetterman (1988). The response is located at *t* = 0, and the Markers are located at the start and end of the intervals. Also shown is the common design of comparing an intermediate duration (X) with well–trained stimuli (A and B).

Examples of prospective timing are concurrent–chain reinforcement schedules and self–control paradigms (e.g., Gibbon, Church, Fairhurst, & Kacelnik, 1988; Logue, 1988). In immediate timing the origin floats from the start of the stimulus to the end, and the subject must continually estimate whether it is in the first or second portion of the interval. Examples of this paradigm are traditional *FI* schedules and the *peak procedure*. In retrospective timing, the ends of the intervals always coincide with the opportunity to estimate, and the subject must judge the relative distance back to the start of the interval. If the experimenter lines up the intervals at the start, then the start is irrelevant; most subjects would choose to be more precise by estimating which interval terminated before the end of the trial. It is possible, however, to lay the intervals end–to–end, placing the point of judgment at the end of the second interval. Subjects must then say whether the first or second interval was longer (e.g., Fetterman & Dreyfus, 1986). This is a retrospective *pair–comparison* task (*MIMIMR*).

Scheduling responses

Durations past, present, and future, as represented in Figure 1, constitute an important dimension of temporal experience. But this scheme does not completely represent various ways in which organisms may respond to their location in the temporal stream. (a) The opportunity to estimate may occur only once, several times, or continuously. In the first case, animals may classify an interval as "short" or "long" by emitting different discrete responses at the beginning (prospective: *RIM*) or end (retrospective: *MIMR*) of an interval. Alternatively, animals may continually estimate an interval by responding to alternatives offered throughout a duration (*MIMR MIMR...MIMR*), or continuously by responding ad libitum (*MRR...RM*), just as students make observing responses to their watches during a boring lecture. (b) In some designs animals are afforded a choice as to whether the interval was short or long (the *choice* procedure: $R \in \{S,L\}$). We signify that the animal has a choice by subscripting the symbol R: R_j. In other designs they must signify times less or greater than a target interval by withholding or emitting a single instrumental response (the *go/no-go* procedure: $R \in \{L,\emptyset\}$). In the following, we briefly describe some standard methodologies in terms of the *discrete* versus *continuous*, and *choice* versus *go/no-go* nomenclatures. In the *choice* designs, the organism is given the opportunity to make a symmetrical yes or no response, whereas in *go/no-go* designs, we infer a "no" response from the absence of a "yes" response, leaving it to the experimenter to determine how long a period of time without a "yes" is tantamount to a "no". The classic procedure of *n-alternative forced choice* ($MIMIMR_j$) has only rarely been used in the study of timing in non-human animals, in part because with time as the stimulus substantial delays must occur between the first and second stimulus, and these may be confounded with the length of the second stimulus. Some nice results have been achieved (e.g., Fetterman & Dreyfus, 1986, 1987), but this paradigm will not be considered further here.

Scheduling reinforcers

Determining the observer's position with respect to the intervals (Figure 1) gives us a better sense of the stimuli that the subject must evaluate. But to understand the consequences of those evaluations we must determine the value the experimenter assigns to correct and erroneous decisions. The best paradigm with which to accomplish this is decision theory as it has been developed for detection and discrimination experiments.

Signal Detection Theory

Two Ways To Be Right; Two Ways To Be Wrong

In any type of discrimination, including temporal discriminations, we ask subjects to choose one or more alternatives corresponding to the state of the world as they sense it. A typical map of this paradigm is given in Figure 2. We have checked off the first cell, indicating that the subject has stated "A" and that the experimenter agrees.

		Experimenter's statement	
		A	−A
Subject's statement	A	X	
	−A		

Figure 2. A 2x2 contingency matrix showing the relation of the subject's categorization of the world to the experimenter's categorization.

Truth propositions are usually treated as binary: Either A or −A, and not both. Such a binary classification is convenient even though the world is usually more complex than two alternatives allow. Even in binary cases there is often an element of arbitrariness in our allocation of data to cells: We conventionally arrogate perfect knowledge of the state of the world to the experimenter, even though the experimenter's knowledge is also a fallible proposition, based on different ("better") data than those available to the subject. For example, in training radiologists to identify pre−cancerous images, the "experimenter"/ trainer may use historical images from patients whose eventual prognosis was cancer or not cancer. But at the time the image was taken it is possible that some of the suspected sites were pre−cancerous and went through remission, leading the trainer to misclassify them. Nonetheless, representation in terms of this matrix is important because it encourages us to generate explicit payoffs for behavior as it falls in each of the cells, and is therefore the first step in analysis of discriminative performance.

Analyses of the Matrix

How do we analyze the data from the matrix? We could use one of the many correlation coefficients that have been designed for categorical data, such as Phi or Chi−square, or employ indices such as the amount of information transmitted between the subject and the experimenter. But all of these statistics reduce the four cells to a single number, and thus reduce the amount of information conveyed from the subject to the analyst and thence to the reader. Because there are two ways to be right in tasks such as that shown in Figure 2, and two ways to be wrong, a single index can never tell the whole story. Reliance on a single index is a common mistake in analyses of discriminations. A test may be 100% accurate when it identifies a disease, a lie, or a Dali Lama. But at the same time it may be completely unable to reject false claimants, with all the attendant costs: If all men are deemed ill, only the drug companies will prosper. No measures of temporal discrimination, nor of any other discrimination, are adequate if they do not communicate the two degrees of freedom inherent in the matrix of Figure 2. There are different ways to accomplish this, but all require complete data. The classic treatment of Figure 2 is provided by the theory of signal detectability. That theory encourages specification of the differential payoffs that may be associated with each of the cells in Figure 2, recognizes the different ways of being correct or incorrect entailed by that matrix, and encourages the plotting of the probability of correct affirmatives (*hits*) against the probability of incorrect affirmatives (*false alarms*) to yield a Receiver Operator Characteristic (*ROC*) curve. As a theory, it provides a context for many specific models that make different assumptions about the representation of percepts and their variability (e.g., Egan, 1975). One standard model represents the signal as a random variable and noise as another random variable with a different mean. The subject evaluates the percept and judges which of the stimuli was most likely to have given rise to the sample on that trial. If that likelihood exceeds a criterion, the subject affirms the presence of a signal, and otherwise negates. If stimuli gave rise to precise and perfectly replicable percepts, and the criterion were likewise invariant, the subject's performance would be perfect. But sometimes a stimulus seems a little more or less than it really is, so that over trials it gives rise to a distribution of percepts. Thurstone (1927; Guilford, 1954) called these distributions "discriminal dispersions". Figure 3 (continuous curves) gives a picture of the discriminal dispersions on a "metathetic" or qualitative continuum (Stevens & Galanter, 1957) such as pitch, where the standard deviation is constant across the continuum. An unbiased subject trained on the 500 and 1000 Hz tones will partition the continuum at 750 Hz, with percepts lower than that called "low" and all others "high".

Three models of discrimination. Rather than replicate portions of standard reviews (see Macmillan and Creelman, 1991), we develop particular models by analyzing the classic timing paradigms in terms of the underlying discriminal dispersions. Although based on "detection" theory, we will be dealing with discrimination, rather than detection, as it is impossible to present a time interval so brief it cannot be detected without confounding the task with presence/absence of the stimulus. Our foundational model is the standard described above, which we apply to discrimination of the training stimuli, and of occasionally–presented test ("probe") stimuli falling somewhere between them. Note that each stimulus has its own dispersion, which we use to evaluate how often any particular stimulus will exceed a criterion boundary. If we look at the equations of prediction, we are always evaluating either a normal function or – for convenience – its logistic approximation, $\{1+\exp[(x-\mu)/(0.55\sigma)]\}^{-1}$. Mu ($\mu$) is the mean of the discriminal dispersion for the test stimulus in question, σ is its standard deviation, and x is the location of the subject's criterion (the *PSE*; in Figure 3, it is located at 750 Hz). This is the basis of the traditional *TSD* treatment of such discriminations.

Notice that this is equivalent to another, quite different model: The stimuli are known exactly, but there is fluctuation in the criterion. In this case, μ becomes the value of the criterion, x the value of the stimulus, and σ the standard deviation of the fluctuations in the criterion. This model is represented by the dashed line in Figure 3. The cumulative distribution function of that dashed line forms the *psychometric function*. This is perhaps the simpler model to think about, and was treated by Thurstone as a model for categorical perception. Notice that, since these dispersions are not observed, the models – variance in the percepts but not the criterion, or variance in the criterion but not the percepts – are empirically indistinguishable on the basis of simple psychometric functions.

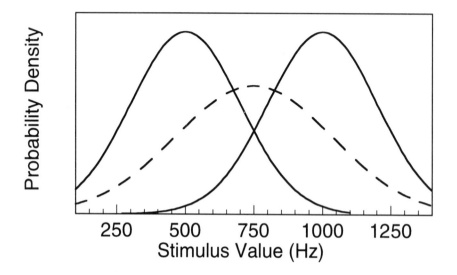

Figure 3. Distributions of perceptual effects (*percepts*) of stimuli located at 500 and 1000 Hz. These distributions are called "discriminal dispersions", and have equal variance on metathetic continua. Also shown is another dispersion, located at 750 and having a variance equal to the sum of the constituent variances.

Both of the above are equivalent to a third model in which there is variation in the subject's perception of both the stimuli and the criterion, with the σ then representing the square root of the sums of those constituent variances (picture Figure 3 with some variance in both the perception of stimuli and in the location of the criterion). *As long as we are dealing with a continuum,* where all standard deviations are constant, *these models are empirically equivalent.* With this preparation, let us review some of the standard paradigms used in the study of timing.

Typical Paradigms

Descriptions of the basic paradigms on these dimensions are

summarized in Figure 4 and Table 1. The key differences among the procedures is whether the opportunities to respond are discrete or continuous, and whether there are explicit opportunities to make symmetric responses (*Choice*, or *Yes/No*) or only the opportunity to make one response (*Go/No-Go*). In the latter, trials end only when a response has collected a reinforcer, and that is the only cell of Figure 2 for which the experimenter explicitly establishes a payoff. Responses are recorded for only half (or less) of the matrix, leaving us in ignorance of the other half. Such half–asked questions leave the experimenter unable to administer symmetric payoffs and thus out of control of the subject's bias. In reviewing the following common timing paradigms refer to Figure 4 and consider the payoffs assigned to each of the cells by the experimenter.

	Full Matrix (Choice)	**Half Matrix (Go/No–Go)**
Discrete	Constant stimuli Pair comparison	Temporal Generalization
Continuous	Free–Operant Psychophysical Choice Time–Left	Fixed–Interval Peak Procedure

Figure 4. Varieties of timing experiments parsed by explicit versus implicit treatment of "Short" responses, and by trial–paced versus continuous response availability.

Discrete Choice Procedures

1. Method of Constant Stimuli. Fetterman and Killeen (1992) used a retrospective task (*MIMR_j*) in which pigeons' responses to one alternative were reinforced following a short duration of keylight offset, and responses to a second alternative were reinforced after a longer period of darkness (also see McCarthy & Davison, 1980). In Figure 2, A becomes $t < t^*$, and −A becomes $t \geq t^*$, where t is the value of the stimulus and t^* is the category boundary established by the experimenter to distinguish the payoffs. Correct classifications (cells A, A and −A, −A) earned the pigeons three seconds access to mixed grain, whereas incorrect classifications (cells A, −A and −A, A) led to a 10 s inter−trial interval.

In one condition of Fetterman and Killeen's (1992) study, the birds were tested with stimuli intermediate to the short and long signal values (*method of constant stimuli*) and in another the value of the long signal was adjusted to maintain accuracy at 75% correct (*staircase method*). In both versions signal durations were varied across a range of three orders of magnitude (0.1 s to 10 s). In the method of constant stimuli the signal pairs consisted of 50 & 100 ms, 500 & 1000 ms, and 5000 & 10000 ms. Under the staircase procedure the standard (short) signal was varied from 25 ms to 5000 ms, with testing at many intermediate values.

Measurement of discriminability (the standard deviation of the discriminations) at multiple points provided a "Weber Function" relating discriminability to time (see Figure 5). These data show that absolute discriminability is constant for very short time intervals, but above 1/5 s the standard deviation becomes proportional to the duration of the intervals. Weber's Law − proportionality of the standard deviation to the mean − holds for moderate to large stimuli here as it does approximately on most other intensive dimensions. The curves through the data are given by the equation:

$$\sigma_t = \sqrt{(w\,t)^2 + pt + c^2} \tag{1}$$

This equation is derived by Killeen and Weiss (1987) as a general model for the Weber Function where that is generated by any sort of pacemaker−counter system. The parameter w is known as the *Weber fraction*, and Killeen and Weiss showed that its value is determined by one form of error in the counter alone. For long durations wt dominates Equation 1, so that changes in variability become proportional to changes in t: $\sigma_t \approx wt$. This is Weber's law, or scalar timing. The Weber fraction for the data from the staircase method shown in Figure 5 was 0.22, whereas for the method of constant stimuli it was 0.15. The constant c in Equation 1 reflects a constant error in the counter, and may also include variance caused by inconsistency in the location of the criterion. For very small durations c dominates, so that variability is approximately constant ($\sigma_t \approx c = 17$ ms in the above figure for both methods). The parameter p reflects the magnitude of pacemaker variability, and also some components of counter variability (e.g., if a constant proportion of pulses to the counter are lost, its signature would be a value of p greater than zero). It was unnecessary to specify a value for p in Fetterman and Killeen's experiment (i.e., p was set to zero with no loss in goodness–of–fit). Data collected with humans using the staircase technique (not shown here) followed Equation 1, with values of 0.10 for w and 31 ms for c.

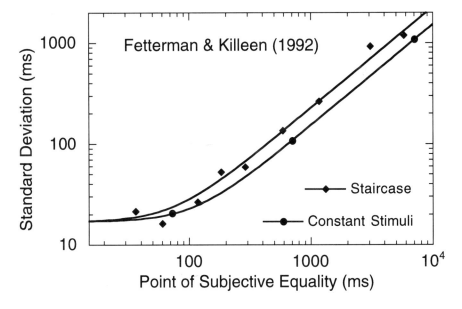

Figure 5. The standard deviation as a function of the *PSE* of the test stimuli, for both psychometric methods. The curves are *Weber Functions* derived from Equation 1.

Let us apply classic psychometric analyses to this paradigm. Figure 3 shows the hypothetical distribution of subjective pitches that a stimulus whose frequency = 500 Hz will engender, and the distribution that a second stimulus at 1000 Hz will engender. To predict the response to an unreinforced probe stimuli such as one at 630 Hz, we ask what is the likelihood that a discriminal dispersion centered at that value will exceed the criterion – that is, we calculate the area under a new curve, centered not at 500 ms but at 630 ms. The area to the right of the criterion gives the probability of saying "high". At any instant, the subject must decide whether the current tone arises from S_1 (500 Hz) or S_2 (1000 Hz). If the *a priori* probabilities of the stimuli are equal, these probabilities shift from favoring S_1 to favoring S_2 when the percept exceeds the criterion. We may apply the same analysis to the temporal data shown in Figure 6. Logistic ogives, $L(\mu,\sigma)$ were fit to these psychometric functions derived from the method of constant stimuli. The standard deviations of these ogives are the data plotted along the lower curve in Figure 5, which runs parallel to that generated by the staircase method.

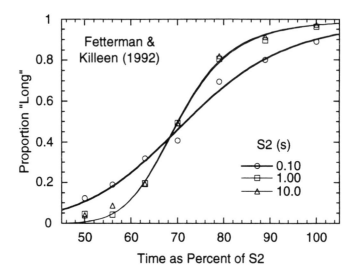

Figure 6. Average data from four pigeons taught to discriminate 0.05 s from 0.1 s; 0.5 s from 1 s; and 5 s from 10 s. Intermediate intervals were also presented for categorization. The ogives are from the pseudo–logistic model described below.

Bootstrapping a fourth model. There is an important inconsistency here: This traditional method assumes that the standard deviation is constant; yet it conclusively demonstrates an approximately linear relationship (Equation 1) between the mean and *SD*. In any one condition, as the probe stimuli move to the right their dispersion increases linearly with time. Only at the PSE is the value of the standard deviation given exactly by the curve in Figure 5.

We have applied a model appropriate for a metathetic continuum to time, which by that analysis is demonstrably a prothetic continuum (as seen by the increasing Weber Functions in Figure 5). The discriminal dispersions are not congruent, as in Figure 3, but affine as in Figure 7. The use of a standard Gaussian or logistic function to fit psychometric data such as those shown in Figure 6 is incoherent. We may achieve a coherent analysis only by fitting dispersions whose standard deviation increases with the stimuli according to Equation 1.

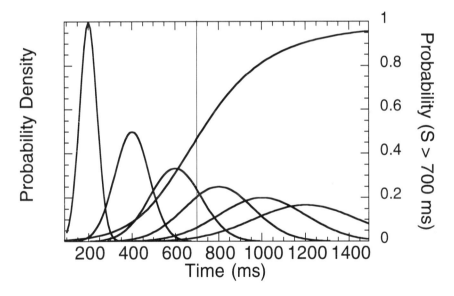

Figure 7. Distributions of perceptual effects of intervals of duration 200, 400, etc., ms. These discriminal dispersions have increasing variance on prothetic continua such as temporal duration. The area under the tail of these densities to the right of the criterion, here placed at 700 ms, gives the probability the subject will say "Long". That probability is given by the rising ogive, which is a pseudo–distribution function (Equation 2).

One claculates the probability corresponding to the deviation $-(S-C)/\sigma_S$ with S the mean of each of the separate dispersions (the value of each of the test stimuli) and C the criterion. That is the logic, and it requires multiple dispersions, one centered over each stimulus. As in Figure 3, however, the equation it generates is equivalent to a single psychometric function centered at C. In this interpretation, one calculates the probability corresponding to the deviation $(C-S)/\sigma_S$, with the mean (μ) of that single ogive now located at C, rather than over each of the stimuli:

$$p = \left[1 + \exp\left(\frac{\mu - t}{0.55\,\sigma_t}\right)\right]^{-1} \quad , \quad \sigma_t > 0. \tag{2}$$

The multiple–densities treatment (Figure 7) is the most fundamental: It asserts that there is variance in the perception of all stimuli, and that variance follows Weber's law. Each of the constituent densities is either the classic Gaussian or its logistic approximation. The single–ogive picture of Equation 2 yields one ogive that is neither the classic Gaussian distribution nor its logistic approximation. It is a bastard curve that is not even a distribution function, as it does not necessarily asymptote at 1 (the area under its "density" does not equal 1). It is not derived from a density. It traces the area to the right of the criterion for each of the test stimuli, and thus rises from 0 to approach 1, following an S–shaped course that reminds one of a distribution function.

Despite the unorthodox nature of this latter treatment of the psychometric function, there is an economy of conceptualization to representing the ogives by a single equation giving rise to a single curve, an equation that looks like the logistic equation, but which has a varying standard deviation given by Equation 1, and has the argument of Equation 1 (t) being the independent variable in Equation 2. We call Equation 2 the *Pseudo–Logistic Function*, and in concert with Equation 1, the *Pseudo–Logistic Model* (which we abbreviate

PLM). Equation 2 is not a logistic function, even though it is derived from one, but rather is significantly skewed. It is not a genuine statistical distribution function because it asymptotes at $F_\infty = [1+\exp(-1.81/w)]^{-1}$ rather than at 1. This is a small deviation (e.g., for $w = 0.3$, Equation 2 eventually asymtpotes at $F_\infty = 0.9976$); more noticeable is the slowness with which it approaches that asymptote (see, e.g., Figure 10; the constant in Equation 2 is more precisely $\sqrt{3}/\pi$, and 1.81 approximates its reciprocal).

The (proper) logistic function may be viewed either as a fundamental choice model in its own right (Luce, 1959, 1963), or as an approximation to a Gaussian function. In the latter case, the equivalent ogive would be the Pseudo–Gaussian Function. In both cases their names are qualified because they are neither logistic (Gaussian), nor statistical distribution functions. However, their ogives legitimately describe the result of well–defined processes and provide a fitting model of psychometric data (see Figs 9–14).

A logarithmic transformation of the abscissae can generate a distribution function (i.e., a lognormal function) that approximates the new model, but it is not the same as the new model (Killeen, Cate, & Tran, 1993). This similarity has been a source of confusion for generations: "How can it be that perceptions of time seem linear with SI time, but we must take logarithms of time to get simple discrimination functions? Is time really logarithmic?". We know that time is not perceived as its logarithm (see, e.g., Allan, 1979; Gibbon, 1981; Gibbon & Church, 1981; Stevens, 1986/1975). A key difference in the models is that under PLM the dispersion of a stimulus is assumed to be symmetric, whereas with the logarithmic transformation it is assumed to be positively skewed. Allan (1983) showed that the dispersions of magnitude estimates are not symmetrized by the log transformation, and concluded "our results do not support the lognormal model for magnitude estimates of temporal intervals" (p.35).

We don't have to make such logarithmic transformations if we use the correct model of the discrimination process, and so the logarithmic rescaling of stimuli need never enter our analysis. We may wish to use a logarithmic transform to condense the representation of the data and restore a more elegant symmetry to the ogive (see, e.g.,

Figure 9); but we should keep the independent variables native and transform the graphs, rather than confusing ourselves and our readers with unnecessary and misleading transforms of the independent variable. Figure 6 shows the result of this consistent analysis. In Figure 6 the criterion is set at the geometric mean of the training stimuli, and the standard deviation is given by Equation 1 with $w = 0.153$ and $c = 17$ ms, the value recovered from Figure 5.

The Pseudo–Logistic function evaluated at the test stimuli predicts the values of all four cells of the matrix in Figure 1. This provides a summary of the subject's accuracy in the task as a whole, and should be very useful when the task is kept constant and the state of the subject is manipulated with drugs, lesions, or motivational variables. If there are additional (probe) stimuli in the experimental task, each deserves its own matrix, which may also be directly inferred from this same function. This is important in situations such as that described by White and Cooney (1996) who showed that bias and detectability may be varied by differential payoffs which themselves vary as a function of time. It is also relevant to the time–left procedure (Gibbon & Church, 1981), where payoff for the time–left response changes continuously with the time of that response due to delay discounting.

The Weber Function (Equation 1) determines the slope of the psychometric function (Equation 2), and thus is our key index of discriminability. Now we turn to the index of bias inferred from that ogive, the location of the criterion, or PSE. (Macmillan & Creelman, 1991, 1996, give measures directly ascertainable from the data, but those indices assume a logistic, rather than pseudo–logistic psychometric function. Comparable indices are yet to be developed for pseudo–distribution functions.)

The PSE. The Point of Subjective Equality (the PSE) is the abscissae of a stimulus that is equally confusable with each of the training stimuli – the point where $p(\text{"}{-}A\text{"}|{-}A)$ equals $p(\text{"}A\text{"}|A)$. For a symmetric curve, the ordinate of the PSE (0.5) is both the mean and the median of the underlying density. But the Pseudo–Logistic is a positively skewed distribution (as long as either w or p in Equation 1 is greater than zero), and therefore its median is less than its mean. The point where the two dispersions in Figure 7 cross locates the stimulus that is equally often confused with those stimuli. We write the equations for those densities and solve them for the abscissa that satisfies both (call it *E*). Unfortunately we can only get so far; for the Pseudo–Gaussian we are stopped at:

$$\left(\frac{E-A}{\sigma_A}\right)^2 = \left(\frac{B-E}{\sigma_B}\right)^2 + 2\ln\left(\frac{\sigma_B}{\sigma_A}\right), \qquad \sigma_A, \sigma_B > 0 \qquad (3)$$

where A and B designate the values of the training stimuli.

For the Pseudo–Logistic we reach a comparable impasse. Notice that if only *c* in Equation 1 is greater than zero, then $\sigma_A = \sigma_B = c$, the logarithm vanishes from Equation 3 (because the log of 1 is zero), and its solution is $E = (A+B)/2$, the arithmetic mean of the stimuli, as expected: *For small values of t* where variability is approximately constant (where *c* dominates), the *PSE falls at the arithmetic mean of the stimuli.* This is the situation for metathetic continua.

Over the range in Figure 3 where Weber's Law holds (*w* dominates because *wt » c*) we may rewrite Equation 3 as:

$$\left(\frac{E-A}{wA}\right)^2 = \left(\frac{B-E}{wB}\right)^2 + \ln\left(\frac{B}{A}\right)^2, \qquad w, A, B > 0$$

This still can't be solved, but we may approximate it as:

$$\left(\frac{E-A}{A}\right) \cong \left(\frac{B-E}{B}\right) + w\ln\left(\frac{B}{A}\right), \qquad w, A, B > 0 \qquad (\sim 3)$$

and solve this approximation for the PSE:

$$E \cong \frac{(1+k)}{\frac{1}{2}\left(\frac{1}{A}+\frac{1}{B}\right)} = HM\,(1+k)\,, \qquad w,A,B>0 \qquad (4)$$

where $k = w\,\ln(B/A)/2$, and HM $\equiv$ the Harmonic Mean.

That is, the PSE lies above the harmonic mean (HM) of the two stimuli by a fraction k, which is proportional to the product of the Weber fraction and the logarithm of the ratio of the stimuli. Thus, we see that the PSE will range from the harmonic mean (in the case of no constant error ($c = 0$ in Equation 1) and a small Weber fraction w) to the arithmetic mean (in the case of a large value for c).

We can find numerical solutions to Equation 3 over the range of typical values. Figure 8 shows the predicted PSEs in an experiment where $S_2 = 3S_1$, for Weber functions corresponding to the smallest obtained by Fetterman and Killeen (1992; circles), one with a Weber fraction w twice as large (squares), and one with a small Weber fraction but a large constant error (triangles). Also drawn are three common measures of central tendency, the arithmetic mean (top line), geometric mean, and harmonic mean. As expected, when c dominates, the PSEs fall close to the arithmetic mean; as c decreases, the PSE moves down to the vicinity of the geometric and harmonic means. This figure validates the accuracy of the approximation, Equation ~3.

This analysis may explain why some research has located the mean of the psychometric function near the arithmetic mean of the stimuli, others place it closer to the harmonic mean. Variables that add a constant source of noise to the task will increase c and elevate the locus of the PSE, as will variables that increase the Weber fraction w. The more precise the discrimination, and the closer the value of the larger stimulus to the smaller one, the closer the data should come to the harmonic mean. As the training stimuli move farther apart, the value of k increases as the logarithm of their ratio. Platt and Davis (1983) studied bisection of widely–spaced temporal intervals, and Equation 4 with $w = 0.3$ falls closer to their reported PSEs than does the geometric mean.

The above analyses concerns the location of the PSE in unbiased subjects. Differential payoffs will shift the value of the PSE, and such a shift over conditions indicates bias. This will be discussed further below, but for now we return to empirical analyses.

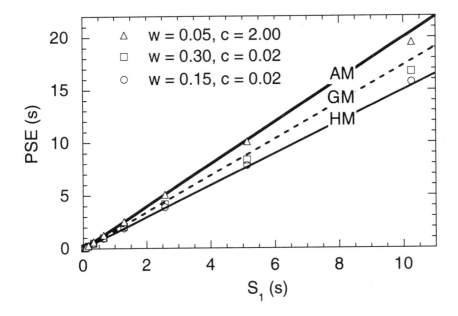

Figure 8. Solutions to Equation 3 for three cases differing in the values of the coefficients in Equation 1. The locus of these solutions fall between the harmonic and arithmetic means, as predicted by the approximations of Equation 3 (the arithmetic mean when $c \gg wt$, and slightly greater than the harmonic mean (Eq. ¯3) when the reverse is true).

Allan and Gibbon (1991) studied human's bisection of temporal intervals using the method of constant stimuli with a range of intervals from 0.75 to 4.0 s. Because they reported the Weber fractions independently of the PSEs, we may use the former to predict the latter. The median Weber fraction (their "sensitivity index") in their second experiment was 0.13, and the median PSEs were 0.87, 1.23, 2.45, 2.92, and 3.30 s. Equation 4 predicts 0.87, 1.23, 2.46, 2.79, and 3.31 s. These points lie very close to the geometric mean.

2. Multiple Stimuli. In the paradigm of Fetterman and Killeen (1992) there are a restricted number of reinforced values for *t* (normally two), each explicitly determined by the experimenter. Stubbs (1968) introduced an important variation of this two–stimulus procedure: Multiple values of stimuli were associated with reinforcement for short and long responses. In one condition, for instance, the short response was correct when the stimulus lasted 1, 2, 3, 4, or 5–s and the long response was correct when the stimulus lasted 6, 7, 8, 9, or 10–s. (This is a MI_jMR_k paradigm, with $j>k$). This many–to–one mapping between stimulus value and choices provides a bridge between traditional psychophysical tasks and free–operant tasks with unlimited values of *t*. Stubbs explicitly stipulated payoffs for all four cells of the matrix, a critical innovation that was little appreciated at the time. His payoff matrix was symmetric. Because we have explicit "short" and "long" responses, we can calculate values of *d'* for various stimulus differences (see Fetterman, 1995, for this analysis). Alternatively, we can treat all durations categorized as "short" by the experimenter as a single stimulus, and arrange differential payoff for the categories of the stimuli, as did Stubbs (1976). A problem with this approach is that the various constituent durations are differentially discriminable from one another, and the ROCs constitute an average over a set and are representative only of some median stimulus. Is there a more precise way to analyze these data? Application of the Pseudo–Logistic model (Equations 1 and 2) is reinforced by its obvious conformity to the median data from Stubb's (1968) Figure 4 (see Figure 9, below).

Rather than treat the value of the additive constant *c* as a free parameter, we assumed that it was comprised primarily of error in estimating the location of the criterion, and subject to the same Weber function as the estimation of real time. According to this assumption its contribution to the standard deviation is simply the Weber fraction *w* times the location of the criterion, μ. There are reasons why this may underestimate the true value of *c* (constant error in the counter – "switch closure" error – is left out), and reasons why it may overestimate the value (because the animal has many trials over which to establish its criterion, its standard error could decrease by the inverse–square rule; see Gibbon and Church, 1990). But in various applications of the PLM (e.g., Killeen, Palombo, Gottlob & Beam, 1997) this assumption has proven parsimonious and close enough to the accuracy achieved with c as a free parameter. Other users of the PLM may wish to keep c free. In the following analyses we have used this Weber model with $c = w$ as the default, fitting the Poisson ($p>0$, $w=0$) only when the goodness of fit of the default model was unacceptable.

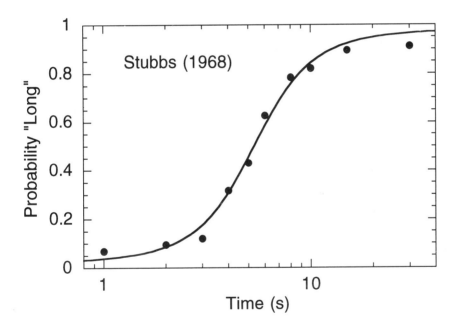

Figure 9. Equations 1 and 2 (the Pseudo–Logistic Model, *PLM*) fit to the data of Stubbs (1968). The parameters took values of $\mu = 5.3$ s, $w = 0.45$, and $c = w\mu$.

Continuous Choice Procedures

3. Free–Operant Choice. This is a modification of a task developed by Stubbs (1979) that combines features of Choice (Yes/No) and schedule procedures. It is the next step in the transition from two stimuli, through multiple stimuli as analysed in the preceding section, to a continuum of stimuli. Responses to one key ("early") are intermittently reinforced during the first half of a trial and responses to a second key ("late") are intermittently reinforced during the second half of the trial. As expected, responses to the early key predominate during the first half of the trial and responses to the late key predominate during the second half. This pattern produces an ogival psychometric function relating the proportion of responses on the "late" key to the time since the trial began (e.g., Bizo & White, 1994a,b). Like the FI schedule, there is a continuous opportunity for responding, but unlike the FI, there is an opportunity for reinforcement for correct "short" responses. This is a MIR_k paradigm, with I being continuous and $k = 1,2$. Representative data are shown in Figure 10, along with the best–fitting Pseudo–Logistic models.

In Figure 10, the effect of decreasing the payoff for "Short" responses was ascertained by changing the schedule from VI 45 s to VI 120 s for responses on the "Short" key. Notice that, even with equal payoffs, the animal's *PSE* is at 18 s, rather than 25 s. Because variance of discriminal dispersions increases with t, stimuli near the middle will be perceived as more likely to have come from the long stimulus. This may be seen graphically in Figure 7, where a percept that appears to be at 700 ms has higher likelihood of being a variant of the 1000 s stimulus than of the 400 ms stimulus.

For two stimuli we may predict the PSE in the analysis given by Equations 3–4. For a continuum of stimuli, we must integrate the discriminal dispersions and find when the likelihood of a sensation having been generated by any of the short stimuli has just decreased to the likelihood of it having been generated by any of the long stimuli. Unfortunately, there is no closed–form solution when the variable of integration also appears as part of the standard deviation. Without further theory development, we can be no more precise than to expect there to be a short bias.

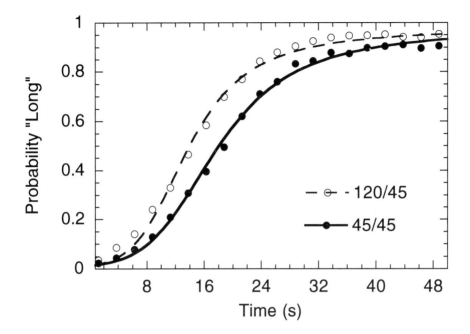

Figure 10. Equations 1 and 2 (the PLM) fit to data from this laboratory using the paradigm of Bizo & White (1994a,b). The filled circles were obtained when responses on both keys were reinforced on VI 45 s schedules; the open circles were collected during the first 5 sessions after the VI schedule in the first half of each trial was changed from 45 s to 120 s. The parameters took values of $\mu = 13.9$ s (open) and 18.1 s (filled), $w = 0.41$, and $c = w\mu$.

It is easier to predict the effect of changing payoff, given knowledge of the unbiassed PSE. If we assume melioration – that the animal shifts its criterion until the expected probability of reinforcement for "long" and "short" responses are equal – this happens when Equation 2 multiplied by the rate of reinforcement during the second half of the trial equals the complement of Equation 2 multiplied by the rate during the first half of the trial. In the simple case when Weber's Law is strictly true (only $w > 0$), this happens when:

$$C = \frac{\mu}{(1 + w \ln [R_L/R_S])} , \qquad (5)$$

where μ is the mean of the unbiased ogive, R_k is the rate of reinforcement for long or short responses, and C is the mean of the biased ogive. Equation 5 predicts a shift of the PSE to 13.2 s for the data in Figure 10.

Another example of free–operant choice is provided by the experiments of Fetterman and Killeen (1995; also see Wearden, 1995), who reinforced pigeon's pecks to one key at *t* s into an interval, to a

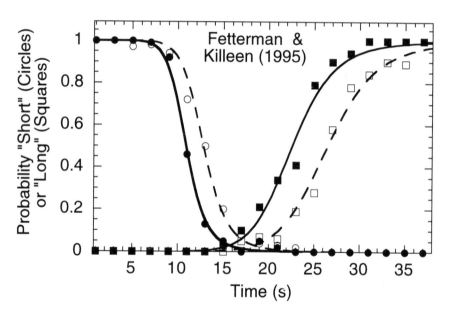

Figure 11. The PLM fit to data from Fetterman and Killeen (1995). The circles depict the probability of responding to a "short" key on which food was available probabilistically after 8 s. The squares depict the probability of responding to a "long" key on which food was probabilistically available after 32 s. The filled symbols are from a condition in which food was available from each key 1/3 of the time; the unfilled symbols are from a condition in which that probability was quartered. The criteria are (from left to right): μ = 10.9, 12.8, 22.7 and 26.6 seconds. The Weber fraction was w = 0.13 for all curves, and $c = w\mu$ for all curves.

second at $2t$ s, and to a third at $4t$ s. This is a MIR_k paradigm, with I being continuous and $k = 1,2,3$. Figure 11 shows response probabilities as a function of time for the first and third keys in the condition where $t = 8$ s.

The curves show the locus of the PLM, with the parameters given in the legend. Notice that under the decreased rate of reinforcement the curves shifted to the right. Because the probability of reinforcement was uniformly decreased, these effects cannot be attributed to melioration–based biases: There was no reason as far as signal detectability theory is concerned that these data should have shifted. The shift is predicted by the Behavioral Theory of timing (BeT). That theory assumes an underlying Poisson process, one in which the speed of the pacemaker is determined by the arousal level of the organism. These data would be equally well fit by a Poisson model, as would most of the other data displayed in this chapter. As the speed of the pacemaker decreases, the value of the parameter p in Equation 1 increases, thus predicting an increase in dispersion as well as mean of these distributions. A more careful consideration of these effects is presented in the last segments of this chapter.

3. Time Left. This is a type of concurrent–chain task. Trials begin with the onset of two initial–link stimuli. Responses during this initial link provide access to mutually exclusive terminal links at variable intervals. One of the terminal links provides reinforcement according to a standard delay of S s timed from the moment of the animal's choice of that option (e.g., $S = 30$ s). The other time–left terminal link provides food after a delay equal to some constant (e.g., $C = 60$ s) minus the time (t) since the beginning of the trial (i.e., $C–t$). Choices favor the standard key early in the trial and the time–left key later in the trial, as they should (e.g., Gibbon & Church, 1981). In this paradigm there are an unlimited number of values for t, each determined by the subject: It is an MIR_k paradigm, with I being continuous and $k = 1,2$. Along with each value of t there exists a unique matrix, with the payoff for an "early" (i.e., standard, S) response equal to the reinforcer devalued by the constant delay of, in this case, 30 s, and the payoff for a response on the comparison ("time–left") key equal to the reinforcer devalued by the delay $C - t$, (not devalued when $t \geq C$). To maximize reward, the animal should switch from C responses to S responses when $t = C/2$. The differential payoff for switching at the correct time is very small around C and continuously grows very large early and late in the interval. Because value is a convex function of delay, the change in payoff is not symmetric around $C/2$:   Early change–overs incur less than proportional losses, and so the animals should switch early, which they do (the average switch–point is 74% of $C/2$). Figure 12 shows the average data and the psychometric functions drawn by the PLM. Note that the data for 60 and 90 s do not approach 1.0 so closely as do the data from the other conditions. Gibbon and Church explained this as due to inattention, and added a parameter to accommodate that.

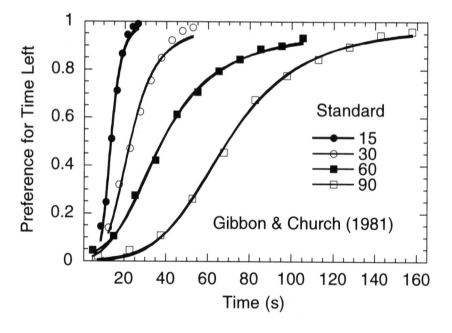

Figure 12. The PLM fit to data from Gibbon & Church (1981). The data depict the probability of responding to the "time–left" key on which food was available at varying delays. The criteria are: $\mu = 13.6, 22.8, 38.2$ and 69.4 seconds. The Weber fractions ranged from $w = 0.23$ to 0.47, and $c = w$ for all curves. Also shown are the ogives corresponding to Poisson timing.

The *PLM* predicts this slower approach to lower asymptotes because of the systematic increase in the variability of temporal judgements at the longer intervals, as given by Equation 1.

Also drawn through the data are the *PLM* psychometric functions generated under the assumption that timing is Poisson (i.e., $w = 0$, $p > 0$, and $c = p\mu$) rather than scalar. They are difficult to discriminate from those drawn under the Weber model. It is clear that data such as these cannot distinguish between those two types of pacemaker– counter systems.

Discrete Go/No–Go Procedures

4. Temporal Generalization. With the temporal generalization task a target response may be reinforced after some particular duration, such as 4–s (the *S*+), but not after other durations (the *S*–), either shorter or longer than the *S*+ value (e.g., 2– s or 8–s). The opportunity to emit the target response occurs after the test durations (retrospective timing). Such studies constitute a $M(IM)_jR$ design. Analyses relate the probability of emitting the response as a function of signal duration, producing a maximum response probability at the *S*+ duration and lower response probabilities at shorter or longer durations (e.g., Church and Gibbon, 1982). These data are essentially discrete versions of those derived from the procedures described in the next section. Figure 13 shows some of the data collected by Wearden and Towse (1994).

Continuous Go/No–Go Procedures

5. Fixed Interval. In this procedure the first response after some interval is reinforced. This schedule typically produces a scalloped pattern of responding, such that the rate of responding increases as the time to reinforcement approaches (e.g., Dews, 1970). It is an *MIR* procedure.

Animals time events whether or not they are explicitly

reinforced for it. Haight and Killeen (1991) observed the various adjunctive behaviors of pigeons given free reinforcements on a probabilistic basis every 15 s. Two of the pigeons showed spontaneous pecking of the front wall of the chamber. Their data are shown in Figure 14. Note that the second curve appears to have a lower asymptote, but that is due to the greater standard deviation of the second ogive that is correlated with its mean ($c = w\mu$); the asymptotic response rates for both pigeons are essentially identical.

Many other adjunctive responses showed an increase then decrease throughout the course of the interval. The PLM also accommodates those data if it is generalized, after the fashion of Fetterman and Killeen (1991), by adding a second temporal judgement concerning when to cease that behavior and move on to another.

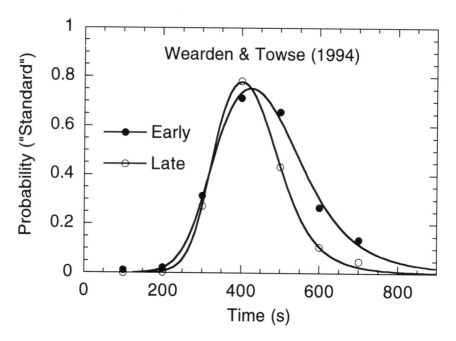

Figure 13. The probability of saying the stimulus was the same as the standard (400 ms) averaged over 20 human observers, from early and late in training. The parameters for the PLMs are: Early: $\mu = 331$ and 545 ms; $w = 0.16$; Late: $\mu = 328$ and 491 ms; $w = 0.12$; $c = w\mu$ for both conditions. Extended training reduces both the Weber fraction and upper criterion. The Poisson model provided a comparable fit to the data.

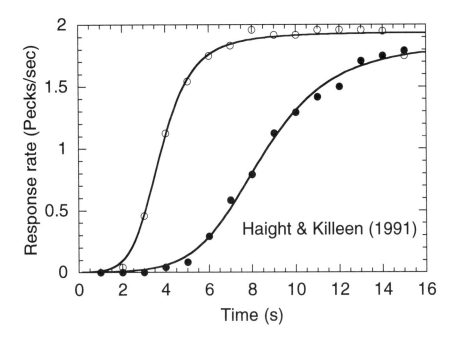

Figure 14. The PLM fit to adjunctive pecking responses of two pigeons. The criteria are: μ = 3.8 and 8.5 seconds; the Weber fraction w = 0.25 and $c = w\mu$ for both subjects. The PLM predicts the probability of being in a state (e.g., "Long"). In the present case where multiple responses are possible in the state, a new parameter – response rate while in the state – is necessary (alternatively, the curves could be divided by that parameter to give the appearance of greater parsimony). In the present case, the rate-in-state was 1.94 responses/s for the first pigeon and 1.87 for the second. The Poisson model provided an equally good fit to the data, but required different Poisson parameters for the two curves (0.23 and 0.52).

6. Peak Procedure. This procedure is a modified FI schedule that includes unreinforced probe trials lasting two or three times the duration of the FI. For example, Catania (1970) trained pigeons on a FI 10 s schedule and scheduled food with a probability of either 0.9 or 0.1. On unreinforced trials the key remained illuminated for 48 s.

Responding increased to a maximum at about 10 s and then decreased, suggesting a temporal generalization gradient about the time of reward. Roberts (1981) christened this task the peak procedure, and identified three important measures, peak rate (the maximum response rate), peak time (the time at which maximum responding occurs), and the standard deviation of the curve passed through the data (Roberts used a Gaussian density).

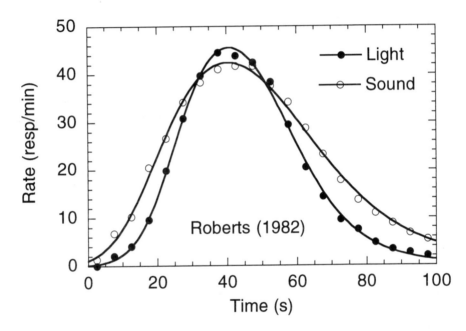

Figure 15. The PLM fit to the average data from 6 rats and 18 days of responding on the peak procedure, reinforcing those responses probabilistically at 40 s. For the light the criteria were μ = 23.2 s to start, 62.3 s to stop, and p = 5.0. For the sound the criteria were μ = 27.7 s to start, 57.7 s to stop, and p = 2.43. The rate–in–state was 60 responses/m and $c = \sqrt{(p\mu)}$ for both conditions.

Figure 15 shows peak data in their classic form, collected by Roberts (1982). Similar data were reported by Church and Broadbent (1990) and Church and Meck (1988). Note that the use of a light, rather than tone, provided a sharper discrimination (despite the lab–lore that rats are more sensitive to tones than lights). This is the first case where the Poisson and scalar (Weber) models make substantially different predictions, with the Poisson providing a clearly superior fit to the data. (The fit of the scalar model could be improved to the same accuracy by treating the additive constant as a free parameter, rather than constraining it to represent criterial error that follows the same rule as real–time error.) Note that in fitting the Poisson, the criterial error is also treated as Poisson, so c is now the root of the Poisson parameter times the mean. The peak data of MacEwen and Killeen (1991) were also fit with the *PLM* and, like Roberts' data, required the Poisson model.

7. Differential Reinforcement of Low Rate. On a DRL procedure reinforcement depends upon some minimum duration separating successive responses (e.g., Bruner and Revusky, 1961; Wilson & Keller, 1953). Responses separated by less than the DRL value are not reinforced. Sometimes a limited hold (LH) specifying an upper bound for reinforcement is used, such that responses that exceed the upper limit are not reinforced. This is a "production" paradigm, which we designate as *RIR*. (In cases where the first response initiates a stimulus, it is RMIR). Jasselette, Lejeune, and Wearden (1990) provide a nice procedure and alternative.

The above constitute a wide array of procedures, but all are amenable to signal detectability analysis using the Pseudo–Logistic model (PLM). This model resides close to the level of the data, as it does not postulate particular underlying processes that generate the data: It does not require scalar or Poisson timing, nor even pacemakers and counters. It is a descriptive, rather than process model. However, in concert with the analysis of Killeen and Weiss (1987), it helps us to understand the relation among various higher–level theories of timing that do specify processes. There remains inadequate room in this chapter to review the various higher–level theories of timing, some of which are represented elsewhere in this volume. Instead, we review the theory we are most familiar with, and show how it dove–tails with the PLM.

The Behavioral Theory of Timing

Killeen and Fetterman's (1988) behavioral theory of timing (BeT) places special emphasis on the role of behavior in mediating timing. The mediating behaviors may include interim, terminal, emitted or elicited behaviors, and these are assumed to act as discriminative stimuli for retrospective estimates – they are the digits on the counter. *BeT* assumes the transitions between behavioral states are caused by pulses from a pacemaker, moving the animal through a sequence of states. These states are of variable duration, and a single behavior may be correlated with just one or with several states. The correlation between a mediating behavior and a timing response evolves with repeated exposures to a duration; behaviors that facilitate timing and thus aid the acquisition of reinforcement will become more strongly conditioned than those behaviors that are less strongly correlated with reinforcement.

Behavioral Mediation of Timing. Many classic analyses of timing in animals have invoked mediating behavior as a causal mechanism. Such behaviors are held to function as the effective stimuli when animals learn to respond to temporal regularities in their environment. For instance, Wilson and Keller (1953) observed systematic patterns of behavior in rats trained on a DRL schedule and concluded that the "collateral" behaviors mediated the temporal spacing of instrumental lever presses demanded by the DRL procedure.

More recently, behavioral explanations of animal timing have been influenced by research on adjunctive behavior. Beginning with the seminal work of Staddon and Simmelhag (1971) it has become clear that behaviors indexed by switch closures constitute only part of the behavioral repertoire engendered by various schedules of reinforcement, especially periodic schedules such as FI. Once researchers moved beyond the recording of switch closures and began to peer inside the operant chamber, it became clear that: a) animals engaged in a variety of activities along with the target response and, b) that these "superfluous" behaviors were highly regular in time and could be described by simple quantitative models (Killeen, 1975;

Figure 14 above). Staddon and Simmelhag observed that there was temporal structure in the patterns of adjunctive behavior, with some behaviors occurring more frequently early in an interval (interim behaviors) and others occurring more frequently later in the interval (terminal behaviors). Such structure affords the possibility of an organism using its behavior to predict its position in time with respect to a forthcoming reinforcement, just as Killeen and Fetterman (1988) hypothesized. Numerous studies have confirmed the observations of Staddon and Simmelhag under a variety of conditions involving periodic and aperiodic schedules, contingent and noncontingent food deliveries, different species, responses, and reinforcers.

Of course, the observation that classes of adjunctive behavior are correlated with temporal locus in the interfood interval does not mean that organisms use these behaviors to anticipate food deliveries. Nonetheless, the temporal regularity of sequences of behavior provides a simple mechanism that might explain how animals time, and in the remainder of this section we evaluate the plausibility of such a mechanism.

One test of this behavioral timing hypothesis involves the use of techniques that disrupt the putative mediators; such disruptions should affect the accuracy of timing while leaving other nontemporal discriminations intact. This maneuver is analogous to manipulations (e.g., distracter tasks) that prevent humans from counting to themselves in order to estimate an interval of time. Humans' temporal estimations are less accurate (and more variable) when counting strategies are precluded, suggesting that chronometric counting strategies mediate temporal estimations. In the case of timing by nonhuman animals, disruption of ongoing activities frequently (but not always) perturbs the accuracy of timing. Laties, Weiss, and Weiss (1969), for example, assessed the efficiency of DRL performance and sometimes allowed subjects to engage in mediative responding (e.g., nibbling the grid floor) and at other times obstructed this activity. Animals timed more accurately when allowed to engage in collateral responding. This literature is ably reviewed by Richelle and Lejeune (1980), who conclude that mediating behaviors support temporal regulation, but that animals also may time accurately without the aid of obvious behavioral strategies.

Our laboratory has applied the observational techniques pioneered by Staddon and Simmelhag (1971) to a retrospective timing task. Pigeons were trained under a discrete retrospective timing procedure to discriminate a short (6–s) signal from a long (12–s) signal. One response was reinforced after the shorter signal and another after the longer ($MIMR_{1,2}$). Some birds were trained on a spatial choice arrangement whereby left–key and right–key choices signified short and long durations. Other birds were trained under a nonspatial arrangement in which the position of the correct choice was identified by key color (e.g., red vs. green), not key location. This arrangement disallows strategies based on spatial position (e.g., standing in front of the "short" key early in the trial and standing in front of the "long" key later in the trial), and may engender different behavioral strategies (e.g., Chatlosh & Wasserman, 1987). Once the temporal discrimination was acquired, sessions were videotaped and observers coded the birds' behaviors on every trial using categories that captured the ongoing repertoires. Behaviors were coded once each second.

Figure 16 shows data from a representative bird trained under the spatial choice procedure. Each panel shows the relative frequencies of different behaviors in one–second bins. Once the original discrimination was learned, the task was made more difficult by reducing the value of the longer stimulus in order to increase error rates; the proportion of correct responses fell from 95% to 70%. The figure shows that this pigeon engaged in differential behaviors that were predictive of the durations of the stimuli. Pecking in front of the left key was predictive of the short signal duration and pecking in front of the right key predicted the long signal duration. The distributions of behavior showed little change in this new condition, so the the "short" behavior remained predictive, but both it and the "long" behavior occurred equally at the 9 s duration. In other words, neither behavior uniquely predicted the occurrence of the long signal.

It is clear that if this pigeon used its ongoing behaviors to discriminate between short and long signals, choice accuracy should be higher on short than on long signal trials (as there would be more

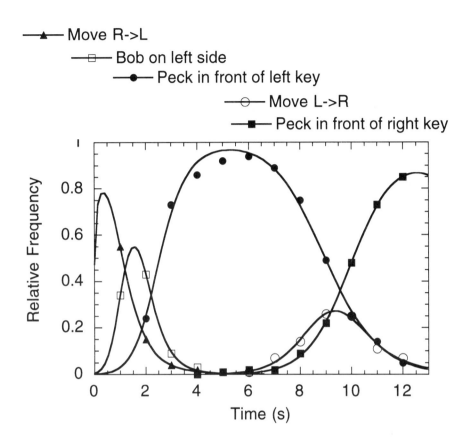

Figure 16. Relative frequency of various collateral behaviors in a retrospective timing experiment. The curves through the data are best–fitting PLMs under the Poisson assumption, with Poisson parameters ranging from 0.12 to 0.20, and $c = \sqrt{(p\mu)}$.

"behavioral confusion" in those instances). This proved to be the case (98% on *short* versus 76% on *long* signal trials). A more powerful test of behavioral timing is provided by a confusion analysis carried out on the data in this figure. Behaviors that immediately preceded choice responses on long trials were classified according to whether the ensuing choice response was correct or incorrect. According to our analysis these antecedent behaviors should be better predictors of choice than the signal duration. If a "short" behavior preceded a choice when a long response was called for, and that behavior provided the cue for choice, the subject should incorrectly classify the signal duration. Conversely, when a "long" behavior preceded choice, subjects would tend to correctly classify the signal. Most of the incorrect responses occurred to the long stimulus, and, of these, 90% occurred while the bird was pecking in front of the left key. Table 1, showing the data from long trials over the last four sessions with this subject demonstrates that this was the case. Whereas this was a very obvious mediating behavior, other subjects showed other mediating behaviors less directly tied to the final response that served as the basis of their final judgement.

Table 1   Mediating behaviors, and correct and incorrect timing choice totals following different mediating behaviors

Category	Behavior
1	Pecking in front of left key
2	Moving from left to right along front wall
3	Pecking in front of right key
4	Bobbing in front of left key
5	Moving right to left along front wall

Category	Correct	Incorrect
1	6	39
2	18	3
3	52	1

BeT thus professes to explain retrospective timing in terms of discrimination of ongoing behavior; but what explains the temporal orderliness of that ongoing behavior? We have called such temporally differentiated behavior "immediate timing", and attributed it to the transition between states as occasioned by pulses from a pacemaker. The pacemaker is a purely hypothetical construct, and can be viewed as the material reification of a formal cause (the stochastic process underlying the state transitions). There are, however, many candidates for neural circuits that function as pacemakers, many of them reviewed elsewhere in this volume. We are confident that many pacemakers will be identified, and that rather than adjust the speed of the pacemaker to optimize timing, animals may simply switch among other pacemakers with different periods.

Just as two mechanisms are posited, the pacemaker and the counter, there are two degrees of freedom in how the animals adapt to different timing demands: They can change the speed of the pacemaker, or change the criterion count. In the following section we examine adjustments in the period of the pacemaker.

Adjusting the Pacemaker. Killeen and Fetterman (1988) showed that the scalar property of temporal behavior could be modeled as a timing system in which the speed of a Poisson pacemaker varied as a consequence of changes in reinforcer frequency. The probability of residence in a state as a function of time is then given by a gamma distribution with mean $\mu = n\tau$ and standard deviation $\sigma = \sqrt{n}\,\tau$, where n is the number of counts and τ (tau) is the period of the pacemaker. Tau is one of the constituents of p in Equation 1. Killeen and Fetterman recognized that pacemakers would be generally somewhat more reliable than the Poisson random emitter (and counters less than perfectly accurate) but the Poisson process constitutes a particularly simple and well–understood system, and it was decided to find its limits before designing more complicated systems.

Whereas most theories of timing assume that the rate of the pacemaker remains approximately constant, Killeen and Fetterman (1988) posited that pacemaker speed varies with the animal's state of arousal in the timing environment (and, *per hypothesis*, with arousal level; see Boltz, 1994; Killeen, 1975; Shurtleff, Raslear, and Simmons, 1990; Wearden & Penton–Voak, 1995). This was not arbitrary: It was necessary to achieve the Weber–timing found in so many studies. In many cases, the rate of reinforcement, R, is confounded with the interval to be timed (I). Situations that deliver high frequencies of reinforcement result in a faster pacemaker than those that deliver lower frequencies. With the period of the pacemaker (τ) proportional to the inter–reinforcement interval, it follows that the standard deviation will increase proportionately with the interval to be timed. This works fine for FI schedules, but has problems elsewhere. When I varies but R remains constant, or vice versa, we find exceptions which do not, unfortunately, prove the rule.

Fetterman and Killeen (1991) assayed the influence of several reinforcement manipulations on pigeons' timing behavior under a discrete retrospective timing task; the birds learned to make one response after a short signal and another after a longer signal. In these experiments, reinforcer density was altered by changing the duration of the intertrial interval (ITI), the probability of reinforcement for correct responding, or the amount of reinforcement. Each of these manipulations would appear to bring about similar changes in the arousal level of the organism, but timing was not influenced equally by the manipulations. Changes in reinforcer probability produced effects consistent with predictions of BeT. Changes in the amount of reinforcement yielded less robust effects, but in the predicted direction. Changes in the ITI had no consistent effect on timing.

A recent study by Fetterman and Killeen (1995) provides a clear illustration of the changes in timing that can be produced by changes in reinforcer density. They trained pigeons on a categorical scaling procedure that reinforced responding to one of three keys according to separate *FI* schedules. Responses to one key could be reinforced after a short period, to a second key after an intermediate period, and to a third key after a long period (e.g., 8–s, 16–s, or 32–s). Reinforcement was assigned to only one key on each trial, producing a pattern of behavior that could be described as "temporal search." The birds responded on the short key early in the trial, moved to the intermediate key somewhat later, and finally switched to the long key if reward was not obtained on the first two keys by the specified times. Responding on each of the keys was recorded as a function of time in a trial, yielding a set of preference curves relating proportion of responses on each key to trial time. Reinforcer frequency was manipulated by varying the probability of a trial ending with reinforcement in an ABA design – 1 (all trials end with food), .25 (1/4 of the trials end with food) – 1. Figure 11 shows some of the results in terms of the proportion of responses on the short and long keys through time. When reinforcer frequency decreased, the response distributions shifted to longer durations and the variance of the distributions increased, consistent with a deceleration of the pacemaker. When reinforcer frequency increased, the response distributions shifted to shorter durations and the variance of the distributions decreased, consistent with an acceleration of the pacemaker. These effects cannot be explained as a shift in bias in order to optimize rate of reinforcement (Equation 5) because the probability of reinforcement was always the same for all responses.

Bizo and White (1994a) reported similar effects. In their experiment, reinforcer frequency was manipulated through changes in the value of a VI schedule. The task involved discriminating the first half of a 50–s trial from the second half of the trial. Responses on the right ("early") key were reinforced according to a VI during the first 25 s since trial onset, and responses on the left ("late") key were reinforced according to the same VI during the second 25 s. Reinforcement rate was manipulated by varying the VI schedule on the two keys. When the VI schedule was made richer (VI 15–s), the response distributions (proportion of responses on the "late" key) shifted to the left; when the VI schedule was made leaner (VI 120–s), the response distributions shifted to the right; the intermediate (VI 30–s) schedule yielded performances between the two extremes. These effects, shown in Figure 17, are consistent with the predicted shifts in pacemaker speed.

Hedging our BeT. When rate of reinforcement and other sources of arousal are constant, the speed of the pacemaker is also constant. Then, according to the classic BeT, error in timing must be due to Poisson–like processes only. But if rate of reinforcement is varied, there is a concomitant change in the speed of the pacemaker. (These speeds are usually inferred by fitting either gamma distributions or Gaussian distributions to psychometric functions, just as we originally did in the analysis of Figures 3 and 11). To achieve pure Weber timing, Killeen and Fetterman (1988) assumed that this change in pacemaker speed would be proportional to the changes in rate of reinforcement. As noted above, this is seldom precisely true. How can we reconcile the flatter linear relation observed between inferred value of pacemaker speed and reinforcement rate?

First, we must remember that pure Weber timing is also the exception; there is usually only a linear relation between the standard deviation and the inter–reinforcement interval (i.e. $c>0$), in which case a linear relation between pacemaker speed and rate of reinforcement is the predicted mechanism.

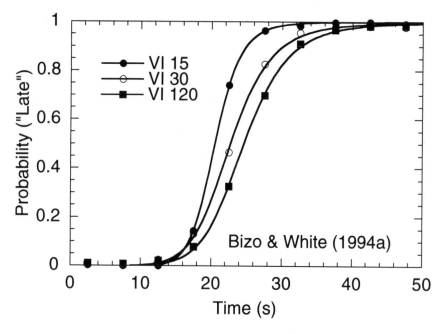

Figure 17. The PLM fit to the average data from 5 pigeons responding on the free–operant choice procedure (Bizo & White, 1994a). The criteria were $\mu = 20.5$, 22.7, and 24.6 s, and $w = 0.11, 0.16$, and 0.16 for the 15, 30, and 120 s VI schedules. In all cases, $c = w\mu$. The Poisson model provided an equivalent goodness–of–fit.

If there is any constant error in timing – for example error in estimating the criterion – we expect variability to grow more slowly than proportionate to rate of reinforcement or to the interval to be estimated. If the current simplification of setting the constant error proportional to the criterion is sustained, this additive constant will be larger when larger intervals are estimated, providing some support for this hypothesis concerning the origin of that error. This source of error – criterial error – is very similar to what Gibbon and associates mean by error in accessing reference memory.

Killeen and Weiss showed that Weber–like error in timing may derive from Weber–like error in the counter. For BeT, this means that sometimes animals must skip a mediating state, and sometimes they persevere in a state for a double count. There are data showing that this is the case (Reid, Bacha, & Moran, 1993). Furthermore, to obtain Weber–like variance in timing, the magnitude of these errors must increase with the number of counts. Alternatively, it is conceivable that as an interval wears on, animals become less excited, leading to a decrease in the pacemaker speed. Fetterman and Killeen (1995) found that response rates were lower for later categories of time than for earlier ones. Poisson changes in timing plus linear changes in arousal within the interval could lead to the PLM parameters found appropriate for these data.

Weber–like error will dominate at intermediate and long intervals. At short to intermediate intervals, the Poisson error will dominate; whereas at very short intervals, there will be constancy in error. This is consistent with the observed Weber functions. However, these functions are so unremarkable that it is rarely necessary to include all sources of error to provide a good fit to the data: With the parameter c governing error at the very shortest intervals, and the Weber fraction w governing it at the longest intervals, there is little work left for the Poisson parameter p to take responsibility for. Evidence for the necessity of Poisson– like processes cannot be found in psychometric functions. But they are found in the data from immediate timing experiments, such as Figures 15 and 16, which cannot be adequately fit with the simple Weber model. Is this a particular feature of immediate timing experiments, or would any experiment which generated bitonic patterns of data also force a decision between process models? Only experiments will tell.

There is very good evidence for the existence of the behavioral processes assumed by BeT (see, e.g., Figure 16 and Table 2). Other processes must also be necessary, such as the changes in arousal level, and error in the counter and criterion. We have a simple model of the Weber Function (Killeen &Weiss, 1987), and now, with PLM, a simple model of psychometric functions that is consistent with it, that gives these various processes expression and permits precise tests among models. Putting finer points on these processes will occupy us, but need not further try your patience.

Authors' note

This chapter would not have been possible without the assistance of NSF Grants IBN 9408022, IBN 9407527 and BNS 9021562, and a sabbatical leave granted by IUPUI to J.G. Fetterman. Ms Josey Chu contributed to the discussions on which this chapter was based.

References

Allan, L. G. (1979). The perception of time. *Perception & Psychophysics, 26,* 340–354.

Allan, L. G. (1983). Magnitude estimation of temporal intervals. *Perception & Psychophysics, 33,* 29–42.

Allan, L. G., & Gibbon, J. (1991). Human bisection at the geometric mean. *Learning and Motivation, 22,* 39–58.

Bizo, L. A., & White, K. G. (1994a). The behavioral theory of timing: Reinforcer rate determines pacemaker rate. *Journal of the Experimental Analysis of Behavior, 61,* 19–33.

Bizo, L. A., & White, K. G. (1994b). Pacemaker rate in the behavioral theory of timing. *Journal of Experimental Psychology: Animal Behavior Processes, 20,* 308–321.

Boltz, M. G. (1994). Changes in internal tempo and effects on the learning and remembering of event durations. *Journal of Experimental Psychology: Learning, Memory, and Cognition, 20,* 1154–1171.

Brown, S. W. (1985). Time perception and attention: The effects of prospective versus retrospective paradigms and task demands on perceived duration. *Perception & Psychophysics, 38,* 115–124.

Bruner, A., & Revusky, S. H. (1961). Collateral behavior in humans. *Journal of the Experimental Analysis of Behavior, 4,* 849–850.

Brunner, D., Kacelnik, A., & Gibbon, J. (1992). Optimal foraging and timing processes in the starling, *Sturnus vulgaris*: Effect of inter–capture interval. *Animal Behaviour, 44,* 597–613.

Catania, A. C. (1970). Reinforcement schedules and psychophysical judgements: A study of some temporal properties of behavior. In W. N. Schoenfeld (Ed.), *The theory of reinforcement schedules* (pp. 1–42). New York: Appleton–Century Crofts.

Chatlosh, D. L., & Wasserman, E. A. (1987). Delayed temporal discrimination in pigeons: A comparison of two procedures. *Journal of the Experimental Analysis of Behavior, 47,* 299–309.

Church, R. M., & Broadbent, H. M. (1990). Alternative representations of time, number and rate. *Cognition, 37,* 55–81.

Church, R. M., Broadbent, H. A., & Gibbon, J. (1992). Biological and psychological description of an internal clock. In I. Gormezano & E. A. Wasserman (Eds.), *Learning and memory: The behavioral and biological substrates,* (pp. 105–128). Hillsdale, NJ: Lawrence Erlbaum Associates.

Church, R. M., & Meck, W. H. (1988). Biological basis of the remembered time of reinforcement. In M. L. Commons, R. M. Church, J. R. Stellar, & A. R. Wagner (Eds.), *Quantitative Analysis of Behavior VII: Biological determinants of reinforcement,* Hillsdale, NJ: Lawrence Erlbaum Associates.

Culler, E. A. (1938). Recent advances in some concepts of conditioning. *Psychological Review, 45,* 134–153.

Dews, P. B. (1970). The theory of fixed–interval responding. In W. N. Schoenfeld (Ed.), *The theory of reinforcement schedules* (pp. 43–61). New York: Appleton–Century–Crofts.

Egan, J. P. (1975). *Signal detection theory and ROC analysis.* New York: Academic Press.

Fetterman, J. G. (1993). Numerosity discrimination: Both time and number matter. *Journal of Experimental Psychology: Animal Behavior Processes, 19,* 149–164.

Fetterman, J. G. (1995). The psychophysics of remembered duration. *Animal Learning & Behavior, 23,* 49–62.

Fetterman, J. G., & Dreyfus, L. R. (1986). Pair comparison of durations. *Behavioural Processes, 12,* 111–123.

Fetterman, J. G., & Dreyfus, L. R. (1987). Duration comparison and the perception of time. In M. L. Commons, J. E. Mazur, J. A. Nevin, & H. Rachlin (Eds.), *Quantitative analyses of behavior: Vol. 5. The effects of delay and intervening events on reinforcement value* (pp. 3–27). Hillsdale, NJ: Erlbaum.

Fetterman, J. G., & Killeen, P. R. (1991). Adjusting the pacemaker. *Learning and Motivation, 22,* 226–252.

Fetterman, J. G., & Killeen, P. R. (1992). Time discrimination in *Columba livia* and *Homo sapiens. Journal of Experimental Psychology: Animal Behavior Processes, 18,* 80–94.

Fetterman, J. G., & Killeen, P. R. (1995). Categorical scaling of time: Implications for clock–counter models. *Journal of Experimental Psychology: Animal Behavior Processes, 21,* 43–63.

Gibbon, J. (1981a). On the form and location of the psychometric bisection function for time. *Journal of Mathematical Psychology, 24,* 58–87.

Gibbon, J. (1981b). Two kinds of ambiguity in the study of time. In M. L. Commons & J. A. Nevin (Eds.), *Quantitative analysis of behavior,* (Vol. 1, pp. 157–189). Cambridge, MA: Ballinger.

Gibbon, J. (1991). Origins of scalar timing. *Learning and Motivation, 22,* 3–38.

Gibbon, J. & Church, R.M. (1981). Time left: Linear versus logarithmic subjective timing. *Journal of Experimental Psychology: Animal Behavior Processes, 7,* 87– 108.

Gibbon, J., & Church, R. M. (1990). Representation of time. *Cognition, 37,* 23–54.

Gibbon, J., Church, R. M., Fairhurst, S., & Kacelnik, A. (1988). Scalar expectancy theory and choice between delayed rewards. *Psychological Review, 95,* 102–114.

Grafe, T. U. (1996). The function of call alternation in the african reed frog (*Hyperolius–marmoratus*) – precise call timing prevents auditory masking. *Behavioral Ecology and Sociobiology, 38,* 149–158.

Grondin, S. (1993). Duration discrimination of empty and filled intervals marked by auditory and visual signals. *Perception & Psychophysics, 54,* 383–394.

Guilford, J. P. (1954). *Psychometric Methods.* New York: McGraw–Hill.

Haight, P. A., & Killeen, P. R. (1991). Adjunctive behavior in multiple schedules of reinforcement. *Animal Learning & Behavior, 19,* 257–263.

Hicks, R. E., Miller, G. W., & Kinsbourne, M. (1976). Prospective and retrospective judgments of time as a function of amount of information processes. *Americal Journal of Psychology, 89,* 719–730.

Hollis, K. L. (1983). Cause and function of animal learning processes. In P. Marler & H. S. Terrace (Eds.), *The biology of learning,* (pp. 357–371). Berlin: Springer–Verlag.

Huxley, J. S. (1914). The courtship habits of the great crested grebe (*Podiceps cristatus*). *Proceedings of the Zoological Society of London, 2,* 491–562.

Jasselette, P., Lejeune, H., & Wearden, J. H. (1990). The perching response and the laws of animal timing. *Journal of Experimental Psychology: Animal Behavior Processes, 16,* 150–161.

Jones, B. M., & Davison, M. (1996). Residence time and choice in concurrent foraging schedules. *Journal of the Experimental Analysis of Behavior, 65,* 423–444.

Kamil, A. C., Krebs, J. R., & Pulliam, H. R. (Eds.). (1987). *Foraging behavior.* New

York: Plenum Press.

Killeen, P. R. (1975). On the temporal control of behavior. *Psychological Review, 82,* 89–115.

Killeen, P. R., Cate, H., & Tran, T. (1993). Scaling pigeons' preference for feeds: Bigger is better. *Journal of the Experimental Analysis of Behavior, 60,* 203–217.

Killeen, P. R., & Fetterman, J. G. (1988). A behavioral theory of timing. *Psychological Review, 95,* 274–295.

Killeen, P. R., Palombo, G.-M., Gottlob, L., & Beam, J. (1997). A Bayesian analysis of foraging by pigeons (*Columba livia*). *Journal of Experimental Psychology: Animal Behavior Processes,* in press.

Killeen, P. R., & Weiss, N. (1987). Optimal timing and the Weber function. *Psychological Review, 94,* 455–468.

Landes, D. S. (1983). *Revolution in time.* Cambridge, MA: Belknap Press.

Laties, V, G., Weiss B., & Weiss, A. B. (1969). Further observation of overt "mediating" behavior and the discrimination of time. *Journal of the Experimental Analysis of Behavior, 12,* 43–57.

Logue, A. W. (1988). Research on self–control: An integrating framework. *Behavioral and Brain Sciences, 11,* 665–679.

Luce, R. D. (1959). *Individual choice behavior.* New York: Wiley.

Luce, R. D. (1963). Detection and recognition. In R. D. Luce, R. R. Bush, & E. Galanter (Eds.), *Handbook of Mathematical Psychology,* (Vol. 1, pp. 103–189). New York: Wiley.

MacEwen, D., & Killeen, P. (1991). The effects of rate and amount of reinforcement on the speed of the pacemaker in pigeons' timing behavior. *Animal Learning & Behavior, 19,* 164–170.

Machado, A. (1996). Learning the temporal dynamics of behavior. *Psychological Review,* in press.

Macmillan, N. A., & Creelman, C. D. (1991). *Detection theory: A user's guide.* Cambridge, England: Cambridge University Press.

Macmillan, N. A., & Creelman, C. D. (1996). Triangles in ROC space: History and theory of "nonparametric measures of sensitivity and bias. *Psychonomic Bulletin & Review, 3,* 164–170.

McCarthy, D., & Davison, M. (1980). On the discriminability of stimulus duration. *Journal of the Experimental Analysis of Behavior, 33,* 187–211.

McNeill, W. H. (1995). *Keeping together in time: Dance and drill in human history.* Cambridge, MA: Harvard University Press.

Ornstein, R. E. (1969). *On the experience of time.* Harmondsworth, England: Penguin.

Pavlov, I. P. (1927). *Conditioned reflexes* (G. V. Anrep, Trans.). London: Oxford University Press.

Platt, J. R., & Davis, E. R. (1983). Bisection of temporal intervals by pigeons. *Journal of Experimental Psychology: Animal Behavior Processes, 9,* 160–170.

Reid, A. K., Bacha, G., & Moran, C. (1993). The temporal organization of behavior on periodic food schedules. *Journal of the Experimental Analysis of Behavior,* *59,* 1–27.

Richelle, M., & Lejeune, H. (1980). *Time in animal behavior.* Oxford, England: Pergamon Press.

Saksida, L. M., & Wilkie, D. M. (1994). Time–of–day discrimination by *Columba livia. Animal Learning & Behavior, 22,* 143–154.

Sawyer, T. F., Meyers, P. J., & Huser, S. J. (1994). Contrasting task demands alter the perceived duration of brief time intervals. *Perception & Psychophysics, 56,* 649–657.

Schmidt, J. M., & Smith, J. J. B. (1987). Short interval time measurement by a parasitoid wasp. *Science, 237,* 903–905.

Selverston, A. I., & Moulins, M. (1985). Oscillatory neural networks. *Annual Review of Physiology,* (Vol. 47, pp. 29–48). Palo Alto, CA: Annual Reviews, Inc.

Shepard, R. N. (1965). Approximation to uniform gradients of generalization by monotone transformations of scale. In D. I. Mostofsky (Ed.), *Stimulus generalization,* (pp. 94–110). Stanford, CA: Stanford University Press.

Shepard, R. N. (1994). Perceptual–cognitive universals as reflections of the world. *Psychonomic Bulletin & Review, 1,* 2–28.

Shettleworth, S. J., & Plowright, C. M. S. (1992). How pigeons estimate rates of prey encounter. *Journal of Experimental Psychology: Animal Behavior Processes, 18,* 219–235.

Shurtleff, D., Raslear, T. G., & Simmons, L. (1990). Circadian variations in time perception in rats. *Physiology & Behavior, 47,* 931–939.

Skinner, B. F. (1989). The origins of cognitive thought. *American Psychologist, 44,* 13–18.

Staddon, J. E. R., & Simmelhag, V. L. (1971). The "superstition" experiment: A reexamination of its implications for the principles of adaptive behavior. *Psychological Review, 78,* 3–43.

Stevens, S. S. (1986/1975). *Psychophysics: Introduction to its perceptual, neural, and social prospects.* (2nd ed.). New Brunswick, NJ.

Stevens, S. S., & Galanter, E. (1957). Ratio scales and category scales for a dozen perceptual continuua. *Journal of Experimental Psychology, 54,* 377–411.

Stubbs, A. (1968). The discrimination of stimulus duration by pigeons. *Journal of the Experimental Analysis of Behavior, 11,* 223–238.

Stubbs, D. A. (1976). Response bias and the discrimination of stimulus duration. *Journal of the Experimental Analysis of Behavior, 25,* 243–250.

Stubbs, D. A. (1979). Temporal discrimination and the psychophysics of time. In M. D. Zeiler & P. Harzem (Eds.), *Advances in the analysis of behavior: Vol. 1. Reinforcement and the organization of behavior* (pp. 341–369). London: Wiley.

Thurstone, L. L. (1927). Psychophysical analysis. *American Journal of Psychology, 38,* 368–389.

Treisman, M. (1963). Temporal discrimination and the indifference interval: Implications for a model of the "internal clock". *Psychological Monographs,*

77 (whole No. 576)

Wearden, J. H., & Towse, J. N. (1994). Temporal generalization in humans: Three further studies. *Behavioral Processes, 32,* 247–264.

Wearden, J. H. (1995). Categorical scaling of stimulus–duration by humans. *Journal of Experimental Psychology – Animal Behavior Processes, 21,* 318–330.

Wearden, J. H., & Penton–Voak, I. S. (1995). Feeling the heat – Body–temperature and the rate of subjective time, revisited. *Quarterly Journal of Experimental Psychology Section B – Comparative and Physiological Psychology, 48,* 129–141.

White, K. G., & Cooney, E. B. (1996). Consequences of remembering: Independence of performance at different retention intervals. *Journal of Experimental Psychology: Animal Behavior Processes, 22,* 51–59.

Wilkie, D. M. (1995). Time–place learning. *Current Directions in Psychological Science, 4,* 85–89.

Wilson, M. P., & Keller, F. S. (1953). On the selective reinforcement of spaced responding. *Journal of Comparative and Physiological Psychology, 46,* 190–193.

Woodrow, H. (1951). Time perception. In S. S. Stevens (Ed.), *Handbook of experimental psychology,* (pp. 1224–1236). New York: Wiley.

Zakay, D. (1993). Time estimation methods – do they influence prospective duration estimates? *Perception, 22,* 91–101.

Zeiler, M. D. (1985). Pure timing in temporal differentiation. *Journal of the Experimental Analysis of Behavior, 43,* 183–193.

Zeiler, M. D. (1991). Ecological influences on timing. *Journal of Experimental Psychology: Animal Behavior Processes, 17,* 13–25.

Zentall, T. R., Sherburne, L. M., & Urcuioli, P. J. (1995). Coding of hedonic and nonhedonic samples by pigeons in many–to–one delayed matching. *Animal Learning & Behavior, 23,* 189–196.

Time and Behaviour: Psychological and Neurobehavioural Analyses
C.M. Bradshaw and E. Szabadi (Editors)
© 1997 Elsevier Science B.V. All rights reserved.

CHAPTER 4

Application of a Mode–Control Model of Temporal Integration to Counting and Timing Behavior

Warren H. Meck

Introduction

An animal's ability to adapt to a changing environment is determined to a large extent by the accuracy and precision of its perception and representation of space, time, and number –– three amodal stimulus attributes that are common to every niche and richly represented by a wide variety of species (Gallistel, 1990). Therefore, it seems reasonable to suppose that these three stimulus attributes have influenced the evolution of mind in a comprehensive and consistent manner. Considering the diversity of animals that use temporally patterned stimuli for purposes of communication, foraging, and reproduction (these stimuli may be chemical, electrical, tactile, visual, or auditory –– see Brown, 1975), it is readily apparent how important computational–representational systems are to behavior. Regular temporal patterns may be represented in one of two ways, either the time between successive pulses is measured (period), or the number of pulses occurring in some specified time is measured (rate). The use of representations of space, time, and number in the computation of other variables such as the rate of return and associated statistics of spatial and temporal uncertainty implies that the internal representations of these physical attributes are rich isomorphisms in a formal sense, not simply nominal ones. As Gallistel (1989, 1990) has indicated, animals do more with their representations of space, time, and number than make simple discriminations based on them. They manipulate these representations in computations whose validity depends on an isomorphism between internal relational and combinatorial operations and the external world. For example, ducklings apparently divide a representation of numerosity by a representation of a temporal interval in order to obtain a representation of rate, thus demonstrating the ability to interpret the meaning of a mallard hen's call (Gaioni & Evans, 1984).

Although a mapping from numerosity to numerons (representatives of numerosity) must be demonstrated if one is to claim that the nervous system of an animal represents number, the demonstration that such a mapping exists is not sufficient to establish the claim that the nervous system has a concept of number. For that claim to be supported, it must be shown that the system employs the representatives of numerosity in a manner consistent with what it is they are intended to represent. Consequently, the major goal of this paper is to review selected behavioral data obtained from rats performing in multi-modal temporal and numerical discrimination procedures that may be used to support a unified theory for the representation of both time and number. Accompanying these behavioral data will be a discussion of the psychological, computational, and physiological basis for these cognitive abilities.

At present, unified theories of timing and counting behavior can be found in two basic forms; a mode-control model of temporal integration (Meck & Church, 1983; Meck, Church, & Gibbon, 1985) and a multiple-oscillator connectionist model (Broadbent, Rakitin, Church, & Meck, 1993; Church & Broadbent, 1990). Each of these theoretical architectures has its own strengths and weaknesses that follow the principles of the parent paradigms of information-processing descriptions and connectionist networks (see Gluck & Rumelhart, 1990 for a review of these ideas). The mode-control model, therefore, concentrates on identifying distinct serial stages of analogical information processing that intervene between stimulus input and response output. The connectionist model attempts to define counting and timing processes within a framework of autoassociation networks that code the digital input from multiple oscillators and form part of a parallel distributed processing system. Since both models identify and account for roughly the same amount of variance in the behavioral data, the models are currently thought of as alternatives rather than as substitutions as one might have inferred from their developmental history. In fact, selected aspects of the two formalizations might yet be profitably combined in order to obtain a more robust theory for the realization of timing and counting processes. Consequently, for reasons of simplicity, the major focus of this review will be on the mode–control model of temporal integration.

Application of a Formal Analysis to Counting Behavior

Clearly animals can be trained to respond differentially to signals that vary along some stimulus dimension. In the case of the training procedures used in this paper, rats must learn which stimuli are relevant for predicting reinforcement and to classify some attribute(s) of these stimuli in order to determine their response. In our previous work on duration and number discrimination, e.g., Meck and Church (1983), we assumed that it was not always necessary to train animals to count or time explicitly, and although we observe them to be timing, they may also be counting and vice versa. In instances where the number of events is a relevant dimension, counting provides animals with representations of numerosity (numerons) for use in arithmetic reasoning (combinatorial and relational operations with representatives of numerosity).

Recently, a number of theorists have reached tentative agreement on a set of definitional principles for the identification of counting and related processes that may serve as the basis for analysis and interpretation of efforts to establish a sense of numerical competence in animals (e.g., Broadbent, Rakitin, Church, & Meck, 1993; Capaldi & Miller, 1988; Davis & Pérusse, 1988; Gallistel, 1990; Meck, 1987). These principles were originally developed in relation to specific instances of counting behavior in human infants (Gelman & Gallistel, 1978), and are now, by extension, used in the general case. This analysis consists of three definitional principles describing the attributes a counting process must have as the core of its representational system. In addition, at least two non-definitional principles specifying constraints which must exist in the application of numerosity judgments have been proposed. This formal analysis will be taken as the basis for a systematic interpretation of the results of a series of experimental studies investigating numerical competence in rats. This analysis will also serve as the basis for defining a concept of number in animals. It is within this context that non-human animals are said to have the ability to "count" using computational-representational mechanisms.

    To meet Gelman and Gallistel's (1978) criteria for counting, a system capable of discriminating stimuli varying in number must first be used to map these numerosities onto an internal representation (i.e., numerons). Thereafter, five axioms or principles of counting may be evaluated. The first of these principles' states that if the system in question is capable of mapping each element of the domain -- defined by elements in the to be counted set -- to one and only one numeron, then the *one to one principle* is observed. Adjunct to this principle is the restriction that the domain of numerosity be functionally limited by the number of numerons. The second principle requires that the order of mapping elements of the domain to states of the range must occur in identical order from one instance of implementation to the next, this is referred to as the *stable-order principle*. This is akin to our common sense notion of avoiding counting a five element sequence as 1-2-3-4-5 on one occasion and 1-3-4-2-5 on the next. One of the reasons that counting must be constrained by principles is to ensure that the representatives of numerosity generated by the behavior of counting can be validly employed in arithmetic reasoning. For example, the stable-order principle is necessary to ensure that when, in the course of arithmetic reasoning, one numeron is reckoned as greater than another, the set whose numerosity is represented by the first numeron is more numerous than the set whose numerosity is represented by the second (Gallistel & Gelman, 1990). The third and final defining principle states that the last numeron assigned as a representor of an element of the domain must represent the numerosity of the domain as a whole rather than simply representing the ordinality of the last element of the domain to be counted. Accordingly, this is referred to as the *cardinality principle*. The fact that these three principles apply only to the mapping process itself and not the mode or format of input gives rise to two additional, non–definitional, principles. The first non–definitional principle, called the *order irrelevance principle*, further defines the stable order principle by stating the irrelevance of the order in which elements of the *domain* are mapped to numerons as long as the one to one principle is obeyed, the only relevant issue in the establishment of counting is the order in which the numerons are assigned. The second non–definitional principle, called the *abstraction principle*, indicates the lack of constraint on what items can be counted by emphasizing that the characteristics of the items in a set are irrelevant to the numerosity of the set. In practice, the abstraction principle has become central to certain arguments pertaining to the establishment of a concept of number in animals owing to its pertinence in the transfer of counting behavior to novel instances (e.g., Davis & Pérusse, 1988). Because the domain is irrelevant to the functioning of the counting process it may consist of any set, as long as that set is finite so as to maintain the existence of some cardinality for that set. In particular, it is assumed in the present case that because the homogeneity of the set is irrelevant to the mapping process, the domain of the set of items to be counted may be multimodal, and therefore direct and immediate cross–modal or cross–procedural transfer to novel stimulus sets is a property inherent to the counting process as delineated by Gelman and Gallistel (1978).

Application of a Mode–Control Model of Temporal Integration to Number Discrimination

Meck and Church (e.g., Church & Meck, 1984; Meck & Church, 1983; Meck, Church & Gibbon, 1985) have presented a substantial body of evidence pointing to the existence of a general–purpose mechanism underlying the quantification of both duration and number. These data and the associated mode–control model of temporal integration for the

discrimination of duration and number are of special significance to the
ongoing definitional debates in the study of numerical competency in
animals. In effect, the mode–control model incorporates the idea that
the nervous system inverts the representational convention whereby
numbers are used to represent linear magnitudes. Instead of using
number to represent magnitude, it is proposed that the nervous system
of animals uses magnitude to represent number. Furthermore, the
mode–control model posits a form of numeron that is robust in the
definitional sense and addresses the general question of whether
duration and number are represented in a like manner for purposes of
computation.

In this fashion the mode–control model provides a unified theory
of duration and number discrimination. In addition, the model is able
to accommodate a relatively rich display of counting behavior on the
part of rats while at the same time offering a plausible mechanism for
the realization of timing and counting in animals that accounts for a
relatively high percentage of the variance in the behavioral data
without producing systematic discrepancies in the residuals (Meck &
Church, 1983). The idea of a single mechanism for both timing and
counting grew out of the heavy emphasis placed on the computational
aspects of modeling the choice behavior of rats performing in temporal
bisection procedures (e.g., Gibbon, 1981). The period of validation for
the information–processing model of animal timing behavior
developed by Gibbon and Church (1984) gave rise to speculation about
the possible application of scalar timing theory to numerical
discriminations. Because the computational model of timing behavior
referred only to the quantitative variables accounting for the form and
distribution of variance in specific data sets, it did not preclude a single
mechanism hypothesis for both timing and counting provided that
animals could base their behavior on the magnitude of a generic
representation, independent of what that representation stood for. If
one assumes the existence of an isomorphism between duration and
number, as first described by Meck and Church (1983), then the
possibility exists for developing a unified theory of duration and
number discrimination.

What follows is a description of the computational model as it pertains to the temporal bisection procedure (Church & Deluty, 1977; Gibbon, 1981; Meck, 1983, 1991). It stands as a fully functional description of timing behavior, and by extension counting behavior, independent of certain definitional (cf., Davis & Pérusse, 1988) and neurophysiological concerns (cf., Amit, 1989). The description is meant to clarify the history of the mode–control model of temporal integration by presenting the definitional variables in the formats in which they are used. In the temporal or numerical bisection procedure subjects are initially trained to discriminate between two anchor values by providing feedback for correct signal classifications. For example, in order to train a discrimination between a 2–sec signal duration and an 8–sec signal duration one of these two signals (e.g., durations of white noise) are randomly selected with equal probability for presentation on each trial. After the signal presentation, two response levers are inserted into the lever–box. If the animal presses the left lever after a 2–sec signal presentation the response is reinforced by the delivery of a food pellet and the trial terminated, if it presses the right lever after an 8–sec signal presentation the response is reinforced and the trial terminated. All other responses simply result in the termination of the trial without the delivery of reinforcement and after a relatively long intertrial interval (e.g., 30 sec) another trial begins. After this initial phase of two–signal training, other signals of intermediate value are presented and the animal is required to classify these signals as being more like the 2–sec signal or more like the 8–sec signal. No feedback is provided for any responses to these intermediate signal values. In addition to the type of response (e.g., left or right), the response latency is recorded following each signal presentation. The function relating the probability of a "long" or right response to signal value typically rises in a sigmoidal fashion from a point near 0% "long" response to a point close to 100% "long" response. The point of subjective equality (PSE –– 50th percentile) indicates the signal value that the subject treats as lying midway between the two anchor values, it serves to evaluate the various models of temporal integration (see Gibbon, 1981). The difference limen (DL) is obtained by calculating the slope of the steepest part of the response function (typically the central region) and using this measure to index the sensitivity of the discrimination processes involved. A Weber fraction (WF) may also be calculated by dividing the PSE by the DL.

Computational Aspects of the Mode–Control Model

The mode–control model proposes a description of duration and number discrimination that is nearly identical to the information-processing model of timing first presented by Gibbon and Church (1984). These investigators outlined the functional units of the timing model and later specified the stimulus transformations that could occur between them (Gibbon, Church, & Meck, 1984). Certain aspects of this computational model of animal timing were later refined by Gallistel (1990) in accordance with his time–of–occurrence model. In general, the mode–control model views the processing of both duration and number in the following way. A pacemaker is proposed that emits a regular stream of pulses at a constant, but alterable frequency (λ). In this model three physical durations are defined: The duration of an external stimulus (T), the latency of an internal accumulation process to begin accumulating pulses when the external stimulus begins (t_1), and the latency of an internal integration process to stop accumulating pulses when the external stimulus ends (t_2). The magnitude of pulses recorded by an internal accumulation process would be $D = T - t_1 + t_2$, or $T - (t_1 - t_2)$. T_0 is defined as $t_1 - t_2$ so that $D = T - T_0$. Under some conditions the value of T_0 has been found to be greater than zero (Meck, 1984; Meck & Church, 1984). These include conditions during which the stimulus onset is unexpected so that the latency to begin the accumulation process (t_1) is longer than the latency to stop accumulating (t_2). In principle, the value of T_0 could be less than zero. For example, if the onset of a signal was expected but the termination was not, t_1 might be shorter than t_2.

At the onset of a relevant stimulus pulses may be gated into an accumulator which integrates the number of pulses over time (Roberts & Holder, 1984). This gating is accomplished by a mode switch which is regarded as closing with some latency after the onset of the stimulus. This switch gates the pulses to the accumulator in one of three different modes depending on the nature of the stimulus, giving this mechanism the ability to act as both a counter and a timer based on the representation of the variable in the accumulator. In the "run" mode the initial stimulus starts an accumulation process that continues until the end of the signal or trial; in the "stop" mode the process occurs whenever the stimulus occurs; in the "event" mode each onset of the stimulus produces a relatively fixed duration of the process regardless of stimulus duration. This is a mechanism that can be used either for the estimation of duration (the run and the stop modes) or for the estimation of number (the event mode).

There is some evidence that the same neural network can be activated in either the run, the stop, or the event mode (Swigert, 1970). In order to function as a counting network the neural system must be shown to have several dimensions: 1) the recognition of the arrival of several identical stimuli requires some mechanism of short–term memory that does not transform into long–term memory, 2) the identification of generic stimulus characteristics in order to provoke counting, 3) the discrimination between different temporal sequences of stimuli according to the abstract property which is their cardinal number, and 4) robustness to variations in the characteristics (e.g., duration, intensity, location, modality, and period) of each of the stimuli. This robustness should be limited, however, in accord with observed cognitive instances of the segregation of items to be counted. It is also clear that the presence of the network in each of the "number states" needs to be long enough for the output neurons to be able to identify the particular cognitive event as a count. Furthermore, when the stimulus sequence stops, the network will spend a significantly longer time in the last state. This again is an identifiable event, which may be biologically distinguished from the numbers encountered on the way, each of which is distinguishable from each other. After a while, significantly longer than is required for identification of the final count, the network should shift from the last "number state" into a state that is uncorrelated with any of the states in the sequence. There it is ready to again start the counting process. Attractor Neural Network (ANN) models are capable of meeting this list of specifications (e.g., Amit, 1989) and may have some relationship to the behavioral evidence for an internal mechanism that can be operated in several modes. For purposes of illustration, a stimulus with varying durations of signal–on and signal–off segments is presented in Figure 1. Included with this stimulus is a diagram of the three modes of operation of the proposed accumulation process that must be mediated by the neural network.

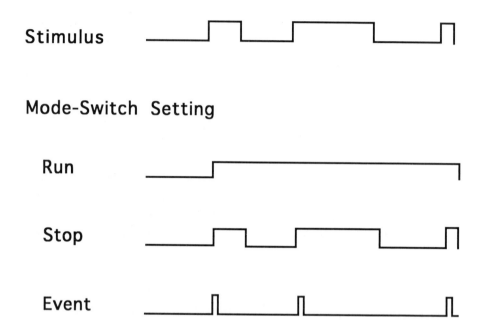

Figure 1. Diagram of three modes of operation of the accumulation process used for timing (run and stop modes) and counting (event mode). Increases in the time line indicate signal–on periods and decreases in the time line indicate signal–off periods. (Adapted from Meck & Church, 1983.)

In all modes of operation the value stored in the accumulator increases as a function of the stimulus duration or segment number, and a comparison is made in real time between the current accumulator reading and values sampled from the distributions of previous signal values associated with reinforcement and stored in long–term memory. A response is made according to a formal response rule establishing a threshold of similarity between the elapsed subjective duration and/or number and the remembered durations and/or numbers associated with the anchor values. When the signal terminates and reward is delivered following a correct response the mode switch opens, stopping the gating of pulses to the accumulator. If a response is reinforced, the current value in the accumulator is multiplied by a constant (scalar distortion -- see Gibbon, Church, & Meck, 1984; Meck, 1983) as it is read into long–term memory. The accumulator is then reset to zero and the animal is considered ready for another trial.

The precise form of the psychophysical function provides evidence for the discrimination processes involved. In the case of the temporal bisection function, the PSE is near the geometric mean of the two anchor values and the WF is fairly constant over a wide range of signal durations (e.g., Church & Deluty, 1977; Meck, 1983). Gibbon has developed several models of time estimation that are compatible with these results (Gibbon, 1981). The version he refers to as the "sample known exactly with a similarity decision rule" provides the best fits and the least systematic deviations from the data (but see Siegel, 1986, for a criticism of the similarity decision rule in temporal bisection). In this version, the pacemaker is assumed to emit pulses at some fixed interpulse interval on any given trial. Across trials the interpulse interval is assumed to be normally distributed with some mean and standard deviation. On the basis of experience the animal learns the accumulator value associated with reinforcement of a left or right response. This information is stored in long–term memory. On any given trial, the animal has a representation of the current accumulator value. In the two–lever signal classification task described above, the animal makes a left response if this value is closer to the value in long–term memory associated with a reinforced left response; it makes the right response if this value is closer to the value in long–term memory associated with a reinforced right response. The measure of distance is a ratio between the accumulator value and the long–term memory values, as described below.

The following account illustrates this process: On a particular trial the signal is presented for t sec, and the animal's pacemaker runs at some rate (λ) that is a normally distributed random variable with a mean (Λ) and some standard deviation (σ). Λ_2, the pacemaker speed at the time of a test, is potentially distinguished from Λ_1, the pacemaker speed at the time of training (Church & Meck, 1988; Meck, 1983). The mode switch is closed for some period of time (d), and pulses from the pacemaker enter the accumulator during this time. The number of pulses in the accumulator is $n_1 = \lambda d$. The ratio of the standard deviation to the mean (γ) is called the coefficient of variation. The lower the ratio, the more sensitive the animal is to time. As γ approaches 0, n_t becomes an increasingly accurate estimate of t.

When the signal value is S, a left response is reinforced. The mean accumulator value at the time of reinforcement is $N_S = \Lambda D_S$, where capital letters are used for expected values. We assume that this value is remembered accurately. By a similar argument, when the signal value is L, a right response is reinforced, the mean accumulator value at the time of reinforcement is $N_L = \Lambda D_L$, and we assume that this value is remembered exactly.

On a given trial, when the signal value is t, the accumulator value is $n_t = \lambda d_t$. The animal compares the distance of the current accumulator value to two samples of accumulator values that were stored in long–term memory, one associated with reinforcement of the "many–long" response m^*_L and the other associated with reinforcement of a "few–short" response m^*_S. The animal responds to the closer one. The specific rule is respond "many–long" if $m^*_L/n_t < n_t/m^*_S$, that is, respond "many–long" if $n_t >$ the square root of $m^*_S \times m^*_L$; which is the geometric mean of the two anchor values. Relative terms (e.g., "few" vs. "many" and "short" vs. "long") are used here in place of the absolute stimulus values in order to provide a generic description of the process. Given the obtained psychophysical functions and the computations required to produce such functions the animal's representations of signal duration undoubtedly incorporate the absolute values of the stimuli.

If the values stored in long–term memory are accurate representations of the values in the accumulator, indifference is predicted at the geometric mean of S and L. This process may be simulated or calculated with the equations derived by Gibbon (1981, Equation 17) that, in the special case of the referents known exactly, are given in Equations 45a and 45b.

Subjects may not always attend to the signal value. On such trials, the process described by the usual decision rule cannot apply. It is assumed that on any particular trial the animal is attending to the signal value with some probability, $p(A)$, and its behavior is described by the process outlined above; however, with probability $1 - p(A)$, the animal is not attending to the signal value, and on these trials its probability of responding "many–long" is some constant bias, $p("R"/\hat{A})$. The term $p(A)$ is called attention and $p("R"/\hat{A})$ is called responsiveness or the probability of a "many–long" response given inattention (cf., Church & Gibbon, 1982).

The formal model can be described with T as a continuous variable identified as the time since stimulus onset, or it can be described with T as a discrete variable identified as the number of events since stimulus onset. In the past we have been able to estimate T_0, but we have not been able to estimate its components (t_1 and t_2) separately. Recent numerical–bisection experiments (e.g., Meck & Church, 1983; Meck, Church, & Gibbon, 1985) have provided a method to make separate estimates of t_1 and t_2. These analyses have indicated that the latency to begin timing or counting (t_1) and the latency to stop timing or counting (t_2) were both close to 200 ms. According to this type of analysis, the latency to begin a temporal integration process and the latency to stop this same process can be measured by the slope and intercept of a linear function that relates the estimated time to the number of segments of a stimulus required to produce a particular estimate of signal duration.

A Review of the Evidence for the Similarity of Duration and Number Discrimination

The experimental history of a unified theory of timing and counting is found primarily in Church and Meck (1984), Meck and Church (1983), and Meck, Church and Gibbon (1985). These experiments are a direct result of the proposal that the psychological processes involved in duration discrimination procedures might be generalizable to other domains. At present, the evidence for a unified theory is derived exclusively from modifications of the bisection procedure. As indicated above, this procedure involves a bifurcated response paradigm used to classify the anchor values of some range of signal durations. Presentation of intermediate signal durations produces a non–linear response function generated by the probability of a response on one of two levers. The sources of variance in this procedure are described by Gibbon's computational model (Gibbon, 1981). What follows is a brief review of the experiments that provide the support for a mode–control model of temporal integration in both timing and counting behavior.

The first experiment in this series (Experiment 1: Meck & Church, 1983) involved a deliberate confounding of duration and number variables by using compound signals composed of repeating cycles of white noise. The one cycle/sec signal had 0.5 sec sound–on periods alternated with 0.5 sec sound–off periods; the number of cycles was equal to the number of sound segments. Consequently, during training the signals differed in two ways: number of sound segments (two or eight) and total signal duration (2 or 8 sec). In this experiment the standard temporal bisection procedure was modified so that a left lever response was reinforced following a discontinuous signal of 2 sec in duration divided into alternating 0.5 sec segments of sound and silence each of equal duration. Right lever responses were reinforced following a signal 8 sec in duration and likewise divided into alternating sound–on and sound–off segments 0.5 sec in duration.

The dual nature of the interpretation of a choice response stems from the imposed discontinuity of the training signals. The cyclic signal presentation precludes the possibility of telling whether the animal is timing the duration of the signal, counting the number of cycles or sound segments in the signal, or both. During testing the conditions of training were maintained except for the addition of probe signals. On half of the trials one of the two training signals was presented with equal probability. On the remaining trials, two sets of probe signals were presented. One set of probe signals held duration constant at 4 sec while the number of cycles was allowed to vary between two and eight. The other set of probe signals had the number of cycles fixed at four while total signal duration was allowed to vary between 2 and 8 sec. Consequently, rats that had been trained to classify signals based on the simultaneous presentation of both duration and number cues were now required to reveal the degree to which these two dimensions independently controlled responding. The choice of 4 sec or 4 cycles as the intermediate control value for both duration and number discrimination stems from the work of Church and Deluty (1977) in establishing the geometric mean of the anchor values as the subjective middle in a range of temporal bisection signals; a point which by extension should also apply to the numerical bisection condition. The experimental design, including the varying lengths of the sound cycles is shown in Table 1.

Table 1: Design of Experiment 1: Meck & Church, 1983: Testing

Total no. of cycles	Total duration (in sec) of signal					
	2	3	4	5	6	8
2	L(1.00)		X(2.00)			
3			X(1.33)			
4	X(.50)	X(.75)	X(1.00)	X(1.25)	X(1.50)	X(2.00)
5			X(.80)			
6			X(.67)			
8			X(.50)			R(1.00)

Note: L = left response reinforced, R = right response reinforced, X = neither response reinforced; each number in parenthesis represents duration in seconds of a single cycle.

The psychophysical functions for both duration and number were fit equally well by the scalar timing theory using a "referents known exactly" response rule (see Gibbon, 1981). Neither the PSE, the DL, nor the Weber fraction differed significantly as function of signal duration or number.

These results indicated that the rats acquired both the duration and the number discrimination during original training with compound signals and, owing to the similarity of the two discriminations as revealed by the associated psychophysical functions, either two separate mechanisms with very similar properties were at work or a single mechanism was used to perform both discriminations.

The single mechanism hypothesis was evaluated pharmacologically by administering methamphetamine to rats trained to produce psychophysical functions for duration and number (Experiment 2 -- Meck & Church, 1983). Establishing that the PSE of a numerical bisection procedure was isomorphic to that of a temporal bisection procedure with a similar range and ratio of signal values provided an easily derived variable with which to compare attempts to selectively modulate the individual discriminations. The stimulation of central dopaminergic activity with methamphetamine is known to produce a proportional leftward horizontal shift in the PSE of psychophysical functions obtained from rats trained with the temporal bisection procedure (e.g., Maricq & Church, 1983; Meck, 1983; 1986; 1988). Neuroleptic drugs that block rather than stimulate dopaminergic receptors (e.g., haloperidol) produce the opposite effect, i.e., a proportional rightward horizontal shift of the psychophysical functions (e.g., Maricq & Church, 1983; Meck, 1983; 1986; 1988). These effects have been interpreted as a modification in the rate of pulse accumulation, i.e., an increase in clock speed for drugs that stimulate dopamine receptors and a decrease in clock speed for drugs that block dopamine receptors. Similar horizontal shifts in the numerical bisection function of equal magnitudes as a function of drug dose would implicate specific neurochemical pathways and therefore support a single mechanism hypothesis, while dissimilar modulations would deny such a hypothesis. To identify the pharmacological substrates involved, subjects were injected on alternating days with saline and then methamphetamine. The results of pharmacological testing with a dose of 1.5 mg/kg of methamphetamine demonstrated a statistically significant drug–induced leftward shift of approximately 10–15% in the PSE of the psychophysical functions compared with functions obtained during saline control sessions. The psychophysical functions for duration and number did not differ from each other as a function of drug treatment. These findings replicate previous reports of the effects of methamphetamine in temporal bisection procedures. The results for the numerical bisection procedure constitute a degree of pharmacological validation for a single mechanism hypothesis. Additional studies employing a greater range of dopaminergic drugs with different receptor profiles, as well as a wider range of drug doses and signal values would be required to firmly establish these pharmacological relationships between timing and counting (See Meck, 1986 for an example of how this has been done in the case of temporal discriminations.)

As discussed above, the literature on concept formation places emphasis on transfer to novel stimuli and procedures as the criteria for determining the presence of a concept of number. Owing to the similarity of the psychophysical functions derived from numerical bisections with those derived from temporal bisections it was assumed that it would be possible to demonstrate cross–modal transfer of numerical discriminations as had been established in the case of temporal discriminations (Church & Meck, 1982; Meck & Church, 1982). Rats were evaluated in the cross–modal transfer of temporal and numerical discriminations from a set of discontinuous auditory signals to a set of discontinuous cutaneous signals mixed with auditory signals (Experiment 3: Meck & Church, 1983). Training involved the presentation of a set of four signals combining duration and number in a manner similar to the one described above. Testing was identical to training except that on half of the trials, two sets of probe signals consisting of alternating auditory and cutaneous segments of equal duration were presented. The "number relevant" set of probe signals held duration constant at 4 sec and had either one or four cycles of combined auditory and cutaneous stimulation, while the "duration relevant" set of probe signals held the number of cycles constant at two and had a total signal duration of either 2 or 8 sec. Note that for both duration and number related sets of probe signals the anchor values were held in a 4:1 ratio with a common high value of eight. This is due to each cycle now being considered as two countable elements instead of one. That is, the previous eight–cycle signal with alternating periods of sound–on and sound–off should generalize to the four–cycle signal composed of combined auditory and cutaneous signals. The results of this cross–modal transfer test showed that the rats learned to discriminate number of segments as well as duration, consistently labeling the four–cycle signal as "long/many" and therefore equivalent to eight cycles in the baseline training. The probe signal containing four auditory segments in combination with four cutaneous segments produced a "long" response on approximately 89% of trials whereas that same signal value for sound–on segments alone had previously been classified "long" on approximately 50% of the trials during baseline training. This result strongly suggests that the cutaneous segments accounted for the change in probe signal classifications. This demonstrates cross–modal transfer of a numerical discrimination where the ability to enumerate auditory stimuli is immediately transferred to cutaneous stimuli without the need of reinforcement –– because the classification of probe signals was never followed by reinforcement.

While the first experiment described above developed the testing paradigm and strengthened the existing links between duration and number discrimination, and the experiments dealing with the pharmacological modulation of the rate of pulse accumulation and the cross–modal transfer of classification rules provided indirect evidence in support of a single mechanism hypothesis, the last two experiments to be described here sought direct evidence in the form of a psychophysical equivalence between the two variables. That is, there should exist some quantitative relation between "counts" and "times" such that a fixed number of seconds could be translated into a fixed number of events and vice versa. Under a single–mechanism hypothesis, the response classification learned for a signal of some fixed number of cycles should, under the appropriate circumstances, show substantial transfer to another signal that meets the specified accumulation regardless of the signal's modality or the mode setting that produced the accumulation. To restate a point that was made earlier, animals should in effect base their behavior on the magnitude of the representation, independent of what that representation stands for (i.e., duration or number). It was obvious from the first experiment that any two values of duration and number could be confounded for the purpose of training a discrimination, but it was not immediately apparent what sets of values would show spontaneous transfer (in either direction) when the animals were initially trained with only one of the variables. The range of possible relations were presumably endless but the experimenters did not start from scratch; there are numerous reports from the physiology literature indicating that dopaminergic "pacemaker cells" are known to fire in bursts every 200 msec (e.g., Bannon & Roth, 1983; Bunney, 1979; Steinfels, Heym, Strecker, & Jacobs, 1983). Based on this electrophysiological evidence our initial hypothesis assumed a range of 100–400 msec per pulse for each count increment. The two experiments each used rats initially trained on a temporal bisection procedure with 2 and 4 sec anchor values serving to define the signal range. In the initial experiment (Experiment 4: Meck & Church, 1983) the animals were trained to respond to different levers as a function of two signal values (e.g., 2 and 4 sec). A second phase of training used five non–reinforced signal values spaced at equal logarithmic intervals between 2 and 4 sec. A second experiment (Experiment 1: Meck, Church, & Gibbon, 1985) had an additional phase of training where rats initially trained with a 2 to 4 sec signal range were transferred to a signal range having equiratio endpoints but a narrower range, i.e., 1 to 2 sec. This manipulation provided for a comparison between absolute vs. relative accuracy of temporal–numerical discriminations for both experiments. The test phase for both experiments involved presenting the seven signals from the original training phase on a random half of the trials along with five probe signals drawn from one of two sets of discontinuous auditory signals on the other half. These auditory probe signals varied in the number of sound–on and sound–off cycles. Each cycle was composed of a 1–sec sound–on segment and a 1–sec sound–off segment such that the shortest probe signal was longer than the longest training signal. Pilot studies indicated that a mapping of a 5:1 ratio of event increments to seconds would produce a range for the probability of a "long" response between 0% and 100%.

For both of the experiments taken together, the median ratio estimate of duration in seconds to number of events was 200 msec ± 10 msec, as determined by fitting linear regression lines to the three central signal values for each of the two psychophysical functions for each rat and then mapping one onto the other and determining the slope of the fitted lines. A direct comparison was also made of the PSE's (signal values at the 50% "long/many" point) for each of the two functions. These analyses yielded an average ratio of 2.84 sec/14.1 events, or approximately 200 msec/event increment. This result fits nicely with both the electrophysiological results described above and the behavioral findings of the various pilot studies conducted in our laboratory. On the whole, these experiments provide the strongest evidence for a direct quantitative relationship between duration and number.

To summarize: The data reviewed here indicate that the discrimination of the number of successive, temporally–spaced events can be accounted for by a mode–control model of temporal integration that is based on the hypothesis that the timing and counting mechanisms used by animals are intimately related, both computationally and neurobiologically. The major hallmarks of this model are that a temporal integrator can be used in several modes: the "event" mode for number discrimination or the "run" and the "stop" modes for duration discrimination. This proposal has guided research showing that (1) rats are equally sensitive to a 4:1 ratio of events and a 4:1 ratio of durations; (2) the PSE between two numbers or durations used as anchor values is at the geometric mean for both attributes; (3) pharmacological manipulations of central dopaminergic systems affect the discrimination of both duration and number in an identical fashion; (4) the magnitude of cross–modal transfer from auditory to cutaneous stimuli is similar for duration and number; (5) and the mapping of number onto duration has demonstrated that a numeron (i.e., "count") is approximately equal to 200 msec. This fifth similarity between duration and number discrimination is the strongest because it claims that there is a measurable quantitative equivalence between a unit of time and a "count".

Application of the Formal Analysis to the Mode–Control Model

Having set forth the mode–control model and the supporting data as it currently stands, the definitional issues of counting may now be addressed. The mode–control model is assumed to be a model of counting when the switch is set to the "event" mode. In this condition discreet stimuli are marked by a fixed increment in the accumulator. It is this temporal integration process that represents the numerosity of events or objects and thus constitutes this model's proposed numeron, just as this same temporal integration process represents duration when pulses are gated through the switch in the run or stop modes. Does this mode–control model constitute a valid model of counting within Gelman and Gallistel's (1978) definitional criteria? Our review of the evidence provided here strongly suggests that the mode–control model of temporal integration does indeed meet the defining characteristics of counting behavior as developed by Gelman and Gallistel (1978). The value in the accumulator after a certain number of events is equal to 200 msec multiplied by the number of events, plus a relatively small amount of pacemaker variance. In this case, the number of events is uniquely represented by the value of the product of temporal integration up to that point, thus fulfilling the *one–one principle* which demands the assignment of one and only one numeron to each item or event. The *stable order principle* which requires that the order of assignment of numerons must be the same from one occasion to the next is obviously fulfilled because in no case can the accumulator operating in the event mode produce anything other than a fixed set of values whose order never varies. The *cardinality principle* is also met because the value of the accumulator at the time of signal termination is taken as the representation of the total numerosity of the stimulus array; the value of which is then remembered for future comparison. The mode–control model imposes no restrictions on which items can be counted thus satisfying the *abstraction principle*, nor does it stipulate some stimulus–specific order of presentation for items to be counted thus honoring the *order–irrelevance principle*.

The implications for a concept of number within the mode-control framework are, as pointed out earlier, contrary to convention. The model suggests that counting is a basic and natural process tied to the regulatory function of a pacemaker system. Numerosity is represented through mediation by the linear magnitude of an internal variable, and it is a representation of this value which is remembered between trials. One could postulate then that the basis for a general concept of number in the rat is the ubiquitous functioning of the pacemaker system in conjunction with a mode switch and appropriate stimulus combination rules for computation. The rat has a general purpose temporal integration system that can be implemented over any serial presentation of events, implying that duration and number are very basic perceptual attributes. The application of this model to simultaneous events would also be straight–forward (see Church & Meck, 1984; Meck & Church, 1984).

The mode–control model successfully accounts for not only the behavior of rats in numerical bisection procedures, but also for the links between timing and counting behaviors. It has accounted for the results of pharmacological manipulations, indicating that it does not stand contrary to the neurophysiological substrates of the processes. Consequently, it has proved a viable interpretation of the computational model.

Some New Evidence Suggesting a Fundamental Equivalence Between Duration and Number Discrimination

General Methods.

Subjects.
Sixty–five experimentally naive female rats selected from 12 litters of timed–pregnant Long–Evans dams obtained from Charles River Laboratories, Stone Ridge, NY were used. Behavioral training began when the subjects reached approximately 4 months of age. One

week prior to this period the rats were placed on a 24-hr food deprivation schedule during which they were given approximately 12 g/day of Purina Rat Chow with continuous access to water in their home cages. Rats were individually housed in transportable polycarbonate cages. A light-dark cycle of 12:12 hr. was maintained in the vivarium with fluorescent lights on from 8:00 - 20:00 EST.

Apparatus.

Lever Boxes. In Experiments 1, 2, and 3 rats worked in 12 similar operant conditioning boxes with two response levers and associated equipment (e.g., pellet/water dispensers, cue lights, and sound generators) that have been described elsewhere in detail (Meck, 1986). The outlet for the pellet/water dispenser was located in the center of the intelligence panel between the two response levers. Immediately above both response levers was a row of three colored cue lights with red, white, and green cue lights arranged from left to right. A speaker panel was centered approximately 20 cm above the outlet for the pellet/water dispenser. Noyes (45 mg) food pellets were used to reinforce appropriate lever presses. A time-shared IBM AT clone computer controlled the experimental equipment and recorded the data.

12-Arm Radial Maze. An elevated radial-arm maze composed of twelve identical alleyways, spreading out from a central elevated platform at equal angles was used in Experiments 2 and 3. The radial-arm maze was constructed such that travel between different maze arms required the subject to return to the central platform prior to each separate selection. Food wells, designed to conceal reinforcers from plain view, were located at the ends of every arm. The maze was placed in a test room that was approximately 4.5 m x 4.8 m. The room was illuminated with fluorescent light and contained a variety of extra maze stimuli in relatively fixed positions which included the cart used to transport the rats and an open doorway in which the experimenter sat and recorded the data. The maze was constructed of plywood and Masonite board and was painted flat gray. Maze dimensions were as follows: central platform = 36.5 cm diameter; height above the floor = 80 cm; arms = 83 cm long x 7.6 cm wide, with edges 1.2 cm high along each side; food wells = 2.5 cm diameter, and 0.6 cm deep. See

Meck, Smith and Williams (1988) for additional details about the test room and apparatus.

Pretraining Procedures.

Lever Box Pretraining. Each rat received at least two sessions of magazine and lever training. During these sessions a pellet of food was delivered once every 30 sec in conjunction with a 1–sec illumination of the white cue light above the lever, and in addition, each lever press produced food. The session continued until the rat had pressed the lever 60 times or until 30 min had passed, whichever came first. If the task required the use of both levers (Experiments 1 and 2) the Pretraining just described was alternated between the two levers in the same session with 30 reinforcements being obtained on each lever.

12–Arm Radial Maze Pretraining. Rats were trained to run to the ends of the arms in order to obtain food. During pretraining the food wells at the end of each arm were baited with approximately 10–15 Noyes (45–mg) food pellets. Rats were placed on the maze in groups of 4–6 after a period of 24 hr. food deprivation and allowed to explore the maze for 25–30 min. Food pellets in each food well were replenished as necessary. Any subject not readily exploring the arms of the maze was physically forced down the arms and permitted to eat from the food wells. After 3–4 days of this shaping procedure all rats readily ran to the ends of the arms and consumed food pellets.

Experiment 1: Training on compound signals with "outside" values: Similarity versus proximity decision rules in the bisection of number and duration.

The similarity decision rule described above and applied to temporal bisection data by Church, Gibbon, and Meck (e.g., Gibbon, 1981; Meck, 1983; Meck & Church, 1983) requires that "outside" signal values (number or duration) less than the lower reinforced anchor value or greater than the upper reinforced anchor value be classified by the rat as being "short" or "long" at least as often as the closest anchor value. A failure of this prediction has been shown to occur for signal duration when rats are consistently trained with "outside" signal values as opposed to the presentation of these signal values on probe trials following the acquisition of a discrimination without the inclusion of "outside" signal values (Siegel, 1986). Based on these findings from maintained discrimination training with "outside" signals durations it has been suggested that a proximity decision rule rather than a similarity decision rule may be more appropriate for the bisection procedure (Siegel & Church, 1984). This modification brings the bisection procedure more into line with the types of decision rules used for the temporal generalization and peak–interval timing procedures (Church, 1984, 1989; Gibbon, Church, & Meck, 1984; Meck & Church, 1984).

The result of the inclusion of "outside" signals in the training of a duration discrimination adds additional support to the claim that animals are not simply performing a "short" vs. "long" discrimination when they are trained in bisection tasks. Rather it appears that they are producing a two–point generalization gradient that is mediated by the absolute discrimination of the two reinforced anchor values. The basic logic of the bisection procedure is still valid under these conditions, i.e., the point of bisection still serves as a measure of the subjective middle between the two trained values, but the decision process is based primarily on the absolute discrimination of signal duration. This raises the question as to whether a similar result would occur for the bisection of number if "outside" signal values were included in the training procedures. In order to answer this question, rats were trained to discriminate between 4 cycles and 8 cycles of an auditory stimulus that could be classified either by number or duration. Rats were trained to press one lever (e.g., left lever) following four sound segments composed of 0.5 sec sound presentations alternating with 0.5 sec silent periods with the number of cycles equaling the number of sound segments and to press the other lever (e.g., right lever) following eight such sound segments.

Method

Subjects. Five experimentally naive female rats identical to those described above (see General Methods) served as subjects.

Apparatus. The lever boxes used in this experiment were identical to those described above (see General Methods).

Procedures

Two–Signal Training (Sessions 1–15). The rats were first trained to press the left lever following a one cycle/sec sound signal of 4–sec duration ("4" response) and to press the right lever following a one cycle/sec sound signal of 8–sec duration ("8" response). The one cycle/sec signal had 0.5 sec sound–on periods alternated with 0.5 sec sound–off periods; the number of cycles was equal to the number of sound segments. These signals differed in two ways: number of sound segments (four or eight) and total signal duration (4 or 8 sec). On each trial one of the two signals was randomly presented with a probability of 0.5. When the signal was turned off, the white cue lights over both levers were illuminated indicating the opportunity for a response to be reinforced. If the rat made the correct response, a pellet of food was immediately delivered; if the rat made the incorrect response, no food was delivered. When either lever was pressed, the cue lights over both levers were turned off. Intertrial intervals (ITI) were 5 sec plus a geometrically distributed duration with a minimum of 0.1 sec and a mean of 25 sec. On Days 1–5, if an incorrect response had been made on the previous trial, the same signal was presented again on the next trial (correction procedure). From Day 6 until the end of the experiment, there were no correction trials. A daily session began within 30 min of the same time each day and lasted 3 hr. A record was kept of the number of left and right responses following each of the two signals and the latency of each response.

Seven–Signal Training (Sessions 16–30). The conditions of two–signal training were maintained except that each of the two training signals were presented with a probability of 0.25 on each trial. On the remaining trials, 5 unreinforced signals were presented that were composed of repeating cycles of sound stimuli similar to those used in two–signal training. These unreinforced signals ranged between 2 – 15 sound segments and 2 – 15 sec (2, 5, 6, 12, 15). Consequently, "outside" signal values that were both lower (2 segment–2 sec signal values) and higher (12 segment–12 sec and 15 segment–15 sec signal values) than those used to maintain the discrimination (4 segment–4 sec and 8 segment–8 sec signal values) were employed during this stage of training.

Testing (Sessions 31–32). The conditions of two–signal training were maintained except that each of the two training signals were presented with a probability of 0.25 on each trial. On the remaining trials, there were two types of unreinforced test signals. One set held the number of cycles constant at 6 while varying the total signal duration between 2 and 15 sec (2, 4, 5, 6, 8, 12, and 15 sec). The other set of test signals held total signal duration constant at approximately 6 sec while varying the number of cycles between 2 and 15 cycles (2, 4, 5, 6, 8, 12, and 15 cycles). These 14 signal combinations were presented with equal probability.

Results and Discussion

The mean psychophysical function for compound signals averaged over the five rats during the last 7 sessions of seven–signal training (Sessions 24–30) is illustrated in Figure 2. During this stage of training the rats could have been using either number and/or duration as a discriminative stimulus to control responding. It can be easily seen that rats formed two separate discriminations, one centered around 4 segments and/or 4 sec and another centered around 8 segments and/or 8 sec. Signal values that were either smaller or larger than the reinforced anchor values tended to be classified as intermediate between the anchor points, presumably reflecting the rats' indifference in classifying these signal values rather than a failure of discrimination.

The point of subjective equality (PSE) during seven–signal training was determined for each rat by fitting a linear regression line to the probability of a right response for the three middle signal values (5, 6 and 8 segments or sec) and calculating the signal value that the rat classified as an "8" on 50% of the trials. The mean ($\pm$ S.E.M.) PSE was 5.82 $\pm$ 0.08 segments – sec. This value is intermediate between the arithmetic mean (6.0) and the geometric mean (5.66) of the two reinforced anchor values. This finding contrasts somewhat with prior reports indicating that the PSE for both number and duration is typically found to be reliably nearer to the geometric mean of the two anchor values (e.g., Meck & Church, 1983). Although this is a relatively small discrepancy between the expected and the observed values, and therefore does not provide a strong test of the prediction made by scalar timing theory (e.g., Gibbon, 1981), it is important to note that with "outside" signal values rats may be more likely to bisect signal number and/or duration at a point intermediate between the arithmetic mean and the geometric mean of the two reinforced anchor values than at the geometric mean. A difference limen (DL) was also estimated from the individual psychophysical functions. From the same linear regression line fit to each rat's data, the signal value that the rat classified as an "8" on 75% of the trials and the signal value that the rat classified as an "8" on 25% of the trials was determined. One half of this range of signal values is defined as the DL. The mean DL was 1.36 $\pm$ 0.04. The mean Weber fraction (WF) was also determined by dividing the DL by the PSE; WF = 0.23 $\pm$ 0.01.

The individual psychophysical functions for each of the five rats and the mean group function during the two test sessions (Sessions 31–32) are shown in Figure 3. The data indicate that the mean psychophysical function is highly representative of the individual response functions both in terms of the general shape of the functions and also in terms of the specifics such as the PSE and the classification of "outside" signal values. The PSE for the number dimension was 5.57 $\pm$ 0.15 and for the duration dimension it was 5.44 $\pm$ 0.12 (segments or sec), a non–significant difference, $t(4) = 0.9$, $p > 0.05$. The DL for the number dimension was 1.9 $\pm$ 0.31 (segments or sec) and for the duration dimension it was 1.56 $\pm$ 0.11 (segments or sec), a non–significant difference, $t(4)$ 1.43, $p > 0.05$. The WF for the number dimension was 0.35 $\pm$ 0.06 and for the duration dimension it was 0.29 $\pm$ 0.02, a non–significant difference $t(4) = 1.2$, $p > 0.05$.

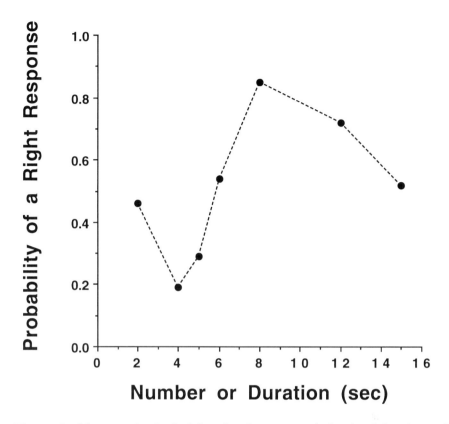

Figure 2. Mean psychophysical function for compound signals of duration and number during the last 7 sessions of Seven–Signal Training with "outside" signals (Sessions 24–30). Reinforced signal values were 4 and 8 seconds or segments; unreinforced signal values were 2, 5, 6, 12, and 15 seconds or segments.

A comparison of the PSE, DL, and WF measures for training and combined testing indicated only the change in the PSE to be significant, $t(4) = 4.76$, $p < 0.01$, $t(4) = 1.8$, $p > 0.05$, and $t(4) = 2.26$, $p > 0.05$, respectively.

Taken together, these data support the claim that when number and duration stimulus attributes are presented together in compound valid number cues associated with stimulus events are not necessarily overshadowed by valid duration cues and vice versa. The data also

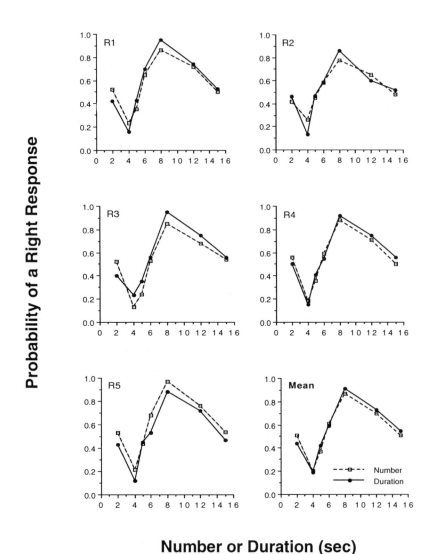

Number or Duration (sec)

Figure 3. Individual and group mean psychophysical functions for duration and number during the two sessions of Testing (Sessions 31–32) following seven–signal training. During this test phase one of the stimulus dimensions was held constant at an intermediate value (e.g., 6 seconds or 6 segments) while the other dimension was allowed to vary.

suggest that a similarity decision rule may not be appropriate in all cases to account for the bisection of number or duration. Furthermore, the generality of this finding may not be restricted to the special case of training with "outside" signal values (see Siegel, 1986). The chance levels of classification for "outside" signals provides new evidence indicating that rats are not simply performing a "few" vs. "many" discrimination and are basing their discrimination, at least in part, on the absolute values of the two reinforced anchor values expressed either in the form of number or duration.

Experiment 2: Cross–procedural transfer of number.

This experiment was designed to address the question of whether rats trained to discriminate between two different numbers of sounds (e.g., 2 and 6) by associating a specific number of sound segments (e.g., 6) with the reinforcement of a specific response (e.g., right lever press) in a bisection–type procedure could transfer the specific numerical aspects of those associations to a different task. In order to answer this question rats were first trained to discriminate 2–cycle signals from 6–cycle signals in a 2–lever bisection task similar to those described above. Following this training, rats were transferred to a 12–arm radial maze procedure where the goal for different groups of rats was to locate 4, 6, or 8 items of food before exiting the maze by approaching the arm of the maze closest to the experimenter. Once they had exited the maze, rats were fed their daily ration of food. The major dependent variable in this experiment was the percentage of trials that rats approached and entered the "final" maze arm as a function of the number of baited arms previously visited.

Consequently, one can examine whether rats trained to discriminate between 2 and 6 sound signals in a numerical bisection–type procedure can abstract the numerical aspects of the task from the modality specific stimulus–response associations. If rats learned that the numbers 2 and 6 were positively associated with reward independent of what was counted and what response was reinforced they should be able to show positive transfer of that knowledge to another task where similar numerical values are also paired with reward. In the present case, it will be determined whether the acquisition of associations between fixed numerical values and the opportunity for reinforcement in a task where the *number* of sound segments in a temporal sequence is the relevant stimulus dimension can be shown to influence performance in a paradigm where the *number* of food items in a spatial array is the relevant stimulus dimension. If positive transfer were evident this would be strong evidence in support of animals having an abstract representation of number that is independent of the modality–specific aspects of the stimulus and the response used to train the discrimination. Furthermore, such evidence would strongly implicate a concept of number that can be transferred across different stimulus contexts (e.g., modalities and procedures). In the present case it was the transfer of the representation for 6 counts from the signal classification task to the radial–arm maze food searching task that was of interest; the representation of 2 counts was irrelevant for this purpose.

Method

Subjects
Thirty–six experimentally naive female rats identical to those described above (see General Methods) served as subjects. These rats were divided into two groups of eighteen rats for behavioral testing.

Apparatus
The lever boxes and 12–arm radial maze used in this experiment were identical to those described above (see General Methods).

Procedures
Phase I: Two–Signal Lever–Box Training (Sessions 1–20). Following pretraining to press the levers (see General Methods) the rats were trained using standard two–lever operant chambers to discriminate between two different types of stimuli. For one group of rats the discriminative stimuli consisted of red and green cue lights located just above the left and right response levers (*visual discrimination* group). The second group of rats was trained on a numerical bisection procedure using non–localizable auditory stimuli

in a manner similar to that described above in Experiment 1 and by Meck and Church (1983: *numerical discrimination* group). All subjects received 20 sessions of two–signal training where half of the rats in the visual discrimination group were trained to press the left lever after a green cue light had been illuminated over both levers and to press the right lever after a red cue light had been illuminated over both levers. The remaining rats in the visual discrimination group had the association between the cue lights and the levers reversed. In contrast, rats in the numerical discrimination group were trained to press the left lever following two sound segments composed of 0.5 sec sound presentations alternating with 0.5 sec silent periods with the number of cycles equaling the number of sound segments and to press the right lever following six such sound segments. The association between the signal number and the correct lever was reversed for the remaining rats.

On each trial one of the two cue lights (red or green) for the visual discrimination group or one of the two signal numbers (2 or 6 segments) for the numerical discrimination group was randomly presented with a probability of 0.5. In the case of the visual discrimination group the cue lights remained on for 4 sec before a choice response was recorded. In the case of the numerical discrimination group the red and green cue lights over both levers were illuminated once the sound signal had terminated, thus signaling the opportunity for a response. In both cases, if the rat made the correct response, a pellet of food was immediately delivered; if it made the incorrect response, no food reinforcement was delivered. When either lever was pressed, the cue lights over both levers were extinguished. ITI's were 5 sec plus a geometrically distributed duration with a minimum of 0.1 sec and a mean of 25 sec. On Days 1–5, if an incorrect response had been made on the previous trial, the same signal was presented again on the next trial (correction procedure). From Day 6 until the end of the experiment, there were no correction trials. A daily session began within 30 min of the same time each day and lasted 3 hr. A record was kept of the number of left and right responses following each of the two signals and the latency of each response.

In this manner, one group of rats was trained to discriminate between signals varying along a visual dimension (e.g., wavelength or intensity) while the numerical and temporal properties of the signals were held relatively constant; whereas, the other group of rats was trained to discriminate between signals varying along numerical and temporal dimensions while the acoustical properties of the signals were held relatively constant.

Phase II: 12–Arm Radial Maze Training (Sessions 1–30). Once the rats had been trained to discriminate between visual or numerical/temporal cues they were pretrained on a 12–arm radial maze as described above in the General Methods section. Following pretraining, the rats were transferred to a maze procedure with a mixed–pattern paradigm of baited (S+) and unbaited (S–) arms similar to that described by Olton and Papas (1979). Rats from the visual discrimination group and the numerical/temporal discrimination group were evenly divided into three separate squads of 12 rats each for maze training. Squads were distinguished by the number of maze arms baited out of the 12 available arms (e.g., 4, 6, or 8 baited arms). Thus, for the *4/12 squad* of rats there were 4 out of 12 arms baited, for the 6/12 squad there were 6 out of 12 arms baited, and for the 8/12 squad there were 8 out of 12 arms baited. One session was given each day, 6 days a week for 30 days.

The maze was baited by placing two Noyes food pellets in the food well at the end of 4, 6, or 8 of the 12 arms. Six different patterns of S+ and S– arms were randomly selected and the rats in each of the three squads were each randomly assigned to one of these 6 patterns with the restriction that rats from the visual discrimination group and rats from the numerical discrimination group be matched for this baiting pattern variable. Once assigned, each rat's baiting pattern was maintained throughout testing. When a rat had completed the maze by locating all of the baited arms it was required to approach and enter the maze arm closest to the experimenter in order to be removed from the maze and returned to its home cage where it was immediately fed its daily ration of food. This particular maze arm was never baited for any of the subjects and thus was designated a S– arm within the confines of the task. Of course, on another level this maze arm served as a S+ arm, conditional upon completion of the maze (i.e., locating all of the food items).

At the beginning of each trial the rat was placed on the central platform and allowed to choose arms in the maze until all pieces of food were obtained. All rats typically completed the maze by finding all pieces of food within 5–10 min. The order of arms chosen was recorded; a choice was defined as a rat advancing more than half of the way down an arm. Early in training the experimenter would stand up after the rat had completed the maze and move towards the "final" arm of the maze in order to pick up and remove the rat from the maze and return it to its home cage for feeding. As training progressed it became unnecessary for the experimenter to display intentional movements toward the maze, as the rats learned to run to and enter the "final" arm when they believed that they had completed the task.

Phase III: Five–Signal Training (Sessions 1–20). For twelve rats randomly selected from those originally trained on the 2 vs. 6 cycle number–duration discrimination the conditions of two–signal training in Phase I were reestablished except that each of the two training signals were presented with a probability of 0.25 on each trial. On the remaining trials, 3 unreinforced signals (3, 4, and 5 cycles) were presented that were composed of repeating cycles of auditory stimuli similar to those used in two–signal training.

Results and Discussion

Phase I. Rats in both the visual discrimination and the numerical/temporal discrimination groups rapidly acquired the discriminations and displayed levels of signal classification well above chance. During the last 10 sessions of training the rats in the visual discrimination group were correct in classifying the visual cues presented on 92 ± 5 % of the trials, whereas the rats in the numerical discrimination group were correct in classifying the sound cycles on 88 ± 4 % of the trials, a non–significant difference. Numbers are means ± the standard deviation.

Phase II. The results from the final seven sessions of the radial–arm maze transfer phase are shown in Figure 4. Here the data for rats originally trained to make a discrimination along the visual dimension are contrasted with data from rats originally trained to make a

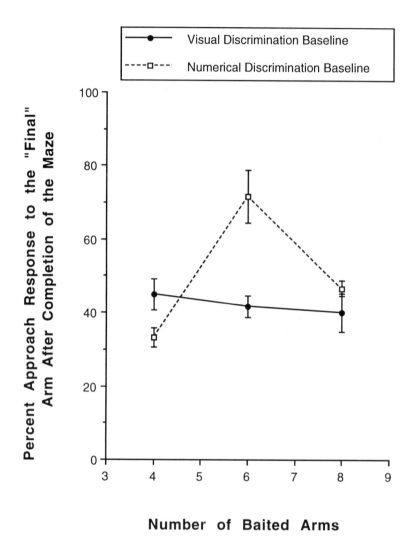

Number of Baited Arms

Figure 4. Mean (± S.E.M) percent approach response to the "final" arm after completion of the maze, (i.e., after visiting all of the baited arms) during the final seven sessions of Phase II: 12–Arm Radial Maze Training (Sessions 24–30). Data are plotted as a function of the number of arms baited for different squads of rats (squads 4/12, 6/12, and 8/12) derived from groups of rats transferred to the maze task after originally receiving either visual discrimination training or numerical discrimination training.

discrimination along a numerical dimension. The percentage of trials during the last 10 sessions during which a rat approached and entered the "final" arm are plotted as a function of the number of baited arms (4, 6, or 8) for different squads. A two–factor analysis of variance with group (visual vs. numerical) and number of baited arms (4, 6, or 8) as factors indicated a significant effect of group, $F(1,30) = 6.98$, $p < .02$; a significant effect of the number of baited arms, $F(2,30) = 5.1$, $p < .02$; and a significant interaction between the two main factors, $F(2,30) = 6.69$, $p < .01$. Positive transfer from a signal classification task to a radial–arm maze task was only demonstrated for those rats originally trained to discriminate 2 sound segments from 6 sound segments and transferred to a situation where the object was to visit 6 baited arms before approaching the "final" arm. A flat transfer gradient was observed for the squads derived from the visual discrimination group and an inverted U–shaped generalization gradient was observed for the squads derived from the numerical discrimination group with the peak of the response function centered at 6 baited arms. As indicated by the analysis of variance reported above, the percent approach response to the "final" arm was significantly higher for the subjects transferred from the numerical discrimination group to the the 6/12 radial maze squad than for *any* other treatment condition.

Phase III. The mean psychophysical functions averaged over the twelve rats during the last 7 sessions of five–signal training are illustrated in Figure 5. The function in the top panel relates the probability of a right response to signal number or duration and the function in the bottom panel relates the choice latency (sec) to signal number or duration. These data were obtained from rats following the radial–arm maze portion of the experiment. The mean PSE of the psychophysical function relating the probability of a right response to signal duration was 3.54 ± 0.07, the mean DL was 0.93 ± 0.06, and the mean WF was 0.23 ± 0.03.

This experiment, in conjunction with the findings reported in Experiment 1, provides evidence that rats acquire an appreciation of the absolute number of sounds comprising the signals that they are trained to discriminate among in numerical bisection procedures. Although it is unclear how such an explanation could account for the

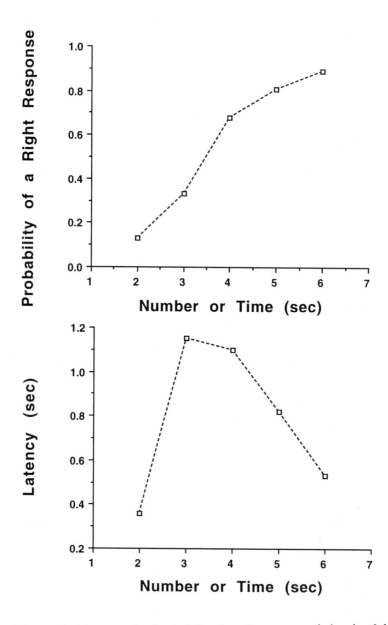

Figure 5. Mean psychophysical functions for compound signals of duration and number during the last 7 sessions of Five–Signal Training (Sessions 14–20). Reinforced signal values were 2 and 6 seconds or segments. The probability of a right response as a function of signal value is shown in the top panel and the choice latency (sec) as a function of signal value is shown in the bottom panel.

observed data, some reviewers have suggested that bisection–type procedures may only require the animal to make "few vs. many" relative numerousness judgments rather than an absolute discrimination of number (e.g., Davis & Albert, 1986; Davis and Pérusse, 1988). Perhaps the present data will help to allay this concern.

It is also important to note that the choice latency function reveals a generalization gradient centered around the midpoint of the two anchor values. This is the first report of a latency function for the bisection of stimulus number and it is very similar to the latency functions reported for the bisection of stimulus duration (e.g., Maricq & Church, 1983; Meck, 1983). The fact that the longest latency is associated with a signal duration near the geometric mean of the two anchor values suggests that rats are the most uncertain about how to classify signal durations in this region and that their uncertainty decreases in a fairly symmetrical fashion as the signals become further away from the midpoint and closer to the anchor values. This sort of generalization gradient for choice latency has been used to infer the nature of the scaling process for stimulus duration and presumably the same analysis could be done for stimulus number with similar results (Meck, 1983). It also seems worth noting that the evidence of transfer from the auditory signal classification task to the radial–arm maze food searching task is in many ways analogous to the transfer of perceived number into performed number (e.g., Seibt, 1988).

Experiment 3: Cross–procedural transfer: Evaluation of the quantitative relationship between the discrimination of duration and number.

The amount of time equal to a single increment of the accumulator used for enumerating stimuli similar to the sound segments used in Experiments 1 and 2 has been shown to be @ 200 msec (Meck & Church, 1983; Meck, Church, & Gibbon, 1985). This is exactly the type of quantitative relationship that the mode–control model of duration and number discrimination has been developed to explain. A straight–forward application of the model to the cross–

procedural transfer observed in Experiment 2 would predict a similar quantitative relationship to hold between the events enumerated in the radial–arm maze procedure and the number of events or durations obtained from the bisection–type procedures employed in this series of experiments. Consequently, rats trained to enumerate a specific number of baited maze arms should, under the appropriate circumstances, transfer a discrimination of a specific quantity from one situation to another. For example, it should be possible to reveal that rats trained to use 8 baited arms as a discriminative stimulus in a radial–arm maze procedure represent that stimulus as being approximately equal to 1.6 sec of time (8 arms x 200 msec/event). The selective facilitation of a specific duration discrimination (e.g., 1.6 sec) in a standard temporal generalization procedure (Church & Gibbon, 1982) by prior maze training with a specific number of arms or rewards as the discriminative stimulus would provide additional evidence for the underlying quantitative equivalence between duration and number as predicted by the mode–control model.

In order to test this prediction, two groups of rats were first trained on a radial–arm maze procedure. One group of rats (Fixed = Number–relevant group) was trained to approach the "final" maze arm after they had located 8 baited arms (see Experiment 2). The other group of rats served as a control group (Random = Number–irrelevant group) that was removed from the maze after they had located a random number of baited arms on each trial that varied between 4 and 12 (mean = 8). Thus, if the mode–control model is correct, the Fixed group should acquire a representation of 8 baited arms that is equivalent to a relatively fixed amount of time, e.g., 1.6 sec. The Random group should not acquire such a discrimination because the number of maze arms visited prior to being removed from the maze varied from trial to trial and hence was unpredictive of removal from the maze.

Method

Subjects
Twenty–four experimentally naive female rats identical to those

described in the General Methods section served as subjects.

Apparatus

The lever boxes and 12–arm radial maze used in this experiment have been described above (see General Methods).

Procedures

Counting procedures

12–Arm Radial Maze Pretraining. Pretraining was identical to that described above (see General Methods).

Phase I: Constrained 12–Arm Radial Maze Training (Sessions 1–20). The rats were evenly divided into two treatment groups of twelve rats each. At the beginning of each session, two pellets of food were placed in the food well at the end of each of the arms. The rat was placed on the central platform and allowed to choose arms in the maze until a randomly determined number of baited arms between 4 and 12 (mean of 8) had been selected (Random = Number–irrelevant group) or 8 baited arms had been selected (Fixed = Number relevant group) at which time the experimenter stood up and approached a specific maze arm ("final" arm) and removed the rat from the maze by guiding them down that arm into their home cage, where the rat was immediately fed its daily ration of food.

Phase II: Unconstrained 12–Arm Radial Maze Training (Sessions 21–30). Training continued as described above except that rats in the Fixed group were now rewarded for independently entering the "final" arm after they had visited exactly 8 other arms and were punished for entering the "final" arm prematurely or belatedly (i.e., prior to selecting 8 other arms or after selecting more than 8 other arms) by leaving them on the maze for 10 additional min before removing them from the maze and placing them in their home cage where they waited an additional 10 min before being fed. One training trial was given per day during which the probability of entering the "final" arm as a function of the number of prior choices was recorded for each rat in the Fixed group.

Timing procedures.

Following the 30 sessions of radial–arm maze training described above, rats were transferred to the timing procedures described below.

Pretraining. Pretraining was identical to that described above in the General Methods section except that only the left lever was used.

Phase III: Temporal Generalization Training (Sessions 1–12). The two groups of rats were divided into four squads of rats as follows: Half of the rats in the Fixed and Random groups were trained to discriminate a 1.6 sec signal from four other signal durations and the remaining rats were trained to discriminate a 6.4 sec signal from four other signal durations. A discrete–trials temporal generalization procedure similar to that described by Church and Gibbon (1982) was used.

For the 1.6 sec discrimination a sound signal was presented for 1.6 sec on 50% of the trials and immediately afterwards the white cue light over the left lever was illuminated. If the rat made a lever press within 4 sec of signal termination a food pellet was delivered, the cue light was turned off, and a 25–s random ITI ensued. If no response occurred within 4 sec of signal termination the cue light was turned off and the ITI was initiated. On the remaining trials, one of four other sound signal durations (0.4, 0.8, 3.2, and 6.4 sec) was randomly presented with equal probability. Following the signal presentation, the white cue light was again illuminated over the left lever and remained lit for 4 sec at which time it was turned off independent of any response.

For the 6.4 sec discrimination a sound signal was presented for 6.4 sec on 50% of the trials and immediately afterwards the white cue light over the left lever was illuminated. If the rat made a lever press within 4 sec of signal termination a food pellet was delivered, the cue light was turned off, and a 25–s random ITI ensued. If no response occurred within 4 sec of signal termination the cue light was turned off and the ITI was initiated. On the remaining trials, one of four other sound signal durations (1.6, 3.2, 12.8, and 25.6 sec) was randomly presented with equal probability. Following the signal presentation, the white cue light was again illuminated over the left lever and remained lit for 4 sec at which time it was turned off independent of any response.

All sessions lasted 2 hr and were conducted approximately 6 days a week. For each trial the signal duration, whether or not a response occurred during the 4-sec response period, and the latency of the response if one occurred was recorded.

Results and Discussion.

The quality of the rat's discrimination of 1.6 sec and 6.4 sec as a function of prior radial-arm maze training was evaluated by examining the probability of a response as a function of signal duration during the last 6 sessions of temporal generalization training. The mean response probability as a function of signal duration for rats in either the Fixed or the Random group trained to discriminate 1.6 sec from other durations is shown in the top panel of Figure 6. A discrimination index (DI) was calculated for the rats in each group by dividing the maximal response probability for any one signal duration by the mean response probability in order to compare the levels of discrimination. The mean ($\pm$ S.E.M.) DI for rats in the Fixed group was 1.25 ± 0.04 and for rats in the Random group it was 1.12 ± 0.02, a significant difference, $F(1,10) = 8.29$, $p < 0.05$.

The mean response probability as a function of signal duration for rats in either the Fixed or the Random group trained to discriminate 6.4 sec from other durations is shown in the bottom panel of Figure 6. The mean ($\pm$ S.E.M.) DI for rats in the Fixed group was 1.14 ± 0.03 and for rats in the Random group it was 1.12 ± 0.02, a non-significant difference, $F(1,10) < 1$.

Temporal generalization procedures in which the quality of the discrimination is indexed by the probability of making a response as a function of signal duration are not easily acquired by rats because they involve learning to inhibit responding following all but one signal duration. The major benefit of such "go – no go" procedures lies mainly in the fact that when a discrimination occurs, as evidenced by a DI measure significantly different from 1.0, it reliably indicates that animals are basing their performance on the absolute characteristics of the stimulus. The degree of control exerted by the discriminative stimulus can be evaluated by the horizontal and vertical placement of the temporal generalization function, as well as by the spread of the function. In the present case, both groups of rats trained to discriminate

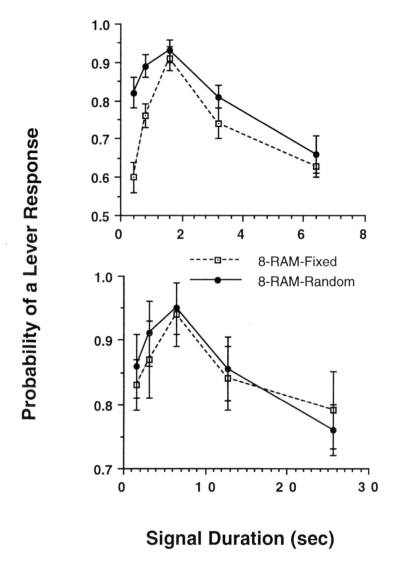

Figure 6. Mean (± S.E.M) duration discrimination functions for different squads of rats initially trained to locate either 8 (Fixed = number–relevant group) or 4–12 (Random = number–irrelevant group) baited arms in a radial–arm maze (RAM) procedure and later transferred to a 1.6–sec (top panel) or a 6.4–sec (bottom panel) temporal generalization procedure. Signal durations were 0.4, 0.8, 1.6, 3.2, and 6.4 sec for the squads trained to discriminate 1.6 sec and 1.6, 3.2, 6.4, 12.8, and 25.6 sec for the squads trained to discriminate 6.4 sec. Data are taken from the last 6 sessions of Temporal Generalization Training (Sessions 7–12).

1.6 sec, not surprisingly, centered their temporal generalization functions around the reinforced signal duration (1.6 sec) and both groups displayed virtually equivalent levels of responding at that value. The two groups differed however, in the spread of their timing functions as revealed by the DI measure. This result indicated that rats originally trained to use 8 baited arms as a discriminative stimulus in the radial–arm maze procedure were more sensitive to signal duration in general and 1.6 sec in particular compared to rats initially trained on the radial–arm maze with number as an irrelevant variable. In contrast, there was no differential effect of prior radial–arm maze training for rats that were transferred to a temporal generalization procedure where the signal durations used were well outside the range of values that could have reasonably been obtained in the radial–arm maze procedure if a mode–control process such as the one described above was functioning.

Thus, although it is difficult at this point to determine whether the observed group differences in the 1.6 sec temporal generalization are due to the Fixed group (number relevant) showing facilitation or the Random group (number irrelevant) showing inhibition of a specific numerical discrimination, or both, it is clear that prior training with number in a radial–arm maze procedure can influence the discrimination of duration in a temporal generalization procedure within a limited range of signal values that may be related to the the number of arms counted in the radial–arm maze task. The pattern of the data suggest that the proposed 200 msec equivalence between an event (i.e., numeron) and a unit of time may have contributed to this result.

Summary and Conclusions

The new data presented here extend the application of a mode–control model of temporal integration for duration and number discrimination from the highly specialized case of a two–choice numerical bisection procedure involving the quantification of an experimenter–controlled sequence of sound segments to a radial–arm maze task requiring the quantification of a subject–controlled sequence of visits to spatial locations in order to obtain a fixed number of food items. This extension of the mode–control model to enumerating items placed in a spatial array is particularly noteworthy, at least from a historical perspective, because it parallels the type of behavioral control used in counting experiments first described in a series of papers by Koehler (1941, 1943, 1950, 1957) and later reintroduced by Davis and Bradford (1991 –– and as described in Davis & Pérusse, 1988).

The cross–species generality of the results and the accompanying theory of preverbal counting presented here is uncertain. The accumulation mechanism proposed by the mode–control model represents numerosities by magnitudes on what some investigators have called a "number line" (e.g., Gallistel & Gelman, 1992). These numerosities are rapidly, but inaccurately generated during the temporal integration process described above and possess a scalar property such that the variability of numerical discrimination increases with the mean of the quantity being enumerated. In this respect, the accumulation process represented by the mode–control model resembles the analogical calculation system found in humans (e.g., Dehaene, Dupoux, & Mehler, 1990). This type of mechanism is specialized in representing magnitudes in an analog form that is inherently more variable than a symbolic or digital representation of number that human subjects might typically use for counting. Research by Dehaene and his colleagues has suggested that human subjects with specific types of brain damage may lose the use of the digital system and may have to revert to the analog system in order to perform a variety of mental calculations (Dehaene & Cohen, 1991). Gallistel and Gelman (1991) have further proposed that the preverbal counting mechanism described herein "is the source of the implicit principles that guide the acquisition of verbal counting." These authors go on to suggest that the preverbal counting system of arithmetric computation required by the decision processes used in the mode–control model "provides the framework for the assimilation of the verbal system" in humans and that, "learning to count involves, in part, learning a mapping from the preverbal numerical magnitude (intervals on the mental number line) to the verbal and written number symbols and the inverse mappings from these symbols to the preverbal magnitudes."

Thus, at least one group of investigators familiar with the development of counting in children considers the present model to be a reasonable foundation on which to base additional numerical abilities.

It has been the position of recent reviewers of numerical competency in animals (e.g., Davis & Pérusse, 1988; Gallistel, 1990; Rilling, 1993) that the identification of a behavior involving the apparent discrimination of number by a member of a particular species is not in itself sufficient evidence to conclude that there exists in that species a robust concept of number related to that known to exist in humans. Davis and Pérusse, authors of one of the most comprehensive and influential reviews of experiments and definitions related to counting in animals, argue this line quite strongly (Davis & Pérusse, 1988). Consequently, use of the term "numerical competency" in the present paper is in line with the general label settled upon in that earlier publication. This term refers to a whole range of behaviors including *relative numerousness judgments* of "greater or less than" –– based on a many versus few dichotomy, *subitizing* –– a category of numerical competency related to rapid tagging behavior in human beings whose general definition and application as descriptor of animal processes is contentious among investigators (e.g., Davis & Pérusse, 1988; Gallistel, 1988; Gallistel & Gelman, 1992), *protocounting* –– a negatively defined classification of behaviors fitting the Gelman and Gallistel rules but not showing evidence of transfer which is described as the distinguishing mark of "true" counting, and *counting* –– following the principles described herein and requiring generality across situations. For Davis and Pérusse, inference of concept formation is proportional to the demonstration of transfer, although it is unclear how this differs from Gelman and Gallistel's own object irrelevance principle other than as a matter of reversing the emphasis in the analysis of the attributes of the representation. Davis and Pérusse also make a number of demands for combinatorial ability in the manipulation of numerons. However, it is Gallistel (1990) who makes the strongest demands of this nature. From Gallistel's perspective, an animal must demonstrate skill with basic arithmetic operators in order for us to accept that animal as possessing a concept of number. This is in line with his assumption that the perception of duration and number are ethologically relevant behaviors because of the advantage that the calculation of rate, defined as division of number by duration, brings to species survival. In contrast, Davis and Pérusse take a more moderate stance in insisting only upon a demonstration of operations conducted between numeron attributes such as symmetry of equality, reflexivity and transitivity; attributes subsumed by Gallistel's standards. It seems correct to assume that these operations stem from an underlying ability to transfer abstract knowledge (i.e., representations) to novel situations.

An alternative proposal is that the criteria for a concept of number should be established in congruence with the literature on other concept formations in animals (see Cerrela, 1979, Herrnstein & de Villiers, 1980), where the emphasis is placed squarely on the issue of transfer to novel stimuli, rather than on some notion of combinatorial ability. However, unlike Davis and Pérusse, we find this issue adequately dealt with by Gelman and Gallistel's abstraction principle. The description of counting solely in terms of the amodal mapping of numerosity and the existence of numerons without reference to the input modality of numerosity implies a strong and natural correlation with the presence of transfer of that process to novel stimuli and procedures and an unnecessary category of numerical competency in the case of "protocounting." The demand for combinatorial skill, while possibly justified on ethological grounds, is not supported as a precondition of concept formation in the literature. It seems an overly robust demand that bears more relation to a mastery of mathematics and symbol manipulation than to a concept of number in a narrow, atomic definition. After all, as one of my colleagues has so aptly stated, a pigeon need not have an understanding of forestry in order to have a concept of a tree.

With these ideas in mind, it will not be argued that the formal analysis of counting by itself provides sufficient criteria to establish the formulation for a concept of number, especially in the case of the numerical bisection procedures described above. It is apparent, however, that when an animal may be said to be counting in the sense that Gelman and Gallistel intended, they are indeed displaying an impressive amount of numerical skill that presumably includes some display of transfer as per the abstraction principle. To establish a concept of number, the quality of the evidence of transfer must be weighed very carefully. The definitional criteria of the formal analysis for the establishment of counting center on the qualities of the numeron. Therefore, the establishment of the closely related concept of number likely hinges on a case by case examination of the nature and implementation of the proposed numeron. If it can be demonstrated that numerons are not rigidly tied to a particular stimulus modality or procedure it can be reasonably concluded that the subject has the ability to abstract certain numerical principles from the stimulus context and to this extent the subject may be said to possess a concept of number. The richness of that concept will obviously depend upon the range of transformations that the subject is able to apply to the numerons in question. At present, it would appear from the evidence for the discrimination of numerosities and its transfer to novel situations that rats possess a relatively well–developed concept of number.

References

Amit, D.J. (1989). *Modeling brain function: The world of attractor neural networks*. Cambridge University Press, Cambridge.

Bannon, M.J., & Roth, R.H. (1983). Pharmacology of mesocortical dopamine neurons. *Pharmacological Review, 35*, 53–68.

Broadbent, H.A., Rakitin, B.C., Church, R.M., & Meck, W.H. (1993). The quantitative relationships between timing and counting. In E.J. Capaldi, & S.T. Boysen (Eds.), *Numerical skills in animals*. (pp. 171–187). Hillsdale, NJ: Erlbaum.

Brown, J.L. (1975). *The evolution of behaviour*. New York: Norton.

Bunney, B.S. (1979). The electrophysiological pharmacology of midbrain dopaminergic systems. In A.S. Horn, J. Korf, & B.H.C. Westerink (Eds.), *The neurobiology of dopamine*, (pp. 417–451). New York: Academic Press.

Capaldi, E.J., Miller, D.J. (1988). Counting in rats: Its functional significance and the independent cognitive processes which comprise it. *Journal of Experimental Psychology: Animal Behavioral Processes, 14*, 3–17.

Cerrela, J. (1979). Visual classes and natural categories in the pigeon. *Journal of Experimental Psychology: Human Perception and Performance, 5*, 68–77.

Church, R.M. (1984). Properties of the internal clock. In J. Gibbon & L.G. Allan

(Eds.), Timing and time perception. *Annals of The New York Academy of Sciences, 423*, 566–582.

Church, R.M. (1989). Theories of timing behavior. In S.B. Klein & R.R. Mower (Eds.), *Contemporary learning theories: Instrumental conditioning theory and the impact of biological constraints on learning*. (pp. 41–71), Hillsdale, NJ: Erlbaum.

Church, R.M., & Broadbent, H.A. (1990). Alternative representations of time, number, and rate. *Cognition, 37*, 55–81.

Church, R.M., & Deluty M.Z. (1977). Bisection of temporal intervals. *Journal of Experimental Psychology: Animal Behavior Processes, 3*, 216–228.

Church, R.M., & Gibbon, J. (1982). Temporal generalization. *Journal of Experimental Psychology: Animal Behavior Processes, 8*, 165–186.

Church, R.M., & Meck, W.H. (1982). Acquisition and cross–modal transfer of classification rules for temporal intervals. In M.L. Commons, A.R. Wagner, & R.J. Hermstein (Eds.), *Quantitative Analysis of Behavior: Discriminative Processes (Vol. 4)*. Cambridge, MA: Ballinger.

Church, R.M., & Meck, W.H. (1984). The numerical attribute of stimuli. In H.L. Roitblat, T.G. Bever, & H.S. Terrace (Eds.), *Animal cognition* (pp. 445–464). Hillsdale, NJ: Erlbaum.

Church, R.M., & Meck, W.H. (1988). Biological basis of the remembered time of reinforcement. In M. L. Commons, R. M. Church, J. R. Stellar, & A. R. Wagner (Eds.), *Quantitative analyses of behavior: Biological determinants of reinforcement. (Vol. 7)*,(pp.103–119). Hillsdale, NJ: Erlbaum.

Davis, H., & Albert, M. (1986). Numerical discrimination by rats using sequential auditory stimuli. *Animal Learning and Behavior, 14*, 57–59.

Davis, H., & Bradford, S.A. (1991). Self-imposed feeding restraint in the rat: A numerical discrimination. Unpublished manuscript.

Davis, H., & Pérusse, R. (1988). Numerical competence in animals: Definitional issues, current evidence, and a new research agenda. *Behavioral and Brain Sciences, 11*, 561–615.

Dehaene, S., & Cohen, L. (1991). Two mental calculation systems: A case study of severe acalculia with preserved approximation. *Neuropsychologia, 29*, 1045–1074.

Dehaene, S., Dupoux, E., & Mehler, J. (1990). Is numerical comparison digital? Analogical and symbolic effects in two–digit number comparison. *Journal of Experimental Psychology: Human Perception and Performance, 16*, 1–16.

Gaioni, S.J., & Evans, C.S. (1984). The use of rate or period to describe temporally patterned stimuli. *Animal Behaviour, 32*, 940–941.

Gallistel, C.R. (1988). Counting versus subitizing versus the sense of number. *Behavioral and Brain Sciences, 11*, 585–586.

Gallistel, C.R. (1989). Animal cognition: The representation of space, time and number. *Annual Review of Psychology, 40*, 155–189.

Gallistel, C.R. (1990). *The organization of learning*. Cambridge, MA: Bradford/MIT Press.

Gallistel, C.R., & Gelman, R. (1992). Preverbal and verbal counting and computation. *Cognition, 44*, 43–74.

Gallistel, C.R., & Gelman, R. (1990). The what and how of counting. *Cognition, 34*, 197–199.

Gelman, R., & Gallistel, C.R. (1978). *The child's understanding of number.* Cambridge, MA: Harvard University Press.

Gibbon, J. (1981). On the form and location of the psychometric bisection function for time. Journal of *Mathematical Psychology, 24*, 58–87.

Gibbon, J., & Church, R.M. (1984).Sources of variance in an information processing model of timing. In H.L. Roitblat, T.G. Bever, & H.S. Terrace (Eds.), *Animal cognition.* Hillsdale, NJ.: Erlbaum.

Gibbon, J., Church, R.M., & Meck, W.H. (1984). Scalar timing in memory. In J. Gibbon, & L. Allan (Eds.), Annals of The New York Academy of Sciences: *Timing and time perception, (Vol.423)*, (pp. 52–77). New York: The New York Academy of Sciences.

Gluck, M.A., & Rumelhart, D.E. (1990). *Neuroscience and connectionist theory.* Hillsdale, NJ: Erlbaum.

Herrnstein, R.J., & de Villiers, P.A. (1980). Fish as a natural category for people and pigeons. *Psychology of Learning and Motivation, 14*, 59–95.

Koehler, O. (1941). Vom Erlernen unbenannter Anzahlen bei Vögeln [Learning about unnamed numbers by birds]. *Die Naturwissenschaften, 29*, 201–218.

Koehler, O. (1943). "Zähl"–versuche an einem Kolkraben und Vergleichsversuche an Menschen ["Counting" experiments in a raven and comparative research with people]. *Zeitschrift für Tierpsychologie, 5*, 575–712.

Koehler, O. (1950). The ability of birds to count. *Bulletin of Animal Behaviour, 9*, 41–45.

Koehler, O. (1957). Thinking without words. *Proceedings of the 14th International Congress of Zoology* (pp. 75–88), Copenhagen.

Maricq, A.V., & Church, R.M. (1983). The differential effects of haloperidol and methamphetamine on time estimation in the rat. *Psychopharmacology, 79*, 10–15.

Meck, W.H. (1983). Selective adjustment of the speed of internal clock and memory storage processes. *Journal of Experimental Psychology: Animal Behavior Processes, 9*, 171–201.

Meck, W.H. (1984). Attentional bias between modalities: Effect on the internal clock, memory, and decision stages used in animal time discrimination. In J. Gibbon & L.G. Allan (Eds.), *Annals of The New York Academy of Sciences: Timing and time perception* (pp. 528–541). New York: New York Academy of Sciences.

Meck, W.H. (1986). Affinity for the dopamine D_2 receptor predicts neuroleptic potency in decreasing the speed of an internal clock. *Pharmacology Biochemistry & Behavior, 25*, 1185–1189.

Meck, W.H. (1987). *Counting behavior in rats: A mode–control model of temporal integration.* Paper presented at the annual meeting of the Midwestern

Psychological Association, Chicago, IL.

Meck, W.H. (1988). Internal clock and reward pathways share physiologically similar information–processing stages. In M.L. Commons, R.M. Church, J.R. Stellar, & A.R. Wagner (Eds.), *Quantitative analyses of behavior: Biological determinants of reinforcement. (Vol. 7)*, (pp. 121–138). Hillsdale, NJ: Erlbaum.

Meck, W.H. (1991). Modality–specific circadian rhythmicities influence mechanisms of attention and memory for interval timing. *Learning and Motivation, 22*, 153–179.

Meck, W.H., & Church, R.M. (1982). Abstraction of temporal attributes. *Journal of Experimental Psychology: Animal Behavior Processes, 8*, 226–243.

Meck, W.H., & Church, R.M. (1983). A mode control model of counting and timing processes. *Journal of Experimental Psychology: Animal Behavior Processes, 9*, 320–334.

Meck, W.H., & Church, R.M. (1984). Simultaneous temporal processing. *Journal of Experimental Psychology: Animal Behavior Processes, 10*, 1–29.

Meck, W.H., Church, R.M., & Gibbon, J. (1985). Temporal integration in duration and number discrimination. *Journal of Experimental Psychology: Animal Behavior Processes, 11*, 591–597.

Meck, W.H., Smith, R.A., & Williams, C.L. (1988). Pre– and postnatal choline supplementation produces long–term facilitation of spatial memory. *Developmental Psychobiology, 21*, 339–353.

Olton, D.S., & Papas, B.C. (1979). Spatial memory and hippocampal function. *Neuropsychologia, 17*, 669–682.

Rilling, M. (1993). Invisible counting animals: A history of contributions from comparative psychology, ethology, and learning theory. In E.J. Capaldi, & S.T. Boysen (Eds.), *Numerical skills in animals*. (pp. 3–37). Hillsdale, NJ: Erlbaum.

Roberts, S., & Holder, M.D. (1984). What starts an internal clock? *Journal of Experimental Psychology: Animal Behavior Processes, 10*, 273–296.

Seibt, U. (1988). Are animals naturally attuned to number? *Behavioral and Brain Sciences, 11*, 609–610.

Siegel, S.F. (1986). A test of the similarity rule model of temporal bisection. *Learning and Motivation, 17*, 59–75.

Siegel, S.F., & Church, R.M. (1984). The decision rule in temporal bisection. In J. Gibbon & L.G. Allan (Eds.), *Annals of The New York Academy of Sciences: Timing and time perception* (pp. 643–645). New York: New York Academy of Sciences.

Steinfels, G.F., Heym, J., Strecker, R.E., & Jacobs, B.L. (1983). Behavioral correlates of dopaminergic unit activity in freely moving cats. *Brain Research, 258*, 217–228.

Swigert, C.J. (1970). A mode control model of a neuron's axon and dendrites. *Kybernetik, 7*, 31–41.

Time and Behaviour: Psychological and Neurobehavioural Analyses
C.M. Bradshaw and E. Szabadi (Editors)
© 1997 Elsevier Science B.V. All rights reserved.

CHAPTER 5

Does A Common Mechanism Account for Timing and Counting Phenomena in The Pigeon?

William A. Roberts

Several lines of evidence indicate that the pigeon and other animals are sensitive to and accurately make discriminations involving the dimensions of time and number. Evidence will be reviewed to support the temporal and numerical abilities of animals. An information processing model of timing and counting behavior will be described, and recent investigations of pigeon performance based on predictions from the model will be presented and discussed. Finally, the implications of data from a time production procedure will be considered in which time and number are first confounded and then dissociated.

Evidence for temporal competence

An early indicator of temporal competence in animals was the discovery of the scallop pattern of the cumulative record shown by animals trained on a a fixed–interval (FI) reinforcement schedule. An animal reinforced for the first response made only after a fixed time interval has elapsed since the last reinforcement develops a pattern of responding in which few responses are made throughout the initial half of the interval but responding at a high rate appears in the latter half of the interval. This finding suggested that animals had learned the time between reinforcements and behaved accordingly by responding only when the moment of reward delivery approached. A very important extension of the FI method, the *peak procedure*, was introduced by S. Roberts (1981) and made it possible to estimate precisely the point in time at which an animal anticipated the delivery of reward. In the peak procedure, an animal that had been trained to respond on an FI schedule was then given periodic *empty trials* on which the reinforcer was omitted, but the opportunity to make free operant responses was extended for some time beyond the point at which the reinforcer normally would be delivered.

Data from empty trials in an experiment using the peak procedure (Roberts, Cheng, & Cohen, 1989) are shown for pigeons in Figure 1. Pigeons had been trained to respond to two FIs, FI 15 s and FI 30 s, each signaled by a different ambient cue; when an overhead red light came on, pigeons were reinforced for the first response after 15 s, but, when a 1325-Hz tone was played, the first response after 30 s was reinforced. On empty trials, the light or tone was presented for 90 s, and no reinforcer was delivered throughout this period. The curves shown in Figure 1 contain response rate averaged over a number of empty trials and plotted over successive 3-s time bins throughout the 90-s period. Three properties of these curves should be emphasized. First, the peak or point of highest rate of response closely estimated the FI, indicating that animals were accurately timing the time interval from signal onset to reinforcer delivery. Second, these response rate curves look like Gaussian functions and approach symmetry on each side of the peak. Third, the degree of error in time estimation indicated by the width of these curves suggests that time estimation obeys Weber's law. That is, error seems to increase in proportion to the length of the interval being timed. All three of these properties must be accounted for in a theoretical treatment of timing behavior.

Experiments that have involved FI training and peak procedure testing may be considered time production experiments, because the animal actually produces its time estimate by the peak of its response rate function. In other experiments, animals have shown temporal

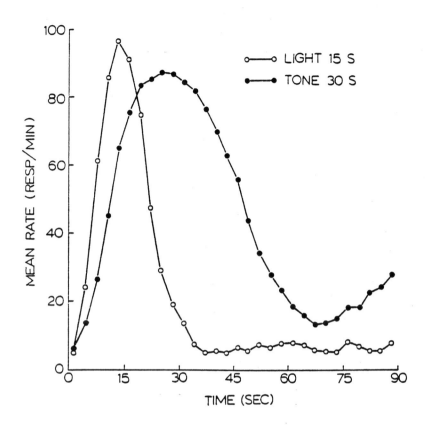

Figure 1. Response rate curves on empty trials for pigeons trained on FI 15 s (light signal) and FI 30 s (tone signal). (From Roberts, Cheng, & Cohen, 1989).

competence by making judgments on time intervals already presented. In a number of experiments, it has been shown that pigeons can learn to make one response after a short signal has been presented and to make another response after a long signal has been presented (Kraemer, Mazmanian, & Roberts, 1985; Spetch & Wilkie, 1983). For example, pigeons given a choice of pecking a red or green key can learn to peck the green key after a 2–s houselight has been presented and to peck the red key after an 8–s houselight has been presented. If birds then are tested with signals varying in duration between short and long, an orderly psychophysical function is obtained with the probability of a long response rising as the signal becomes longer and the point where there is indifference or a 50–50 split between short and long responses closely approximating the geometric mean of the short and long durations (Stubbs, 1968). In another discrimination procedure, Dreyfus, Fetterman, Stubbs, and Montello (1992) have shown that pigeons can make relative temporal judgments. When pigeons have been presented with two signals in succession, for example a red light followed by a green light, they have learned to make one response if the first signal exceeds the second and a different response if the second signal exceeds the first. In fact, pigeons can be trained to make such a discrimination based on a particular ratio difference between the two signals. Thus, a pigeon can learn to make a particular response only if the second signal is twice as long or longer than the first signal.

Evidence for numerical coompetence

In a major review of numerical competence in animals, Davis and Perusse (1988) concluded that animals readily discriminate between different numbers of stimuli. Experiments which have examined numerical discrimination in animals involve either the simultaneous presentation of a number of objects or the successive occurrence of a number of discrete events. Such experiments are fraught with possibilities of confounding from other cues. When an animal must discriminate between two cards containing different numbers of geometric patterns, for example, the discrimination might be made not only on the basis of number but also on the basis of the area occupied by the patterns, the area occupied by the background, the length of contour of the patterns, or perhaps differences in the overall brightness of the cards. All of these alternative sources of discrimination can be controlled through random variation. Several primate studies have shown that monkeys can discriminate number of objects when the size and shape of the objects discriminated changes from trial to trial (Hicks, 1956; Thomas, Fowlkes, & Vickery. 1980; Thomas & Chase,

1980). In a recent experiment, it was found that pigeons also could discriminate between projected pictures that contained 1 or 2 patterns and 6 or 7 patterns (Emmerton, Lohmann, & Niemann, in press). These pictures were presented successively, and pigeons had to peck one key for a "small" number response and another key for a "large" number response. This discrimination was maintained when the size, shape, amount of contour, and brightness of the patterns was varied, showing that pigeons indeed were discriminating number. When probe tests were given with pictures that contained 3, 4, and 5 patterns, the probability of a "large" response increased monotonically as the number of patterns increased. This important finding suggests that pigeons respond to the dimension of number as they do to other stimulus dimensions such as brightness or wavelength.

When number of sequential events is used as a discriminative stimulus, time becomes a major confounding variable. In a three–key operant chamber, Rilling (1967) trained pigeons to peck one side key after completing a small number of pecks on the center key and to peck the other side key after completing a larger number of pecks on the center key. Pigeons could have based the discrimination either on number of pecks made to the center key or on the amount of time it took to complete the required pecks on the center key. Further analyses of Rilling's data and other research (Laties, 1972; Wilkie, Webster, & Leader, 1979) indicate that number of pecks and not time spent pecking controlled choice behavior. In other experiments, the experimenter may control the sequences of events to be discriminated by presenting a series of discrete visual or auditory stimuli to an animal prior to a discriminative response. Roberts and Mitchell (1994) presented pigeons with the task of discriminating between 2 and 8 light flashes. After a pigeon had seen 2 or 8 flashes, each lasting 200 ms, it was presented with two keys, one of which yielded reward when pecked after 2 flashes and the other of which yielded reward when pecked after 8 flashes. To control for the potential confounding effects of time, each number of light flashes (F) was presented at three durations, 2, 4, or 8 s. The results of this experiment are shown in Figure 2. Although pigeons took longer to learn to respond accurately to some sequences than others, based on transfer from a preceding experiment, they clearly learned to respond accurately to number regardless of the time period over which the number of flashes was presented.

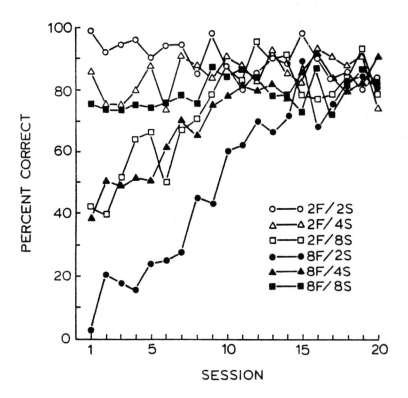

Figure 2. Acquisition of a discrimination by pigeons between 2 flashes and 8 flashes presented at durations of 2, 4, and 8 s. (From Roberts & Mitchell, 1994).

Evidence from experiments in which visual stimuli were presented both simultaneously and successively indicates that pigeons can discriminate along the dimension of number. The claim that pigeons and other animals can count, however, has met with some controversy. Although Davis and Perusse (1988) concluded that animals had shown numerical competence by their ability to make numerical discrimations, they argued strongly that animals failed to show evidence of the counting abilities found in humans. Specifically, humans use a set of cardinal number tags which they assign to objects counted in a fixed order (Gelman & Gallistel, 1978). Animals showed only some primitive precursors of counting, what Davis and Perusse called *protocounting*, but failed to show evidence for the use of a number system with the properties of cardinality and ordinality.

In contrast to the view of Davis and Perusse, both Gallistel (1990, 1993) and Capaldi and Miller (1988) have argued that animals count and do so routinely and automatically. Gallistel and Gelman (1992) have presented the interesting idea that numerical competence in animals is based on automatic and innate neural processes that record the frequency of events as quantities they refer to as *numerons*. Numerons have the properties of cardinality and ordinality because these neural quantities cumulate in a fixed order as events counted are presented successively. Gallistel and Gelman further suggest that the numerical operations of addition, subtraction, multiplication, and division are performed upon numerons by animals. For example, animals may make foraging decisions about which patch to enter or when to leave a patch by adding numerical estimates of food quanities or dividing food quantities by time or spatial area to determine measures of rate of intake or food density. Although it is held that animals and preverbal children can count and perform operations with numerons, the use of verbal symbols for counting operations only appears in people after language ability has developed to the point at which verbal representations can be mapped onto numerons.

A model of counting and timing

Given that ample evidence exists for both accurate timing and numerical discrimination in animals, a model will be considered which provides an integration of these processes and suggests that a common mechanism underlies them. The model was presented in a paper by Meck, Church, and Gibbon (1985) and represented an extension of the scalar expectancy theory earlier introduced by Gibbon (1977). As shown in Figure 3, the model depicts the course of information processing during the process of timing or counting initiated by an external signal. The three rows of boxes shown represent clock, memory, and decision components of the model. In the top row, a pacemaker is shown which emits pulses at a constant rate. These pulses may enter an accumulator only if a switch is opened. The closing and opening of a switch is controlled by the presentation of an external signal. This model is called a *dual-mode model* for the important reason that two switches operate independently and allow pulses to accumulate in separate accumulators. The time switch operates in the *time mode* and closes at the initial presentation of a signal to be timed and only opens when the signal is terminated. The number of pulses in the time accumulator then serves as a measure of the amount of time that has gone by. The count switch is held to operate in the *event mode*; it closes and opens briefly each time an event occurs and allows a fixed number of pulses to enter the count accumulator. The number of pulses residing in the count accumulator then serves as an indicator of how many events were presented. It should be noted that this mechanism of counting by quantity of neural pulses collected in an accumulator is one means of depicting the numerons postulated by Gallistel and Gelman (1992).

In order to store the time or number of events recorded in an accumulator, numbers of pulses are sent from the accumulators to short-term or working memory; values stored in working memory at the time of reward are transmitted to the reference memory for long-term storage. Finally, response decisions are made in a comparator. The comparator is fed information from both working and reference memory. From the working memory, it continuously receives a record of the number of pulses stored in the accumulators. From reference memory, it retrieves criterion quantities of pulses for comparison with the current values in working memory. Comparison is based on a ratio of the absolute difference between the criterion value from reference memory (RM) and the current value in working memory (WM) divided by the RM value. As the ratio of $|RM-WM|/RM$ approaches zero, the ratio will drop below a threshold for a particular response, triggering choice of say a left key or a right key or simply triggering the beginning of a run of responses in FI training. Since the decision is based on a ratio with the absolute value of the RM–WM difference in the numerator, the model captures all of the important properties of timing depicted in Figure 1. It predicts that timing should be symmetrical around the peak of response rate functions and that these functions should have the scalar property that error increases in proportion to the length of the period of time to be estimated.

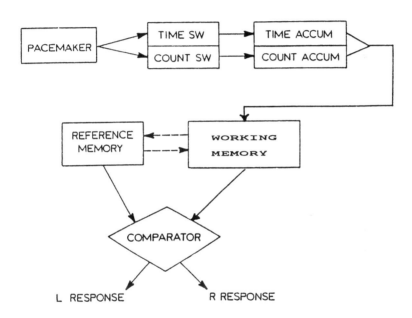

Figure 3. A depiction of the dual–mode information processing model introduced by Meck, Church, & Gibbon (1985).

An important experiment that supported the dual–mode model was carried out by Meck and Church (1983). They trained rats to press one bar following a sequence of two noise bursts and another bar following a sequence of eight noise bursts. Since the noise bursts occurred at the rate of 1/s, two bursts lasted 2 s, and eight bursts lasted

8 s, and rats could base their discrimination of these sequences on either time or number. To test for control by each of these dimensions, rats were given probe test trials on which either the time or number dimension was held constant and the other dimension was varied. The results revealed almost identical psychophysical functions for time and number, both of which showed strong control by the dimension varied. These findings indicate that rats were both timing and counting the sequences of noise bursts, as the notion of dual accumulators for timing and counting suggests.

Tests of the dual–mode model using pigeons

In a series of experiments carried out in my laboratory, we have examined the applicability of the dual–mode model to pigeons discriminating between sequences of light flashes that vary in time and number. As will be shown, the outcomes of these experiments generally support the idea that pigeons are both counting and timing sequences of light flashes and also lead to some modifications and extensions of the dual–mode model.

Testing for control by time and number

Our first order of business was to perform an experiment with pigeons parallel in nature to the Meck and Church (1983) experiment with rats. Roberts and Mitchell (1994, Experiments 1 and 3) trained pigeons to discriminate between sequences of 200–ms light flashes presented at the rate of 1/s that lasted for either 2 s or 8 s. After 2 or 8 flashes of the overhead red light had occurred, side keys were illuminated; a peck on the left key produced reward after 2 flashes, and a peck on the right key produced reward after 8 flashes. Once pigeons were making this discrimination at over 80% accuracy, probe test trials were introduced to examine control by the time and number dimensions. On time–control tests, the number of flashes was held

constant at the intermediate value of 4, and the length of the sequences
of flashes was varied from 2 to 8 s in 1–s intervals. On number–
control tests, the length of the sequence of flashes was held constant
at 4 s, and the number of flashes was varied from 2 to 8. The resulting
psychophysical curves showing the percentage of choice of the "long"
or right key after presentation of test stimuli can be seen in Figure 4.
The rising curves for time– and number–control tests show an
equivalent increase in "long" responses as either time or number was
increased. The pigeon data then largely yielded the same results as the
rat data of Meck and Church and implicated mechanisms that were
simultaneously keeping track of time and number.

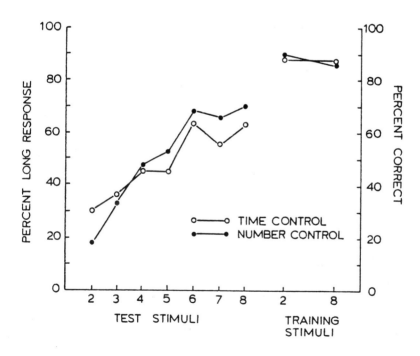

Figure 4. Curves showing equivalent stimulus control by time and number variation
in pigeons. (From Roberts & Mitchell, 1994).

Evidence for The Selective Retrieval of Time and Number Information

A question which interested us was what a pigeon would do if presented with an ambiguous sequence of light flashes, that is a sequence in which time and number dictated different responses. Pigeons continued to be tested on 2f/2s and 8f/8s sequences, in which time and number indicated the same response. In addition, the ambiguous sequences of 2f/8s and 8f/2s were presented on half the trials. Notice that after 2f/8s, 2 flashes directs the animal to choice of the left key, but the 8–s duration directs it to choice of the right key. Similarly, following 8f/2s, 8 flashes should cause the pigeon to choose the right key while 2 s should cause the pigeon to choose the left key. In order to disambiguate these sequences, different colors were presented on the side keys to indicate whether the pigeon should respond on the basis of time or number. Thus, red keys indicated response on the basis of time, and green keys indicated response on the basis of number. For example, if 2f/8s had been presented and followed by red keys, then a choice of the right key based on the 8–s duration of the sequence was rewarded. If the 2f/8s sequence was followed by green side keys, however, then choice of the left key based on 2 flashes was rewarded. After about 40 days of training, pigeons learned to make these discriminations at over 80% accuracy (Roberts & Mitchell, 1994, Experiment 4). It should particularly be noted that pigeons were cued to respond on the basis of time or number only *after* the sequence of flashes had been presented. It cannot be argued, therefore, that birds selectively attended to one dimension over the other during the sequence presentation. Rather, these findings suggest that the side–key color directed the selective retrieval of time or number information.

As a further test of this idea, Roberts and Mitchell (1994; Experiment 5) examined stimulus control when pigeons were directed to retrieve either time or number information. While birds continued to be trained on trials that presented 2f/2s and 8f/8s, as well as the ambiguous 2f/8s and 8f/2s, periodic time and number control probe trials were introduced. As was the case for the experiment depicted in Figure 4, number was held constant at 4 flashes while time was varied from 2 to 8 s, and time was held constant at 4 s while the number of flashes was varied from 2 to 8. An important further aspect of this experiment was that the red and green side keys each were presented on half the number–control probe tests and half the time–control tests. Therefore, the appropriate retrieval cue for number information (green keys) appeared on half the number–control tests, but the inappropriate cue for retrieval of time information (red keys) appeared on the other half of the number–control tests. Similarly, red keys signaling retrieval of time information were an appropriate cue on half the time–control trials, and green keys signaling retrieval of number information were an inappropriate cue on the other half of the time–control trials.

As shown in Figure 5, each dimension showed stimulus control when the appropriate cue appeared on the side keys. Percentage of "long" responses rose monotonically with number of flashes when green keys signalled counting, and a similar rise in the time–control curve from 2 to 8 s occurred when red keys signalled timing. When the key color signalled retrieval of the inappropriate dimension, however, essentially flat curves were found. Since the non–tested dimension was held constant at 4 flashes or 4 s, retrieval of information about the non–tested dimension should have led to a flat curve near the indifference point of 50%.

These findings require some modification of the dual–mode model depicted in Figure 3. The original model made no provision for the separate storage and retrieval of time and number information. Accumulated pulse totals from both the time and number accumulators could simultaneously be compared with reference memory criteria, with whichever value yielded the closest match to the criterion determining the choice response. This model provided no mechanism for accurate choice when ambiguous sequences were presented. The pigeon results argue that birds had independent access to time or number information. In a revised model (Roberts, 1995), the working memory contains separate stores for number and time information, and either type of information may be retrieved based upon an external cue, such as key color, for comparison with a reference memory criterion in the comparator.

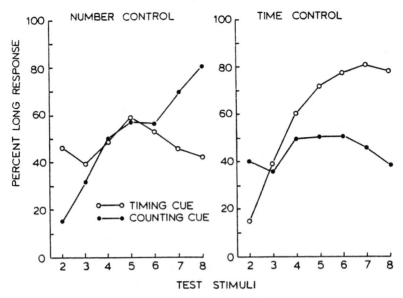

Figure 5. The left panel shows stimulus control by number when the counting cue was presented, and the right panel shows stimulus control by time when the timing cue was presented. The flat lines show no stimulus control by presentation of inappropriate cues. (From Roberts & Mitchell, 1994).

Memory for Time and Number

Short–term or working memory for time has been investigated in several experiments with pigeons using the delayed matching–to–sample procedure. Pigeons initially were trained to match different key colors to sample stimuli that consist of different lengths of time. Since time duration and key color are different dimensions, the relationship learned is symbolic matching–to–sample. For example, a houselight would be presented for a 2–s duration or an 8–s duration and be followed immediately by the appearance of red and green side keys. Only a peck on the red key delivers reward after the 2–s houselight, and only a peck on the green key delivers reward after the 8–s houselight. After pigeons have learned to discriminate time duration accurately by pecking the correct color of side key, memory tests are introduced by delaying the onset of the test keys for a variable number of seconds after the presentation of the time sample stimulus. Retention curves showing the percentage of correct choices at successive delays typically yield a phenomenon called the *choose-short effect* (Kraemer, et al., 1985; Spetch & Wilkie, 1982). After the 2-s or short sample duration, the retention curve declines only gradually over the retention interval. After the 8-s or long sample duration, however, forgetting takes place far more rapidly, and the retention curve falls below the 50% level of choice. The choose-short effect then refers to the fact that pigeons show an increasing preference for the test stimulus (red key in this example) associated with the short-sample duration as the retention interval becomes longer.

The leading theoretical interpretation of the choose-short effect is *subjective shortening theory*. The theory suggests that memory for sample stimulus duration weakens in the sense that it becomes progressively shorter as the retention interval becomes longer. Thus, after several seconds have gone by, an 8-s sample duration may be remembered as only a 4-s or 2-s sample duration. As memory of a long or 8-s sample shortens, it will become more often judged as a short sample and give rise to choice of the red test key. Although memory of the short 2-s sample also may subjectively shorten, its memory should still be more similar to that of a 2-s sample than that of an 8-s sample. Several lines of evidence now support the notion that the choose-short effect arises from subjective shortening (Spetch & Wilkie, 1983; Spetch & Sinha, 1989; Wilkie & Willson, 1990).

It is of interest to consider the choose-short effect within the context of the dual-mode theory of timing and counting already outlined. Since this model conceives of counting as an accumulation of pulses which are sent to a working memory, the notion of subjective shortening may be conceived of as the loss of pulses from working memory as time elapses. If a pigeon's pacemaker emits pulses at the rate of 1/msec, 2000 pulses would be collected by the time accumulator and sent to the working memory when a 2-s houselight duration occurs, and 8000 pulses would collect when an 8-s houselight duration occurs. If the number of pulses lost over a 4-s retention interval is half of the initial contents of the working memory, the working memory should contain 4000 pulses on a long-sample trial and 1000 pulses on a short–sample trial. Both of these quantities are closer to the criterion of 2000 pulses for a red–key choice than to the criterion of 8000 pulses for a green–key choice. The choose–short effect then is readily understandable within the framework of the timing component of the dual–mode model.

The interesting prediction which follows from this exercise is that a *choose–small effect* should appear that parallels the choose–short effect. That is, if memory for small and large numbers of events is measured at increasingly long retention intervals, the retention curve for the large number should fall at a faster rate than the retention curve for the small number. Let us assume that 200 pulses accumulate through the brief closing and opening of the event switch each time a light flash occurs. After 8 light flashes (large number), 1600 pulses should have accumulated in the count accumulator and been sent to working memory, whereas, after 2 flashes (small number), only 400 pulses should have gone to working memory. If we assume that half these pulses are lost over 4 s, there should then be 800 pulses left on a large– number sample trial and 200 pulses left on a small–number trial. Just as in the temporal example given above, memory for both the large and small samples will most closely match the criterion of 400 pulses for choice of the red key associated with the small sample number.

We tested this prediction by training pigeons to choose one test key after 8 light flashes and another test key after 2 light flashes (Roberts, Macuda, & Brodbeck, 1995). To insure that matching–to–sample was not based on time, both numbers of flashes were presented over a 4–s period of time. Pigeons learned to peck the correct test keys after 2 and 8 flashes at over 90% accuracy within 20 sessions of training. Retention tests then were given by inserting delays of 0, 2, 5, and 10 s between the sample stimulus and the test stimuli. The retention curves from these tests are shown in Figure 6. These curves show a striking choose–small effect. The retention curve for 2f/4s is virtually flat, but the curve for 8f/4s declines rapidly and drops below the 30% level of correct choice by the 10–s delay. These data clearly show a choose–small effect for memory of number that parallels the choose–short effect found for memory of time. They further support the ideas that time and number are represented by the common mechanism of accumulated pulses and that pulses are progressively lost from both time and number accumulators over a retention interval.

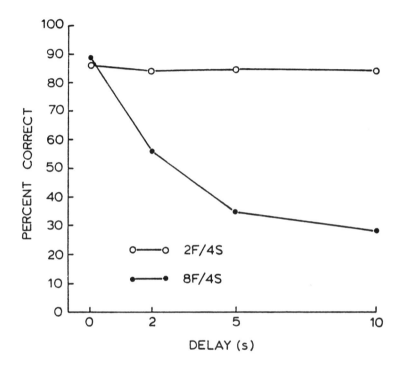

Figure 6. A choose–small effect in pigeons is revealed by the increase over the delay in the tendency to choose the comparison stimulus associated with 2 flashes after both 2 flashes and 8 flashes. (From Roberts, Macuda, & Brodbeck, 1995).

What Happens When Time and Number are Dissociated?

In several of the experiments thus far discussed, pigeons have been trained to respond to sequences of light flashes in which food reward appeared after a fixed time duration and fixed number of flashes. If food always appears after 20 s of a light that flashes at the rate of 1/s, then two sets of criteria for expecting food are established. Once again assuming that pulses are emitted at the rate of 1/msec, the distribution of pulses for the time criterion in reference memory should center around 20,000 pulses. If the event mode allows 200 pulses to enter the counting accumulator each time a flash occurs, the distribution of pulses for the number criterion should center around 4,000 pulses. As long as the signal continues to present flashes at the rate of 1/s, these criteria are redundant in the sense that they both predict the delivery of food after 20 s.

In a recent experiment, we asked what would happen if time and number were dissociated. That is, suppose that the time and number pulse totals accumulating and being sent to working memory meet the criteria in reference memory at different points in time. In our experiment, pigeons initially were trained to complete an FI 20–s schedule by pecking on the center key of an operant chamber. A green field was projected on the center key, but that green field flashed red for 200 msec at the rate of once every second. After pigeons had developed a clear FI scallop, showing strong control of pecking by the FI schedule, empty trials were inserted among FI training trials on 25% of the total daily trials. Empty trials lasted 100 s, and no reward was delivered on empty trials. Among the empty trials, the rate at which the center key flashed from green to red varied. The flash rate on different empty trials was fast (2/s), the medium training rate (1/s), or slow (.5/s). Both the distribution of responses over time bins during empty trials and the points at which runs of responses started and stopped were examined.

In order to consider possible predictions from the dual–mode theory about this experiment, Figure 7 presents a theoretical depiction of changes over time in the |RM–WM|/RM ratio. The solid line depicts changes in this ratio in the training condition in which flashes are presented at 1/s. Notice that the ratio drops to zero at 20 s and rises at the same rate as it declined. This line represents the change in |RM–WM|/RM for both the time and number accumulators; computations within the two modes are completely redundant with

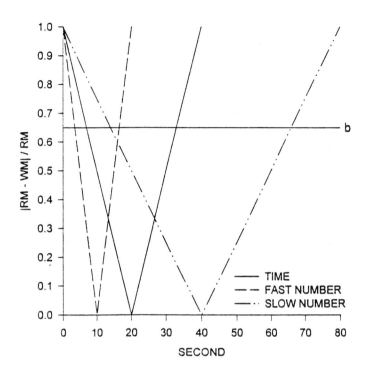

Figure 7. Theoretical curves derived from the dual–mode model showing the decline to zero and rise of |RM − WM|/RM for output of the time accumulator and the count accumulator when the flash rate is fast (2/s) or slow (.5/s). The horizontal line *b* is the threshold for starting and stopping responding.

respect to the decision ratio output. The horizontal line (*b*) represents the threshold for responding. Although there may be different thresholds for time and counter output, for ease of theoretical exposition, only a single threshold is shown. A run of responses should begin when the |RM−WM|/RM ratio drops below *b* and should end when the |RM−WM|/RM ratio rises above *b*. On all empty trials, regardless of the rate at which the center key is flashing, decisions about time should be based on changes in the |RM−WM|/RM ratio shown by the solid line. The broken lines, on the other hand, depict changes in the |RM−WM|/RM ratio over time based on the number accumulator when the flash rate is either fast or slow. At the fast rate, the number accumulator will indicate that reward should occur after 10 s, and, at the slow rate, the number accumulator will indicate that reward should occur after 40 s. Time and number are now dissociated, and the question is how will the dual–mode system respond to this circumstance.

Several hypotheses can be derived from the theoretical situation depicted in Figure 7. One possibility is that the system would use only one mode and ignore the other. If a pigeon only used time output, its behavior on fast– and slow–rate tests should look exactly like that on medium–rate tests, since only the time curve shown in Figure 7 would be used. If only the counter output were used, on the other hand, on fast–rate trials the pigeon should start earlier and end earlier than on medium–rate trials, as shown by the fast number curve; on slow–rate trials, the pigeon should start later and end later than on medium–rate trials.

Another possibility is that the pigeon would use both modes together. There are two ways in which this might occur. According to a *conservative hypothesis*, the pigeon would use both modes by letting the first mode to drop below *b* dictate the start of responding and the last mode to rise above *b* to dictate the end of responding. According to this hypothesis, a pigeon should start earlier on fast–rate trials than on medium–rate trials but should stop at the same time on both fast– and medium–rate trials. Conversely, the pigeon should start at the same time on slow– and medium–rate trials and end later on slow–rate trials than on medium–rate trials. In other words, as long as the |RM–WM|/RM ratio is below *b* for either mode, the pigeon will respond. An alternative is the *strategic hypothesis*, which suggests that the pigeon responds only to the first mode to drop below *b* and the first mode to rise above *b*. This strategy would save the pigeon the extra effort of further responding on empty trials by telling it to stop as soon as the output of either mode rises above threshold. The empirical predictions from the strategic hypothesis are that the pigeon should both start and stop earlier on fast–rate trials than on medium–rate trials. On slow–rate trials, however, the pigeon should not differ in either start or stop time from the medium–rate trials. Since the time curve in Figure 6 drops below *b* before the slow number curve and rises above *b* before the slow number curve, decisions about starting and stopping on slow–rate trials should be based on only the output of the time mode. Finally, it should be noted from Figure 6 that the model in general suggests that if differences appear between the flash–rate conditions, these differences should be larger for stop times than for start times, since the temporal distances between the points where the curves rise above *b* are greater than those between the points where the curves drop below *b*.

Data from two pigeons tested on empty trials over 10 sessions are shown in Figures 8 and 9. The curves shown are based on pecks made in 2–s time bins and have been transformed in two ways. The curves have been smoothed by averaging each point with the points preceding and following it, and the curves have been normalized by expressing the pecks in each bin as a proportion of the number in the peak bin. The curves for Pigeon 10 in Figure 8 and the curves for Pigeon 204 in Figure 9 tell a similar story. The peak of the fast–rate curve is shifted to the left of the peaks for the medium– and slow–rate curves, and the fast–rate curve drops faster than the other two curves. Although pigeon 204's fast–rate curve rises somewhat earlier than its medium– and slow–rate curves, little difference in the rise of the curves is seen for Pigeon 10. Both birds show little difference between the medium– and slow–rate curves, with the major effect being the earlier decline in the fast–rate curve.

These data also were examined on a trial–by–trial basis to identify runs of responding. For the initial run on each empty trial, we determined its start time, stop time, width, and midpoint. Using the method of analysis suggested by Cheng and Westwood (1993), the start of a run was identified as the time at which the pigeon made at least one response in the first of two consecutive 1–s time bins. The stop time was the moment at which the first of two consecutive time bins appeared with no responses. The width of the run was the stop time minus the start time. The midpoint between the start and stop time was used as an estimate of the peak time on each trial. There were four empty trials at each rate in each session. The daily estimates of start, stop, width and midpoint were the median times over these four trials.

In Table 1, the means and standard deviations of these daily medians are shown.

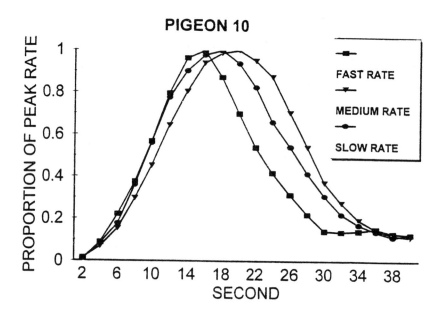

Figure 8. Smoothed and normalized response rate curves for Pigeon 10 on empty trials when the flash rate was fast, medium, or slow.

One clear finding seen in Table 1 is that start time did not vary across flash rates. Pigeon 10 started between 8 and 9 s at every rate, and Pigeon 204 started between 7 and 8 s at every rate. The results of the F-tests failed to suggest any difference between the start time means for either pigeon. Stop time, on the other hand, differed significantly among the flash rates for both birds. Both pigeons stopped around 18 s at the fast rate but took considerably longer to stop at the medium and slow rates; Pigeon 10 stopped between 26 and 27 s at both rates, and Pigeon 204 took between 35 and 37 s to stop at both rates. Similar patterns of differences are seen for width and midpoint. In particular, the midpoint, which is an estimate of peak time, was around 13 s for both birds at the fast rate. At the medium and slow rates, Pigeon 10's peak times were between 17 and 18 s, showing a slight underestimation of the 20-s FI, whereas Pigeon 204's peak times somewhat overestimated the 20-s FI, at values between 21 and 22 s. Stop time, width, and midpoint all differed significantly across rates, as shown by the F-tests for each pigeon. Post-hoc t-tests between the stop time, width, and midpoint means showed in all cases that the means for medium and slow rates did not differ significantly but both were significantly higher than the mean for the fast rate.

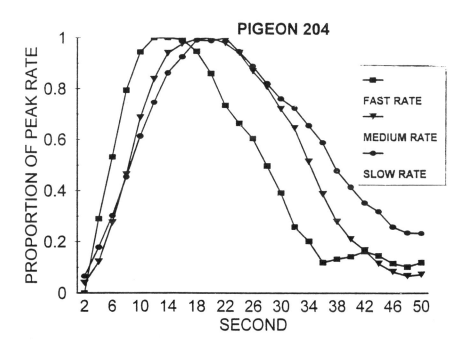

Figure 9. Smoothed and normalized response rate curves for Pigeon 204 on empty trials when the flash rate was fast, medium, or slow.

Table 1.   **Means, Standard Deviations, and F–Tests for Start Time, Stop Time, Width, and Midpoint (sec) of Response Runs made at Fast, Medium, and Slow Flash Rates by Two Pigeons**

	Fast Rate		Medium Rate		Slow Rate		
	M	*SD*	*M*	*SD*	*M*	*SD*	*F*–Test
Pigeon 10							
Start Time	8.85	1.58	8.30	1.12	8.30	1.21	0.63
Stop Time	18.30	3.39	26.80	4.18	26.30	4.20	20.68
Width	9.45	3.56	18.50	4.39	18.00	4.44	17.43
Midpoint	13.58	2.08	17.55	2.21	17.30	2.62	13.89
Pigeon 204							
Start Time	7.65	2.24	7.20	1.55	7.35	2.81	0.16
Stop Time	18.05	6.52	35.55	2.55	36.70	6.75	53.28
Width	10.40	4.54	28.35	2.00	29.35	7.88	48.62
Midpoint	12.85	4.59	21.38	1.96	22.02	3.39	30.07

The findings shown in the response rate curves and in Table 1 tell the same story. Speeding up the rate at which the center key flashed had no effect on start time but caused both pigeons to stop pecking sooner than they did at the slower flash rates. Slowing down the flash rate had no effect on any of the temporal components when compared to the medium flash rate. In terms of the hypotheses earlier discussed, these data seem to come closest to the strategic hypothesis. That hypothesis suggested that pigeons should start when the $|RM-WM|/RM$ ratio for either mode dropped below the threshold (b) and that they should stop when the ratio calculated for either mode rose above the threshold. This hypothesis predicted that pigeons would stop sooner at the fast rate than at the slower rates but would not differ in stop time between the medium and slow rates. It also predicted that pigeons should start sooner at the fast rate than at the slower rates, and this difference was not found. The model shown in Figure 6 does suggest that differences in stop time should be larger than those in start time. Nevertheless, it is a problem for the strategic hypothesis that pigeons did not start sooner under the fast flash rate. One possibility is that it was difficult for pigeons to overcome a response intertia learned under training at the medium rate. Training and testing pigeons at longer fixed intervals might reveal evidence for a difference in start time between the fast and medium rates.

Summary and Conclusions

Evidence was reviewed suggesting that pigeons and other animals have temporal and numerical competency. Animals readily discriminate different magnitudes of time and number. Some timing experiments have examined the production of a time estimate by measuring the peak time on empty trials following FI training. In other experiments, animals have been shown to discriminate readily between successively presented short and long durations of a signal. Numerical competency experiments typically have used discrimination procedures in which a number of objects or events have been presented simultaneously or successively. When various other dimensions that may be correlated with number are controlled, animals have consistently shown sensitivity to number. The question of whether animals can count using the principles of cardinality and ordinality remains a controversial issue.

The dual-mode model introduced by Meck, et al. (1985) offers a parsimonious approach to timing and counting by suggesting that a common mechanism underlies both operations. Both operations involve the accumulation of pulses in accumulators, with separate switches and accumulators for measuring the passage of time and the frequency of events. Evidence for this model has been revealed with both rats and pigeons in experiments in which it was shown that both time and number controlled behavior when animals discriminated between confounded time–number stimuli. The data presented in Figure 4 suggest that pigeons presented with 2f/2s or 8f/8s are both counting the number of flashes and keeping track of the time over which the flashes are presented.

Our further experiments with pigeons indicate some revisions and extensions of the dual–mode model. When memory for number was investigated, a choose–small effect was found that parallels the choose–short effect found for memory of time. Time appears to subjectively shorten, as pigeons judge an 8–s duration sample stimulus to be a 2–s sample stimulus at longer retention intervals in a working memory experiment. As seen in Figure 6, working memory for 8 flashes becomes judged as 2 flashes as the retention interval lengthens by a few seconds. Both of these effects can be explained by the assumption that pulses transmitted to time and number accumulators are lost as time passes during a retention interval.

Some other findings with pigeons suggest more flexibility in the use of time and number information than was envisioned in the original formulation of the dual–mode model. Evidence from experiments in which pigeons were presented with ambiguous sequences indicates that birds can selectively retrieve time or number information. Color cues presented on choice keys after the flash sequence has been presented control the selective retrieval of the duration or frequency of the preceding light flashes. When time and number were dissociated, the data from empty trials supported the strategic hypothesis. This hypothesis suggests that pigeons used information from the output of time and number working memories to stop responding at a point that minimized the response effort made on empty trials. Once either time or number sufficiently exceeded the reward criterion, pigeons stopped responding, although the alternate mode might indicate that reward was still to come.

It is interesting to note that the time–number dissociation experiment has much in common with human timing experiments in which the effects of events during a time interval on the subjective duration of the interval are measured. A popular experiment in human time estimation has been to ask a person to judge the length of an interval when either nothing happens during the interval or the interval is filled with sensory events or tasks the subject must perform (Brown, 1995; Burnside, 1971; Zakay, Nitzan, & Glicksohn, 1983). In retrospective judgments of an interval of time already gone by, people judge filled intervals as longer than empty intervals. If subjects are asked to produce intervals of time, they usually stop sooner if the interval is filled than if it is empty. This effect is often referred to as the *filled–duration illusion* (Thomas & Brown, 1974).

One theoretical account of the filled–duration illusion is that time is based only on the number of events that occur during a time period, and subjective time then is determined by the number of events stored in memory (Ornstein, 1969) or the number of changes in stimulation remembered over an interval (Block, 1978; Fraisse, 1963; Poynter, 1989). In another account, Thomas and Weaver (1975) suggested that people use two processors, a visual information processor and a timer. The timer keeps track of the number of internal or biologically generated signals, while the visual information processor counts external visual events. A final time estimate is based on a weighted average of these two processors. The Thomas and Weaver model is interesting here because it bears a strong similarity to the dual–mode model which is the focus of this essay. A comparison of the models suggests that the dual–mode model offers a more precise and testable model for examining the interaction of time and number on performance. For example, no mechanism is offered to specify exactly how much weight is to be given to the timer and to the visual processor in the Thomas and Weaver model. In the dual–mode model, separate accumulators and working memories are given to time and number, and the relative weight given to one mode over the other may depend upon prior training. Roberts and Mitchell (1994), for example, found that initally weaker number control than time control was eliminated by giving pigeons number discrimination training.

The fact that peak time was shortened by doubling the flash rate, as shown in Table 1, may be seen as evidence that a filled-duration illusion has occurred in the pigeon. That is, doubling the number of flashes that occur in each second causes birds to perceive time as passing more rapidly and thus lowers their peak time. However, the fact that peak time was not lengthened in the slow–rate condition relative to the medium–rate condition fails to support the notion that time is based simply on the number of events occurring during a time interval. An analysis of the pigeons' behavior within the framework of the dual–mode model suggests that pigeons are *both* timing and counting. The apparent filled–duration illusion seen under the fast rate can be better understood by the notion that pigeons are using a strategy of responding to the output of the first mode that falls below or rises above its response threshold. This model has the advantage of accounting for behavior under all three of the flash rates tested.

The results of research thus far carried out are encouraging with respect to the use of the dual–mode model as a device for understanding and investigating temporal and numerical competence in animals. It has the parsimony of suggesting that a common mechanism is responsible for timing and counting, a mechanism which may be linked to preverbal information processing in both animals and people (Gallistel & Gelman, 1992). New invesigations may allow us to further see how flexible and how hard–wired the use of time and number accumulators may be.

Acknowledgment

Preparation of this chapter and research reported in it were supported by a Research Grant to W.A. Roberts from the Natural Sciences and Engineering Research Council of Canada.

References

Block, R.A. (1978). Remembered duration: Effects of event and sequence complexity. *Memory & Cognition, 6,* 320–326.

Brown, S.W. (1995). Time, change, and motion: The effects of stimulus movement on temporal perception. *Perception & Psychophysics, 57,* 105–116.

Burnside, W. (1971). Judgment of short time intervals while performing mathematical tasks. *Perception & Psychophysics, 9,* 404–406.

Capaldi, E.J., & Miller, D.J. (1988). Counting in rats: Its functional significance and the independent cognitive processes that constitute it. *Journal of Experimental Psychology: Animal Behavior Processes, 14,* 3–17.

Cheng, K., & Westwood, R. (1993). Analysis of single trials in pigeons' timing performance. *Journal of Experimental Psychology: Animal Behavior Processes, 19,* 56–67.

Davis, H., & Perusse, R. (1988). Numerical competence in animals: Definitional issues, current evidence and a new research agenda. *The Behavioral and Brain Sciences, 11,* 561–579.

Dreyfus, L.R., Fetterman, J.G., Stubbs, D.A., & Montello, S. (1992). On discriminating temporal relations: Is it relational? *Animal Learning & Behavior, 20,* 135–145.

Emmerton, J., Lohmann, A., & Niemann, J. (in press). Pigeons' serial ordering of numerosity with visual arrays. *Animal Learning & Behavior.*

Fraisse, P. (1963). *The psychology of time.* New York: Harper and Row.

Gallistel, C.R. (1990). *The organization of learning.* Cambridge, MA: Bradford Books/MIT Press.

Gallistel, C.R. (1993). A conceptual framework for the study of numerical estimation and arithmetic reasoning in animals. In S.T. Boysen, & E.J. Capaldi (Eds.), *The development of numerical competence: Animal and human models* (pp. 211–223). Hillsdale, NJ: Erlbaum.

Gallistel, C.R., & Gelman, R. (1992). Preverbal and verbal counting and computation. *Cognition, 44,* 43–74.

Gelman, R., & Gallistel, C.R. (1978). *The child's understanding of number.* Cambridge, MA: Harvard University Press.

Gibbon, J. (1977). Scalar expectancy theory and Weber's law in animal timing. *Psychological Review, 84,* 279–325.

Hicks, L.H. (1956). An analysis of number–concept formation in the rhesus monkey. *Journal of Comparative and Physiological Psychology, 49,* 212–218.

Kraemer, P.J., Mazmanian, D.S., & Roberts, W.A. (1985). The choose–short effect in pigeon memory for stimulus duration: Subjective shortening versus coding models. *Animal Learning & Behavior, 13,* 349–354.

Laties, V.G. (1972). The modification of drug effects on behavior by external discriminative stimuli. *Journal of Pharmacology and Experimental Therapeutics, 183,* 1–13.

Meck, W.H., & Church, R.M. (1983). A mode control model of counting and timing processes. *Journal of Experimental Psychology: Animal Behavior Processes, 9,* 320–334.

Meck, W.H., Church, R.M., & Gibbon, J. (1985). Temporal integration in duration and number discrimination. *Journal of Experimental Psychology: Animal Behavior Processes, 11,* 591–597.

Ornstein, R.E. (1969). *On the experience of time.* Middlesex, England: Penguin Books.

Poynter, D. (1989). Judging the duration of time intervals: A process of remembering segments of experience. In I. Levin, & D. Zakay (Eds.), *Time and human cognition: A life–span perspective* (pp. 305–331). Amsterdam: Elsevier Science Publishers.

Rilling, M. (1967). Number of responses as a stimulus in fixed interval and fixed ratio schedules. *Journal of Comparative and Physiological Psychology, 63,* 60–65.

Roberts, S. (1981). Isolation of an internal clock. *Journal of Experimental Psychology: Animal Behavior Processes, 7,* 242–268.

Roberts, W.A. (1995). Simultaneous numerical and temporal processing in the pigeon. *Current Directions in Psychological Science, 4,* 47–51.

Roberts, W.A., Cheng, K., & Cohen, J.S. (1989). Timing light and tone signals in pigeons. *Journal of Experimental Psychology: Animal Behavior Processes, 15,* 23–35.

Roberts, W.A., Macuda, T., & Brodbeck, D.R. (1995). Memory for number of light flashes in the pigeon. *Animal Learning & Behavior, 23,* 182–188.

Roberts, W.A., & Mitchell, S. (1994). Can a pigeon simultaneously process temporal and numerical information? *Journal of Experimental Psychology: Animal Behavior Processes, 20,* 66–78.

Spetch, M.L., & Sinha, S.S. (1989). Proactive effects in pigeons' memory for event duration: Evidence for analogical retention. *Journal of Experimental Psychology: Animal Behavior Processes, 15,* 347–357.

Spetch, M.L., & Wilkie, D.M. (1982). A systematic bias in pigeons' memory for food and light durations. *Behavior Analysis Letters, 2,* 267–274.

Spetch, M.L., & Wilkie, D.M. (1983). Subjective shortening: A model of pigeons' memory for event durations. *Journal of Experimental Psychology: Animal Behavior Processes, 9,* 14–30.

Stubbs, A. (1968). The discrimination of stimulus duration by pigeons. *Journal of The Experimental Analysis of Behavior, 11,* 223–238.

Thomas, E.A.C., & Brown, I. Jr. (1974). Time perception and the filled–duration illusion. *Perception & Psychophysics, 16,* 449–458.

Thomas, E.A.C., & Weaver, W.B. (1975). Cognitive processing and time perception. *Perception & Psychophysics, 17,* 363–367.

Thomas, R.K., & Chase, L. (1980). Relative numerousness judgments by squirrel monkeys. *Bulletin of the Psychonomic Society, 16,* 79–82

Thomas, R.K., Fowlkes, D., & Vickery, J.D. (1980). Conceptual numerousness judgments by squirrel monkeys. *American Journal of Psychology, 93*, 247–257.

Wilkie, D.M., Webster, J.B., & Leader, L.G. (1979). Unconfounding time and number discrimination in a Mechner counting schedule. *Bulletin of the Psychonomic Society, 13*, 390–392.

Wilkie, D.M., & Willson, R.J. (1990). Discriminal distance analysis supports the hypothesis that pigeons retrospectively encode event duration. *Animal Learning & Behavior, 18*, 124–132.

Zakay, D., Nitzan, D., & Glicksohn, J. (1983). The influence of task difficulty and external tempo on subjective time estimation. *Perception & Psychophysics, 34*, 451–456.

Time and Behaviour: Psychological and Neurobehavioural Analyses
C.M. Bradshaw and E. Szabadi (Editors)
© 1997 Elsevier Science B.V. All rights reserved.

CHAPTER 6

Pigeons' Coding of Event Duration in Delayed Matching–To–Sample

Douglas S. Grant, Marcia L. Spetch
& Ronald Kelly

A fundamental issue in the analysis of working memory is the nature of the representation, or code, that mediates accurate performance across a retention interval. The delayed matching–to–sample (DMTS) task is the primary analytical tool used to investigate working memory processes in pigeons. The DMTS task is typically implemented in a conditioning chamber containing a row of three pecking keys which can be illuminated by various stimuli (colors, geometric forms, lines in various orientations). A houselight, located above the center key, can be activated to provide general illumination. Reinforcement is provided by timed access to a raised and illuminated grain feeder located below the center key.

In the simplest version of DMTS, trials begin with presentation of one of two sample stimuli (e.g., red on some trials and green on other trials) on the center key as the to–be–remembered event. Following a delay of a few seconds, two comparison stimuli are presented on the outer keys for a choice. On each trial, one of the comparisons is designated correct and a single peck to that stimulus results in reinforcement, and the alternative comparison is designated incorrect and a single peck to that stimulus is followed only by the intertrial interval. Which comparison is correct (and hence which comparison is incorrect) is dependent on which sample was presented at the onset of that trial. The selection of the stimulus that serves as the sample is determined quasi–randomly, and the spatial position of the correct choice stimulus is equally often left and right.

In the identity version of DMTS, the correct comparison is a physical match to the sample (e.g., red sample, red comparison correct; green sample, green comparison correct). In the symbolic version of DMTS, there is an arbitrary relation between sample and correct comparison (e.g., red sample, vertical line comparison correct; green sample, horizontal line comparison correct). The symbolic version of DMTS extends the scope of analysis by affording an opportunity to study working memory for events which could not easily be used as comparison stimuli (e.g., tones differing in pitch, the occurrence or nonoccurrence of food). Symbolic and/or identity versions of DMTS have been used to assess working memory for auditory and visual stimuli (e.g., colors, lines in different orientations, geometric forms), number of responses emitted, number of events presented (e.g., light flashes), rate of alternation between two stimuli, the spatial position of a stimulus, the occurrence or nonoccurrence of an event (e.g., food), and the duration of an event.

At a conceptual level, consensus has emerged that DMTS involves processes of both working and reference memory. As suggested by Honig (1978), working memory may be conceptualized as a repository for dynamic information of temporary relevance, whereas reference memory may be conceptualized as a repository for information of more enduring relevance. According to this view, training in a DMTS task results in representations of the reinforcement contingencies being established in reference memory (e.g., if the sample is red, then pecking the vertical comparison is reinforced; if the sample is green, then pecking the horizontal comparison is reinforced). Reference memory representations of the contingencies are not, however, sufficient to mediate accurate performance in the DMTS task. In particular, a working memory representation, or code, is required to bridge the delay that intervenes between sample termination and presentation of the comparison stimuli.

At a general level, the information maintained in working memory during the delay in DMTS could be either retrospective or prospective. In a retrospective (or "backward–looking") coding strategy, working memory is held to contain a representation of sample stimulus information. Reference memory is then queried at the time of comparison choice, and performance is controlled by the combination of the retrospective code in working memory and representations of the contingencies in reference memory. To illustrate, if a vertical line is the correct comparison on red–sample trials, then presentation of a red sample is held to establish a representation of redness in working memory. The code "red" is not, however, sufficient to generate accurate choice between vertical and horizontal comparisons. Thus, reference memory, which in the present example represents the contingency "if red then pecking vertical is reinforced", must be queried to generate choice of the vertical comparison.

In a prospective (or "forward–looking") coding strategy, on the other hand, a sample stimulus is held to activate, with some probability, the reference memory representation of the associated comparison stimulus. In contrast to retrospective codes, prospective codes contain information sufficient to generate accurate choice between the comparisons without need to query reference memory at the time of test. To illustrate, again suppose that a vertical line is the correct comparison on red–sample trials. In a prospective coding scheme, presentation of a red sample would tend to activate the reference memory representation of the vertical line. The code "vertical", or perhaps "peck vertical", is sufficient to generate accurate choice between the vertical and horizontal comparisons.

The focus in the present chapter is the analysis of working memory processes used by pigeons to remember event duration (e.g., 2 or 10 s of light presentation) in symbolic DMTS. Of primary interest is the nature of the working memory representation or code employed to mediate accurate performance following a delay. The research reviewed leads to two main conclusions. Evidence reviewed in the first major section leads to the conclusion that pigeons code event duration retrospectively and analogically in the standard choice version of DMTS. Specifically, we maintain that presentation of a duration sample establishes a representation of that duration, and hence that the representation is retrospective. We maintain further that the representation has properties comparable to those of duration, and hence that the representation is analogical. Also considered in that section is the issue of what factor or factors promote such a coding strategy.

In distinguishing between analogical and nonanalogical representations, it may be useful to provide some examples of each. Representing different sample durations in terms of (a) the number of counts generated by a pacemaker during that duration, (b) images of lines that change in length as a function of duration, or (c) images that change in intensity as a function of duration, all constitute analogical representations. These representations are analogical in that each can change in a continuous and cumulative fashion, as does event duration. Nonanalogical representations, on the other hand, do not share with duration the property of changing in a continuous and cumulative fashion. Thus, representing sample duration categorically would constitute a nonanalogical coding process. Such categorical codes could, for example, be retrospective (e.g., "short" and "long") or prospective (e.g., "peck red" and "peck green").

Evidence reviewed in the second major section of this article leads to the conclusion that pigeons code event duration nonanalogically in at least some "nonstandard" versions of DMTS. In particular, evidence suggests that pigeons code event duration nonanalogically in two instances: (1) in the successive version of DMTS in which retention is indexed by a go/no–go measure rather than by a choice measure and (2) when more than one to–be–remembered event is associated with each comparison stimulus in the choice version of DMTS (a many–to–one, MTO, sample–to–comparison mapping arrangement).

Coding in the Standard Choice Version of DMTS

The Choose–Short Effect

A number of studies have employed the standard choice version of DMTS to investigate memory for event duration in pigeons (e.g.,

Grant & Kelly, 1996c; Grant & Robinson, 1993; Grant & Spetch, 1991, 1993a,b; Kraemer, Mazmanian, & Roberts, 1985; Spetch & Rusak, 1989, 1992a; Spetch & Wilkie, 1982, 1983). In a typical experiment, trials begin with presentation of an overhead houselight for either a short (e.g., 2 s) or long (e.g., 10 s) duration. Termination of the sample is followed immediately (0-s delay) by illumination of two comparison stimuli for a choice. A single peck to one comparison (e.g., green) is reinforced if the sample on that trial had been short, and choice of the alternative comparison (e.g., red) is reinforced if the sample had been long.

Following acquisition, memory for duration is assessed by interpolating a delay of varying length between termination of the sample and onset of the comparison stimuli. Several studies have revealed a pronounced asymmetry in the retention functions on short- and long-sample trials (e.g., Fetterman, 1995; Grant & Spetch, 1991, 1993a; Kraemer et al., 1985; Santi, Ducharme, & Bridson, 1992; Spetch, 1987; Spetch & Rusak, 1989, 1992a; Spetch & Wilkie, 1982, 1983). The data plotted in Fig. 1 illustrate the asymmetry. On trials initiated by a short sample, accuracy declines very little as delay is lengthened. In contrast, on trials initiated by a long sample, accuracy decreases markedly as delay is lengthened, and frequently drops significantly below chance at long delays. Thus, pigeons tend to respond to the comparison stimulus associated with the short sample at longer delays, a phenomenon referred to as the *choose–short effect*.

Spetch and Wilkie (1982), who discovered the choose–short effect, argued that it could be interpreted as reflecting a process of subjective shortening of an analogical representation of sample duration. According to this view, testing a pigeon immediately following termination of a long sample (i.e., at a 0-s delay) results in a high proportion of correct choices because the analogical representation of the sample in working memory corresponds closely to the reference memory representation of a long sample. As time passes since termination of a long sample, the working memory representation is held to subjectively shorten. At longer delays (e.g., 10 and 20 s), therefore, the working memory representation of a long sample corresponds less closely to the reference memory

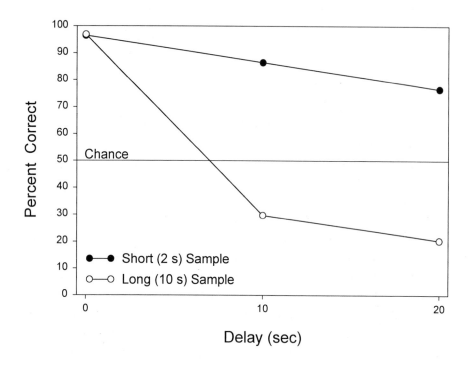

Figure 1. Mean percentage of correct responses on trials initiated by a short (2 s) and long (10 s) sample as a function of delay. The data are from the control group in Spetch and Grant (1993a, Exp. 1).

representation of a long sample and corresponds more closely to the reference memory representation of a short sample, thereby leading to an increased tendency to choose the short–associated comparison as delay increases. Evidence consistent with the notion that remembered duration may undergo a process of subjective shortening has also been obtained using other types of timing procedures in both pigeons (Cabeza de Vaca, Brown, & Hemmes, 1994) and humans (Wearden & Ferrara, 1993).

Notice that the subjective–shortening account of the choose–short effect is predicated on duration being represented analogically, rather than nonanalogically (e.g., categorically). To render the discussion less abstract, suppose that pigeons represent duration analogically in terms of the intensity of an image of white light. Suppose further that as time spent in the presence of the sample increases, the intensity of the image also increases. Thus, long durations are represented by intense images and short durations by dim images. In such a scheme, subjective shortening would occur if the image lost intensity as a function of the passage of time since termination of the sample. A nonanalogical coding process in which, for example, samples are coded categorically as "short" and "long" or "peck green comparison" and "peck red comparison", precludes operation of a process of subjective shortening.

Other investigators have argued, however, that the choose–short effect can be accounted for without assuming a process of subjective shortening and that, therefore, the choose–short effect does not provide definitive evidence of analogical coding of sample duration. In particular, two accounts of the choose–short effect have been offered in which the coding process is not necessarily held to be analogical. According to the account developed by Kraemer et al. (1985), pigeons code samples differing in duration categorically, either in terms of a retrospective ("short" and "long") or prospective ("peck red" and "peck green") representation. To account for the choose–short effect, Kraemer et al. assumed a bias to respond to the short-associated comparison in the absence of a remembered code. They suggested that the bias to respond short in the absence of a remembered code arises because the animal judges the content of working memory (i.e., "nothing") to more closely resemble a short event than a long event. As Killeen and Fetterman (1988) have noted, the plausibility of the biased–guessing assumption, and hence the viability of Kraemer et al.'s account of the choose–short effect, may be questioned. More importantly, evidence presented in the next section permits rejection of this account.

Grant (1993) and Dougherty and Wixted (1996) have noted the empirical similarity between the choose–short effect obtained with short– and long–duration samples and the *choose–no–sample effect* obtained with absence versus presence samples (e.g., nonoccurrence versus occurrence of food, nonoccurrence versus occurrence of a triangle). The choose–no–sample effect refers to the finding that, as delay is lengthened, there is marked forgetting on sample–present trials, but little if any forgetting on sample–absent trials (Colwill, 1984; Dougherty & Wixted, 1996; Grant, 1991; Maki, 1981; Sherburne & Zentall, 1993a,b; Wilson & Boakes, 1985; Wixted, 1993). In contrast to the choose–short effect, the choose–no–sample effect has typically been interpreted in terms of asymmetrical sample coding (Colwill, 1984; Grant, 1991, 1993; Maki, 1981; Sherburne & Zentall, 1993a,b; Wilson & Boakes, 1985). According to this view, the animal responds to the sample–associated comparison if it remembers a presence sample, and to the no–sample–associated comparison if it does not remember a presence sample.

Because duration samples differ in salience, although not as markedly as absence versus presence samples, duration samples might also be coded asymmetrically. That is, the pigeon might code only the long sample, and respond to the short–associated comparison whenever a long–sample code is not present in working memory. Notice that an asymmetrical coding process would result in a choose–short effect (providing, of course, that the occurrence of a long sample was often forgotten during long delays) regardless of the content of the code used to represent a long sample. That is, a choose–short effect would be anticipated whether the long code was an analogical (e.g., bright image) or nonanalogical (e.g., "long", "peck green") representation of duration. Although plausible, evidence reviewed below also permits rejection of the asymmetrical–coding account of the choose–short effect.

Evidence for Analogical Coding

In this section we review four lines of research that, taken collectively, provide convincing evidence that pigeons code duration samples retrospectively and analogically in the standard choice version of DMTS. Each of these lines of research has revealed findings problematic for one or both of the alternative accounts of the choose–

short effect discussed previously. Perhaps more importantly, all of the findings can be readily accommodated by an analogical–coding model, and many of the findings were predicted by that model.

Error patterns. Wilkie and Willson (1990) adapted the procedure and analytical technique developed by Roitblat (1980) to identify the nature of the representation in a memory–for–duration task. Three durations of houselight (2, 8 and 10 s) were used as samples and three colors (red, orange, and green) were used as comparisons. Red was correct after a 2–s sample, orange was correct after an 8–s sample, and green was correct after a 10–s sample. Note that an easy sample discrimination (2 versus 8 s) was mapped onto a difficult choice discrimination (red versus orange), and a difficult sample discrimination (8 versus 10 s) was mapped onto an easy choice discrimination (orange versus green). In one experiment, delays of 0, 5, and 10 s intervened between termination of the sample and presentation of all three comparison stimuli for a choice. As predicted by the analogical–coding model, confusion among similar samples increased more rapidly than did confusion among similar comparisons as delay was lengthened.

Wilkie and Willson's (1990) results are inconsistent with a prospective–coding model which predicts that confusion among similar comparisons should increase more rapidly than confusion among similar samples. A retrospective–categorical model could, however, accommodate this result by assuming that the codes representing 8– and 10–s samples (e.g., "long", "very long") are more confusable than either is with the code representing a 2–s sample (e.g., "short"). It is unclear whether the asymmetrical–coding model can be applied in instances in which more than two sample durations are employed. The fact that Wilkie and Willson did obtain a robust choose–short effect when three durations were used as samples, a result which has also been reported by Spetch and Wilkie (1983), suggests that, at the least, not all instances of the choose–short effect can be explained in terms of asymmetrical coding.

The choose–long effect. A second phenomenon predicted by
the analogical–coding model is that, following training at a longer
delay (e.g., 10 s), testing at a shorter delay (e.g., 0 s) should reveal the
opposite of the choose–short effect – – a choose–long effect. As
shown in Fig. 2, Spetch (1987) confirmed this prediction. At the
training delay (10 s) there was no bias to choose either the short– or
the long–associated comparison. At delays longer than the training
delay (15 and 20 s) the typical choose–short effect occurred. Of
primary interest is the finding that at delays shorter than the training
delay (0 and 5 s) a choose–long effect occurred.

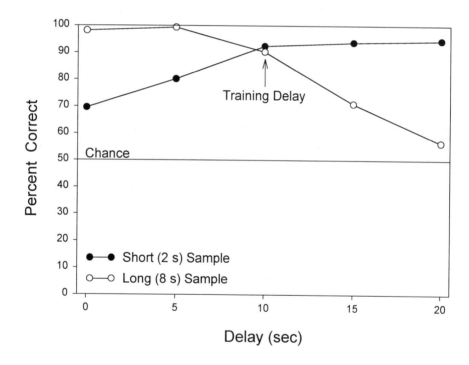

Figure 2. Mean percentage of correct responses on trials initiated by a short (2 s)
and long (8 s) sample as a function of delay following training with a delay of 10 s.
The data were reported by Spetch (1987).

As Spetch (1987) argued, an analogical coding process (in which both short and long samples are coded in terms of perceived duration) combined with a process of subjective shortening (in which remembered duration becomes subjectively shorter as a function of time since the termination of the sample) correctly anticipates both choose–short and choose–long errors. Specifically, training at a fixed nonzero delay should result in foreshortened representations of event duration being associated with particular comparison responses in reference memory. Choose–long errors would predominate at delays shorter than that employed in training because the durations represented in working memory would not have sufficiently foreshortened, and therefore would be subjectively longer than the durations represented and associated with particular comparison responses in reference memory.

The choose–long effect appears particularly damaging to the view that pigeons code only the long sample. Although an asymmetrical–coding model correctly anticipates that testing at delays shorter than that used in training would support high levels of accuracy on long–sample trials (because it is unlikely that the code would be forgotten), it provides no obvious mechanism for decreased accuracy on short–sample trials. Although categorical–coding models make the same erroneous prediction as the asymmetrical–coding model, the addition of certain ad hoc assumptions (see Kraemer et al., 1985) may permit the choose–long effect to be explained adequately.

Forget–cue effect. A number of studies, using samples differing in visual properties (e.g., red and green), have examined the effect of cues to remember (R cue) and forget (F cue) on DMTS performance (see Grant, in press, for a review). In a typical experiment, red and green sample stimuli are followed by an R cue (e.g., horizontal line) on some trials and by an F cue (e.g., vertical line) on other trials. Trials involving the R cue terminate in a test for sample memory, whereas trials involving the F cue terminate without a test for sample memory. Following training, DMTS performance is assessed on R–cued trials identical to those in training and on F–probe trials which, contrary to training, terminate in the same memory test as that employed on R–cued trials. The results of such testing reveal markedly lower accuracy on F–probe than on R–cued trials (the *F–cue effect*) and a larger F–cue effect (a) at longer than at shorter delays and (b) when the cues are presented early rather than late in the delay. These findings have led to the view that pigeons typically postperceptually process, or rehearse, codes activated by visual samples, and that training on F–cued trials endows the F cue with the ability to reduce or terminate that processing (e.g., Grant, 1981, in press; Maki, 1981).

Grant and Kelly (1996a) recently examined the effects of forget cuing in a matching–to–duration task. The samples were 2 and 10 s of houselight, the 2–s cues were horizontal (R cue) and vertical (F cue) lines, and the comparisons were red and green. R–cued trials terminated in the standard choice test, whereas F–cued trials terminated without a test. Following extensive training, a probe–testing phase was conducted in which red and green comparison stimuli appeared on infrequent F–probe trials.

The results of this experiment are shown in Fig. 3. Although accuracy was reliably lower on F–probe trials than on R–cued trials, the magnitude of that reduction was not large (approximately 6 percentage points, collapsed across delay). Moreover, in contrast to when visually differentiated samples are employed, the F–cue effect was not larger at longer than at shorter delays; in fact, a significant trend in the opposite direction was obtained. In a subsequent experiment, Grant and Kelly (1996a) presented the cues during the first 2–s (beginning) or final 2–s (end) of a 7–s delay. As shown in Fig. 4, and again in contrast to when visually differentiated samples are employed, the F–cue effect was not reliably greater when the cues were presented at the beginning rather than at the end of the delay.

Thus, the F–cue effect obtained with duration samples differs markedly from that obtained with visually differentiated samples in that, with duration samples, the F–cue effect is weak and its magnitude does not increase with delay and is not affected by point of cue interpolation. Grant and Kelly (1996a) concluded that pigeons do not postperceptually process when duration samples are employed in the standard choice version of DMTS. We speculated that whether or not pigeons postperceptually process in a DMTS task is a function of the content of the working memory representation. We argued that an analogical representation of duration may be inherently difficult, or perhaps impossible, to rehearse. In contrast, the representation employed in tasks involving visually differentiated samples, which presumably consists of images of colored fields or line orientations, is more conducive to postperceptual processing.

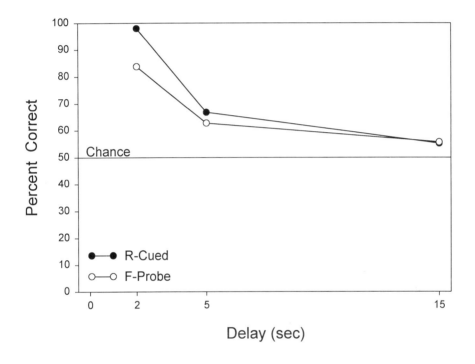

Figure 3. Mean percentage of correct responses on R–Cued and F–Probe trials as a function of delay. The samples were short (2 s) and long (10 s) presentations of an overhead light. The data were reported by Grant and Kelly (1996a, Exp. 2).

Although the analogical–coding view certainly does not predict the differences in the F–cue effect with visual versus duration samples, it can, in our view, interpret those differences sensibly and without additional assumptions. On the other hand, the differences in the F–cue effect as a function of sample dimension are particularly problematic for a prospective–coding view. According to that view, the codes employed in matching with visual and duration samples are equivalent in content. But if so, then why do pigeons demonstrate different F–cue effects in the two cases? Similar, although perhaps less severe, implications follow for the retrospective–categorical coding view. Although Kraemer et al. (1985) refer to the codes as "short" and "long", the codes must have some alternative, nonanalogical content. But if the actual content of the code is a nonanalogical representation of duration, as Kraemer et al. suggest, we see no reason why the content of such a code would discourage, or perhaps preclude, a rehearsal process.

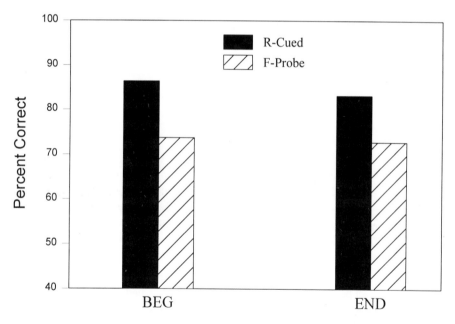

Figure 4. Mean percentage of correct responses on R–Cued and F–Probe trials as a function of whether the cue was presented during the first 2 s (BEG) or last 2 s (END) of a 7–s delay. The data were reported by Grant and Kelly (1996a, Exp. 4).

Aspects of Grant and Kelly's (1996a) results also present serious difficulties for the asymmetrical view of coding in the matching–to–duration task. As noted previously, there is general consensus among researchers that explicit absence versus presence samples are coded in terms of whether the presence sample had occurred. Consistent with this view is the finding of Maki and his associates, illustrated in the top half of Table 1, that presentation of an F–cue reduces accuracy markedly on trials initiated by the presence (food) sample, and reduces accuracy to a considerably lesser extent on trials initiated by the absence (no food) sample (Maki, Olson, & Rego, 1981; see also Maki, 1981; Maki & Hegvik, 1980).

Table 1. **Mean Percentage of Correct Responses on R–Cued and F–Probe Trials as a Function of Sample Type**

	No–Food Sample	Food Sample
R–Cued	92.9	92.9
F–Probe	79.6	60.5
Difference	13.3	32.4
	Short Sample	Long Sample
R–Cued	86.0	75.3
F–Probe	76.0	66.0
Difference	10.0	9.3

Note. Data using no–food and food samples were reported by Maki et al. (1981, Exp. 1, Omission Group). Data using short and long samples were reported by Grant and Kelly (1996a, Exp. 3). In both cases, data were collapsed across retention interval.

If the matching–to–duration task also involves asymmetrical coding, then one would anticipate an asymmetry in F–cue effectiveness in the duration task similar in magnitude to that obtained in the absence/presence task. Specifically, a large F–cue effect should be apparent on long–sample trials (functionally sample–present trials) and little or no F–cue effect should be apparent on short–sample trials (functionally sample–absent trials). As illustrated in the bottom half of Table 1, Grant and Kelly (1996a) did not obtain a differential F–cue effect on short– and long–sample trials. Instead, the F cue was equally effective regardless of the duration of the preceding sample. This finding, combined with our previously discussed results from research involving F–cue effects with duration samples raises serious doubt concerning the viability of the view that duration samples are coded categorically (either retrospectively or prospectively) or asymmetrically in terms of whether the long sample was presented.

Mixed–choice tests. As argued previously in this chapter, the occurrence of choose–long errors is not readily explained by an asymmetrical–coding model. In the present section, we present direct tests of the notion that samples differing in duration are typically coded in terms of whether or not a long sample was presented.

Grant and Spetch (1994b) trained pigeons in a duration DMTS procedure in which asymmetrical coding could mediate accurate performance (see left column in Table 2). Two matching-to-duration tasks were employed; one involved 2- and 10-s samples and the other involved 4.5- and 22.5-s samples. Trials initiated by either the 2- or 10-s sample terminated in a choice between two colors, and trials initiated by either the 4.5- or 22.5-s sample terminated in a choice between two line orientations. Accurate performance in this task could be achieved by coding only the longer sample in each set (i.e., 10 and 22.5 s) and by responding to the choice stimulus associated with either of the short samples (i.e., 2 and 4.5 s) when no code is active in memory.

Following acquisition, a retention test was administered. As anticipated, this test revealed a robust choose-short effect with both sets of samples, a result which can be explained by theories postulating either symmetrical and analogical sample coding or asymmetrical sample coding in which only the long sample is coded. In an effort to determine

Table 2. **Trial Types Employed During Training and Probe Testing by Grant and Spetch (1994b, Exp. 1)**

Training Trials		Mixed–Choice Probe Trials	
2 s:	G+ / R–	2 s:	G+ / H–
10 s:	R+ / G–	4.5 s:	H+ / G–
	and		and
4.5 s:	H+ / V–	10 s:	R+ / V–
22.5 s:	V+ / H–	22.5 s:	V+ / R–

Note. Numbers followed by "s" indicate the duration of the overhead houselight sample. Letters following the colon refer to the choice stimuli, + denotes correct and – denotes incorrect choice stimulus. Choice responses on mixed–choice probe trials were not reinforced. G=green, R=red, H=horizontal line, V=vertical line. Balancing of correct choice stimulus within each set of samples is not shown.

whether the choose–short effect was mediated by symmetrical or asymmetrical sample coding, pigeons next received testing on infrequent, nonreinforced "mixed–choice" probe trials (see right column in Table 2). On some mixed–choice probes, presentation of a short sample (either 2 or 4.5 s) was followed immediately by presentation of the two short–associated choice stimuli (the color associated with the 2–s sample and the line associated with the 4.5–s sample). On the remaining mixed–choice probes, presentation of a long sample (either 10 or 22.5 s) was followed immediately by presentation of the two long–associated choice stimuli (the color associated with the 10–s sample and the line associated with the 22.5–s sample).

Both the symmetrical and the asymmetrical sample coding views anticipate above chance accuracy on mixed–choice probes involving the two long–associated choice stimuli because both views maintain that each of the long samples are coded, and would therefore exert control over choice behavior. The two views make different predictions, however, concerning performance on mixed–choice probes involving the two short–associated choice stimuli. If asymmetrical coding mediated acquisition, then accuracy on mixed–choice probes involving the two short–associated choice stimuli should be at chance (50%) because (a) no code would be present in working memory to direct responding and (b) the absence of a code requires that a response be directed toward each of the choice stimuli. If, on the other hand, each of the four samples was coded, then accuracy on mixed–choice probes involving the two short–associated choice stimuli should be above chance.

Accuracy on mixed–choice probe trials involving both the short–associated (M = 62.6%) and the long–associated (M = 64.3%) choice stimuli was significantly above chance. Statistical analysis also revealed that accuracy on probe trials involving the short–associated choice stimuli did not differ from that on probe trials involving the long–associated choice stimuli. These results are inconsistent with the view that choice responding is controlled primarily by whether the animal remembers the presentation of a long sample. Although that view correctly predicted that accuracy would be above chance on probe trials involving the long–associated choice stimuli, it incorrectly predicted that accuracy would be at chance on probe trials involving the short–associated choice stimuli.

Why are Durations Coded Analogically?

The evidence reviewed previously in this chapter leads us to conclude that pigeons code duration samples analogically in the typical choice version of DMTS. In this section, we discuss research which attempted to identify the factor or factors which promote this form of coding.

DeLong and Wasserman (1985) noted that in contrast to nontemporal events which can be distinguished almost immediately, a subject must wait at least until the shorter duration has elapsed in order to distinguish between events differing in duration – – a period which may be referred to as *minimum wait time*. In the typical instance in which the short sample is 2 s in duration, the minimum wait time is 2 s because it is impossible to determine whether the sample is short or long until at least two seconds in the presence of the sample have elapsed. Spetch and Rusak (1992b) and Spetch and Sinha (1989) noted that a nonanalogical code of a temporal sample cannot be formed quickly following perceptual analysis of the sample, a fact that may encourage analogical representation, and/or discourage nonanalogical representation, of duration.

Grant and Kelly (1996c) assessed the role of minimum wait time in promoting analogical coding of temporal samples in the standard choice–matching task. This was accomplished by reducing the minimum wait time required to identify a temporal sample as long or short. The short sample, and hence minimum wait time, was .5 s in duration. To maintain comparability with other studies employing temporal samples in pigeons, a 1:4 short–to–long sample ratio was maintained. Thus, pigeons were trained with .5– (short) and 2–s (long) durations of houselight as the samples. If pigeons employ categorical codes, either retrospective or prospective in content, when a sample can be identified rapidly following onset, then retention testing should reveal little or no evidence of a choose–short effect. Contrary to this prediction, as shown in Fig. 5, a robust choose–short effect was obtained.

Grant and Kelly (1996c) also examined the possibility that a high degree of sample discriminability is responsible for temporal samples being coded in an analogical form in the standard choice–matching procedure. Several authors have noted that coding processes in pigeons are flexible in the sense that the process that is employed is determined, at least in part, by the relative demand that alternative coding processes place on the memory system (e.g., Grant, 1993; Honig & Thompson, 1982; Zentall, Urcuioli, Jackson–Smith, & Steirn, 1991). Zentall, Urcuioli et al. (1991), for example, have argued that when samples are highly discriminable, pigeons code retrospectively, whereas when samples are difficult to discriminate, pigeons are more sensitive to comparison discriminability and are more likely to code prospectively. On this view, retrospective coding would be anticipated to the extent that temporal samples are highly discriminable.

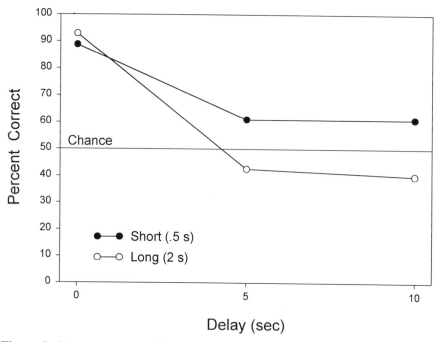

Figure 5. Mean percentage of correct responses on trials initiated by a short (.5 s) and long (2 s) sample as a function of delay. The data were reported by Grant and Kelly (1996c, Exp. 1).

Unpublished data from the first author's laboratory suggests that the temporal samples typically employed (e.g., 2 and 8 s, or 2 and 10 s) are indeed highly discriminable. As parts of separate investigations, two groups of pigeons were trained with different samples mapped onto red and green comparisons. The samples were 2– and 10–s durations of houselight in one group, and 5–s presentations of red and green keylight in the second group. Pigeons trained with temporal samples reached a criterion of 80% accuracy in considerably fewer trials (M = 406) than did the group trained with color samples (M = 704). This marked difference in the rate of acquisition suggests that 2– and 10–s durations are more discriminable as samples than are red and green keylights. This high discriminability may contribute to the coding of sample duration in an analogical form.

Grant and Kelly (1996c) trained pigeons with temporal samples which we anticipated would be difficult to discriminate, 2– and 3–s durations of houselight. The duration of the short sample was set at 2 s to maintain comparability with other studies of memory for event duration in which the short sample has typically been a 2–s event. The duration of the long sample was set at 3 s on the view that a 1–s difference between short and long samples would render the samples difficult to discriminate. To the extent that these samples are indeed difficult to discriminate, Zentall, Urcuioli et al.'s (1991) conclusion that less discriminable samples are more likely to be coded prospectively leads to the expectation of prospective, and hence nonanalogical, coding of the 2– and 3–s samples. If so, little or no evidence of a choose–short effect should be obtained.

The data from acquisition indicated that the temporal samples were indeed difficult to discriminate. In particular, the mean number of trials required to reached the 80% acquisition criterion was over 6,000, more than eight times the number of trials typically required to reach the same criterion using visually differentiated samples. Although the samples were difficult to discriminate, a robust choose–short effect was nonetheless obtained (see Fig. 6). Thus, use of highly discriminable samples is not necessary to the occurrence of analogical coding of event duration in pigeons.

Recently, Grant and Kelly (1996b) trained naive pigeons with samples which we viewed as being highly conducive to categorical coding: 1 and 30 s for some birds and 1 and 60 s for other birds. In addition, all training trials involved a short, variable delay (M = 2 s, range 1 to 3 s). The purpose of the delay was to afford an opportunity for the duration sample to be categorically coded, either retrospectively (e.g., "short" and "long") or prospectively ("peck red" and "peck green"). Variable delay training contrasts with the typical DMTS training regime, in which onset of the comparison stimuli immediately follows termination of the sample (0–s delay on all trials). In the typical procedure, the animal has an opportunity to categorically code

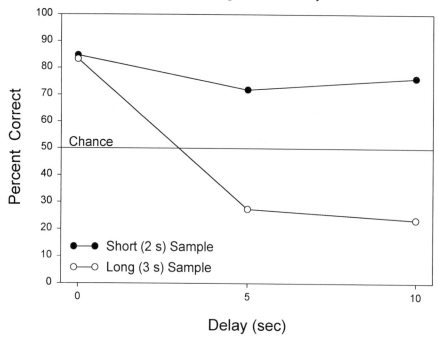

Figure 6. Mean percentage of correct responses on trials initiated by a short (2 s) and long (3 s) sample as a function of delay. The data were reported by Grant and Kelly (1996c, Exp. 2).

on long–sample trials because at some point the duration of the sample will exceed that required to identify it as long. Once identified as long, the categorical code could be activated during the remaining, redundant portion of the long sample. On short–sample trials, however, the animal cannot identify the sample as short until the sample terminates, and thus the representation is necessarily in an analogical form at the time of comparison onset. Employing a delay interval on all training trials allows both short and long samples to be identified well in advance of comparison onset and might, therefore, encourage use of categorical coding. As shown in Fig. 7, however, use of readily categorizable samples and a variable delay on all training trials did not eliminate the choose–short effect and hence did not result in nonanalogical coding of the samples. Spetch and Rusak (1992a), using the more conventional durations of 2 and 8 s as samples, also obtained a choose–short effect after training in which a delay of 2, 4, 6, or 8 s occurred on each trial.

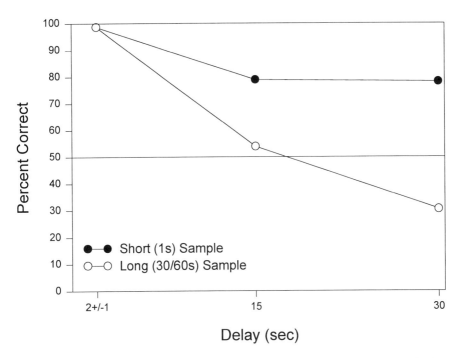

Figure 7. Mean percentage of correct responses on trials initiated by a short (1 s) or long (30 s for half the pigeons and 60 s for the remaining pigeons) sample as a function of delay following training in which a variable delay ($M = 2$ s, range 1 to 3 s) occurred on all trials. Data from these variable–delay trials collected during testing are plotted at the point labelled "2±1". The data were reported by Grant and Kelly (1996b).

The experiments reviewed in this section suggest that neither high sample discriminability nor use of substantial minimum wait time is necessary to the occurrence of analogical coding of event duration. In addition, providing pigeons an opportunity to identify both types of samples, and hence form or activate a categorical code well in advance of onset of the comparisons, failed to induce nonanalogical coding of event duration. This failure to induce nonanalogical coding occurred even though the samples employed (1 *versus* 30s and 1 *versus* 60s) appear from an anthropocentric perspective to be highly conducive to categorical coding. Collectively, these results demonstrate that naive pigeons are strongly disposed to code duration samples analogically in the standard choice version of DMTS. We believe that the fact that duration is inherently temporally extended and cumulative may be critical in promoting the analogical representation of event duration.

Coding in Nonstandard Versions of DMTS

The research reviewed previously in this chapter constitutes, in our view, compelling evidence that pigeons code duration samples analogically in the standard choice version of DMTS. However, research reviewed in the present section indicates that analogical coding is not an inevitable result of employing duration samples in a DMTS task. Research described below has revealed two circumstances in which samples of different durations are coded nonanalogically. One circumstance is the successive version of DMTS in which rate of responding to successively presented comparisons is used as the performance index. The second involves some instances in which an MTO mapping is employed in the choice version of DMTS.

Successive Version of DMTS

In the successive task, there are two different test stimuli, but only one of them is presented following the sample on each trial. Pecks to one test stimulus are reinforced following short samples but not following long samples, whereas pecks to the other test stimulus are reinforced following long samples only. Memory for the sample is assessed by the discrimination ratio, which is the proportion of all responses that are emitted to positive (i.e., reinforced) test stimuli. Even though the choice– and successive–matching tasks appear to involve similar memory requirements (e.g., Nelson & Wasserman, 1978), evidence reviewed below suggests that pigeons code samples of event duration differently in the two tasks.

One experiment involved both a between–subjects and a within–subjects comparison of the effect of assessment task on coding processes (Grant & Spetch, 1991). The between–subjects comparison involved two groups, one assigned to the choice task and a second assigned to the successive task. In both groups, 2– and 8–s presentations of houselight served as the short and long samples, and red and green colors served as the test stimuli. Following acquisition, a retention test was conducted. As shown in Fig. 8, pigeons trained and tested in the choice task demonstrated asymmetrical retention on short– and long–sample trials, indicative of the choose–short effect. In contrast, equivalent rates of forgetting on short– and long–sample trials were obtained in pigeons trained and tested in the successive task.

Next, the birds originally assigned to the successive task were transferred to the choice task, and the birds originally assigned to the choice task were transferred to the successive task. The transfer was arranged such that the reinforcement contingencies were consistent across tasks for each bird. Thus, for birds transferred to the choice task, the former positive test stimulus on short–sample trials became the correct comparison stimulus on short–sample trials, and the former positive test stimulus on long–sample trials became the correct comparison stimulus on long–sample trials. Similarly, for birds transferred to the successive task, the former correct comparison on short–sample trials became the positive test stimulus on short–sample trials, and the former correct comparison on long–sample trials became the positive test stimulus on long–sample trials.

Following acquisition, a retention test was conducted. As shown in Fig. 9, birds that were first trained in the choice task demonstrated equivalent rates of forgetting on trials initiated by short and long samples when subsequently trained and tested in the successive task. As shown in Fig. 10, birds that were first trained in the successive task did not demonstrate differential rates of forgetting on short– and long–sample trials when subsequently trained and tested in the choice task. Instead, accuracy declined at an approximately equivalent rate as a function of delay on trials initiated by short and long samples.

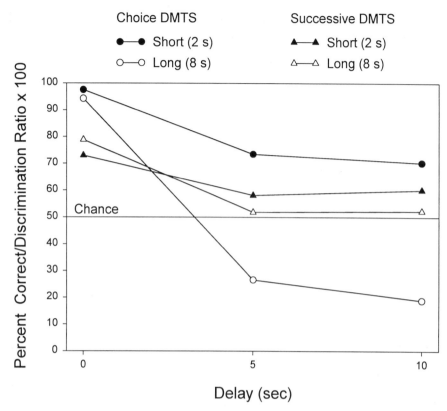

Figure 8. Mean matching accuracy on trials initiated by a short (2 s) and long (8 s) sample as a function of delay. Subjects in the choice group were trained and tested in a choice DMTS task, and subjects in the successive group were trained and tested in a successive DMTS task. The dependent measure is percent correct in the choice task and discrimination ratio, multiplied by 100, in the successive task. The data were reported by Grant and Spetch (1991, Exp. 2).

Thus, differential rates of forgetting as a function of sample duration are not obtained when pigeons are trained and tested in the successive task (see also Wasserman, DeLong, & Larew, 1984), regardless of whether or not they had been previously trained in the choice task. On the other hand, differential rates of forgetting as a function of sample duration are obtained when naive pigeons are trained and tested in the choice task. However, previous training in a successive task, in which the contingencies are consistent with those in a subsequent choice task, is sufficient to eliminate the differential rates of forgetting which would otherwise obtain.

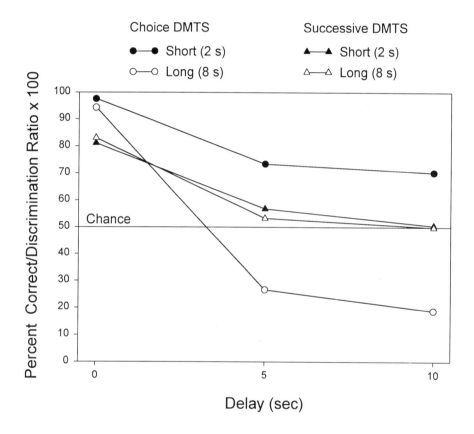

Figure 9. Mean matching accuracy on trials initiated by a short (2 s) and long (8 s) sample as a function of delay. Subjects were first trained and tested in a choice DMTS task, and were subsequently trained and tested in a successive DMTS task. The dependent measure is percent correct in the choice task and discrimination ratio, multiplied by 100, in the successive task. The data were reported by Grant and Spetch (1991, Exp. 2).

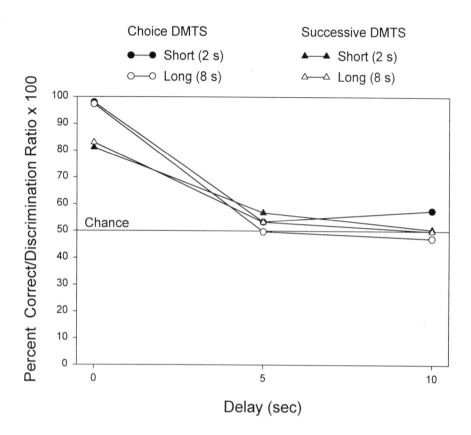

Figure 10. Mean matching accuracy on trials initiated by a short (2 s) and long (8 s) sample as a function of delay. Subjects were first trained and tested in a successive DMTS task, and were subsequently trained and tested in a choice DMTS task. The dependent measure is percent correct in the choice task and discrimination ratio, multiplied by 100, in the successive task. The data were reported by Grant and Spetch (1991, Exp. 2).

The different effects of delay in the choice and successive tasks may be interpreted as reflecting different coding strategies which naive pigeons employ in the two tasks. As argued previously in this chapter, naive pigeons code samples of different durations retrospectively in terms of perceived duration in the choice task. A process of subjective shortening operating during the delay would then lead to differential rates of forgetting on the two types of trials. In the successive task, naive pigeons might code the samples prospectively in terms of an instruction to respond and/or not to respond to a particular test stimulus. If the appropriate code is forgotten during the delay, responding should be nonsystematic with respect to sample duration, resulting in nondifferential rates of forgetting as a function of sample type.

The view that naive pigeons employ a retrospective and analogical coding strategy in the choice task and a prospective coding strategy in the successive task, combined with a consideration of the extent to which each strategy is applicable to the alternative procedure, allows one to provide a plausible account of the results from both transfer and delay testing. Specifically, the degree of transfer should reflect the extent to which accurate performance in the two tasks is readily mediated by coding processes acquired in the initial task. A prospective–coding strategy of the form "respond to test stimulus A" and/or "do not respond to test stimulus B", acquired initially in the successive task, is sufficient to generate accurate performance in the choice task. Hence, animals transferred from the successive task to the choice task should, and did, demonstrate substantial positive transfer (M=78.1% correct on the first session of choice matching). Moreover, because these animals are employing a prospective–coding strategy in the choice task, they should not, and did not, demonstrate a choose–short tendency during delay testing in the choice task.

Consider next animals trained initially in the choice task and transferred to the successive task. A retrospective and analogical coding strategy of the form "if remembered duration is X, then choose comparison A instead of comparison B", acquired in the initial choice task, is not sufficient to generate accurate performance in the successive task. In particular, retrieval of a rule of the form "choose comparison A instead of comparison B" would not necessarily lead to either (a) a high rate of keypecking if test stimulus A was present and/or (b) a low rate of keypecking if test stimulus B was present.

Hence, animals transferred from the choice task to the successive task might, and did, demonstrate little positive transfer (the mean discrimination ratio on the first session of successive matching was .512). Moreover, to the extent that these animals abandon the retrospective and analogical strategy and adopt a prospective one in acquiring the successive task, subsequent delay testing should, and did, fail to reveal a tendency to respond at longer delays as if the long sample was short (i.e., they failed to show a *respond–short effect*, analogous to the choose–short effect obtained in the choice task).

Subsequent research has supported our assumption that the differential delay effects seen in successive and choice tasks reflect, specifically, differences in the way in which duration information is retained and not differences in timing processes (Spetch & Grant, 1993). In these studies, the effects of manipulations that should alter the perceived duration of the sample were compared in the two tasks. In contrast to the differential effects of varying the delay interval, manipulating either the duration of the sample, the number of sample presentations, or the duration of the preceding intertrial interval, had comparable effects on performance in the two tasks. Thus, the difference between the two tasks seems specific to the effects of delay.

The results discussed to this point suggest that coding processes differ in choice and successive tasks with duration samples, but they leave open the question of why these tasks would induce different coding processes. One approach to this question is to isolate the specific differential feature(s) of the two tasks that critically determines the form of coding. Although the choice and successive tasks differ only in the test component of the trial, their test components differ in several potentially important ways. First, the test components of the two tasks are perceptually different in that two stimuli are present during the test phase of the choice task, whereas only a single stimulus is present during the test phase of the successive task. Second, the spatial location of the correct test stimulus varies across trials in the choice task, whereas test stimuli are always presented in the same spatial location in the typical successive task. Third, the test components of the two tasks are temporally different. In the standard choice task, the test stimuli are presented only until a single peck occurs, typically less than 1 s, whereas in the successive task the test stimulus is presented for a fixed period of time, typically 5 s. In addition to any effects due to the difference in duration of the test component, the use of a fixed interval (FI) in successive tasks might cause reactivation of the timing system during the test component of the trial. Fourth, different performance indexes are used in the two tasks. The choice task uses accuracy of a single peck per trial as the performance index, whereas the successive task uses differential rates of responding to the positive and negative test stimuli as the performance index. Finally, the test components of the two tasks are functionally different in that: (a) responding inaccurately in the choice task reduces the obtained rate of reinforcement whereas responding inaccurately in the successive task has no effect on obtained rates of reinforcement, and (b) the choice task permits subjects to select a stimulus that will lead to reinforcement, whereas the successive task provides only an opportunity to respond or not to respond to a stimulus.

To determine which of these differences might underlie the differences in coding strategies with duration samples, we developed a set of "hybrid" tasks that incorporated some features of the typical choice task and some features of the typical successive task. The use of such hybrid tasks has indicated that spatial and perceptual factors do not underlie the use of nonanalogical codes in the successive task. For example, in one condition the successive task was altered by varying the spatial location of the test stimulus between two pecking keys. In a second condition, the variously–located test stimulus was accompanied by an ineffective distracter stimulus at the alternative location. Neither condition resulted in retention functions typical of the choice task (Kelly & Spetch, 1996). Presumably then, these spatial and perceptual features are not responsible for the occurrence of analogical coding of duration.

Temporal factors also do not seem to underlie the differences in coding of duration samples in the choice and successive tasks. Adding a 5–s FI to the test component of a choice task, thereby making it temporally similar to a successive task, did not eliminate the choose–short effect (Spetch, Grant & Kelly, 1996). Apparently, reactivation of the timing system by an FI during the test component is not the reason that pigeons code duration samples nonanalogically in the successive task.

The between–task difference in coding of sample duration also cannot be due to the use of rate versus choice measures as the performance index. First, as discussed previously, birds initially trained on a successive task and then transferred to the choice task do not show a choose–short effect (see Fig. 10). Thus, symmetrical functions typically obtained with the rate measure may also be obtained with the choice measure. Second, in choice tasks with an FI requirement during the test phase (Spetch et al., 1996), asymmetrical retention functions were obtained with both choice and rate measures.

Differential consequences of inaccurate responding also do not appear to be responsible for the between–task difference in coding with duration samples. In an unpublished study, Spetch (1996) modified the choice task to eliminate the strong relationship between choice accuracy and rate of reinforcement. In this task, pecks to the incorrect comparison were recorded but had no effect: Both the correct and the incorrect comparisons stayed on until the correct comparison was pecked. Thus, as in the successive task, pecks to the incorrect (negative) stimulus expended energy but did not reduce the amount of food that could be obtained in the session. The performance index used was accuracy of the first peck made on each trial. Even though the consequence of responding inaccurately was similar to that in the successive task, a significant choose–short effect occurred nevertheless.

Thus, it appears that the critical determinant of coding processes with duration samples must be whether the test component allows subjects to select a correct stimulus or instead permits only a go/no–go response. Evidence that this functional difference is indeed responsible for the difference in coding processes was provided by Spetch et al. (1996) through the use of a "successive–option" hybrid procedure. In this task, the test component involved an FI 5–s schedule during which two stimuli were presented: a red or green test stimulus on the left pecking key and a yellow "option" stimulus on the right pecking key. The test stimulus was either positive or negative depending on which duration sample had been presented. The "option" stimulus was positive when a negative test stimulus was presented and was negative when a positive test stimulus was presented. The pigeons could obtain reinforcement if they completed the FI schedule by choosing the test stimulus if it was positive, or by choosing the option stimulus if the test stimulus was negative. Therefore, the valence of each test stimulus depended on which sample was presented, whereas the valence of the option stimulus was not dependent on which sample was presented but instead depended on the combination of sample and test stimulus. Basically then, this procedure was a successive task with an added option stimulus that allowed subjects to *select* a correct stimulus on each trial. Consistent with the notion that the opportunity to select (choose) is responsible for the analogical coding of sample duration, the successive–option task generated asymmetrical retention functions during testing, comparable to those obtained in standard choice tasks. Furthermore, when the option stimulus was subsequently removed, rendering the procedure identical to a standard successive task, the respond–short effect disappeared (see Fig. 11).

We conclude that an opportunity to select a correct test stimulus is the critical difference between choice and successive tasks that determines the form of coding with duration samples. Retention functions suggestive of analogical coding were obtained when naive birds were tested on any of the tasks that permitted subjects to select the correct stimulus during the test component, whereas retention functions suggestive of nonanalogical coding were obtained whenever the test component required a go/no–go response. This conclusion corroborates our earlier speculations regarding the coding strategies used in each task. Specifically, we speculated that the choice task might encourage a retrospective and analogical strategy such as "if remembered duration is X, then choose comparison A instead of comparison B", whereas the successive task might encourage a prospective strategy such as "respond to stimulus A and/or do not respond to stimulus B". It follows from these speculations that opportunity to choose would be the critical differentiating feature of the choice and successive tasks that encourages different forms of sample duration coding.

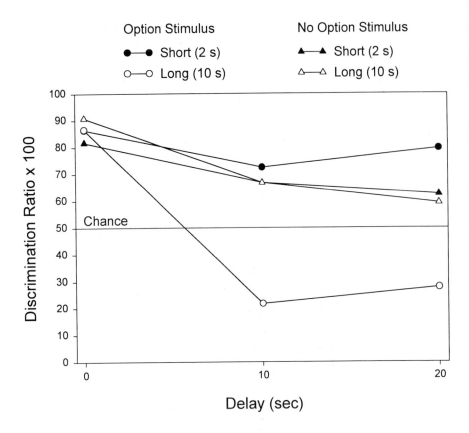

Figure 11. Mean discrimination ratio, multiplied by 100, on trials initiated by a short (2 s) and long (10 s) sample as a function of delay. The birds were first trained and tested in the successive/option procedure and were then trained and tested in the standard (no option stimulus) successive procedure. The data were reported by Spetch et al. (1996, Exps. 2a and 2b).

Use of an MTO Mapping in Choice DMTS

As noted above, naive pigeons trained and tested in the choice–matching task demonstrate differential rates of forgetting on trials initiated by short and long samples. Recall that this result is interpretable within the subjective–shortening account, which maintains that naive pigeons code durations retrospectively and analogically in the choice–matching task. The research described in the preceding section suggests that pigeons code durations nonanalogically, and probably prospectively, in the successive–matching task. The research described in this section asked whether, under some conditions, naive pigeons might code durations in a nonanalogical form in the choice–matching task.

Grant and Spetch (1993a) manipulated the sample–to–comparison mapping arrangement in an attempt to induce different forms of sample coding in a choice–matching task. In one experiment, the sample on any particular trial was either a short (2 s) or a long (10 s) presentation of either houselight or keylight (crossed white lines on the center key). Red and green keylights were presented as comparison stimuli on each trial. Groups of pigeons received one of two MTO sample–to–comparison mapping arrangements. In the consistent mapping arrangement, the two short samples were associated with one comparison and the two long samples were associated with the alternative comparison. In this arrangement, the event which carried the duration information, houselight or keylight, was irrelevant, and only duration was relevant to problem solution. In the inconsistent mapping arrangement, in contrast, both duration and carrier were relevant to problem solution. This was the case because one comparison was correct following a short presentation of one carrier and a long presentation of the alternative carrier (2–s houselight and 10–s keylight), and the alternative comparison was correct following either of the two remaining samples (10–s houselight and 2–s keylight).

It was anticipated that the inconsistent mapping arrangement would be particularly likely to result in nonanalogical coding. This anticipation was based, in part, on the fact that only two codes would be required to mediate accurate performance in a nonanalogical–coding process (e.g., "peck red" and "peck green", or "event A" and "event B"), whereas an analogical–coding process sufficient to mediate accurate performance would require the use of four codes (one for each of the four samples). Moreover, analogical codes would be rather complex because information about both duration and carrier would need to be preserved in each of the four codes. Honig and Thompson's (1982) suggestion that the coding process which requires less information will operate leads to the expectation of nonanalogical coding in the inconsistent group.

In the consistent group, in which only duration is relevant, accurate performance could be mediated by a coding process in which each of the four samples would be represented analogically as one of two values; one value associated with a 2–s duration and the other value associated with a 10–s duration. However, given that the houselight is brighter than the keylight, Wilkie's (1987) finding that bright events are perceived as longer than equivalent durations of dim events suggests that an analogical–coding process might not be particularly effective. Likely candidates for nonanalogical–coding processes that might operate in the consistent mapping are prospective coding ("peck red" and "peck green") and intermediate coding. In an intermediate–coding process, samples associated with the same comparison stimulus activate a common code which is in turn associated with a particular comparison response (for similar suggestions see, for example, Grant, 1982; Maki, Moe, & Bierley, 1977; Roitblat, 1980; Urcuioli, Zentall, Jackson–Smith, & Steirn, 1989). In the most general form of this coding process, samples associated with one comparison activate an "event A" code and samples associated with the alternative comparison activate an "event B" code. At the time of testing, the active code would be used to query reference memory to generate the appropriate comparison response. The intermediate code might be related arbitrarily to the samples in the consistent mapping condition (as is the case for the codes "event A" and "event B"), or they might represent a feature common to each of the samples associated with that code, in which case the intermediate codes would be "short" and "long".

A delay test was employed to determine whether either or both of these MTO mappings would result in nonanalogical coding of event duration. Nonanalogical coding, whether prospective ("peck red" and "peck green") or intermediate (either "event A" and "event B", or "short" and "long") would be implicated to the extent that accuracy on both short– and long–sample trials decreased at approximately the same rate as a function of increases in delay. This prediction is predicated on the notion that there is no reason to anticipate that two prospective codes, or two intermediate codes, should be forgotten at different rates. As shown in Fig. 12, neither mapping arrangement produced any evidence of a choose–short effect. The equivalent rates of forgetting on short– and long–sample trials suggests that the pigeons in both mapping arrangements employed a nonanalogical–coding

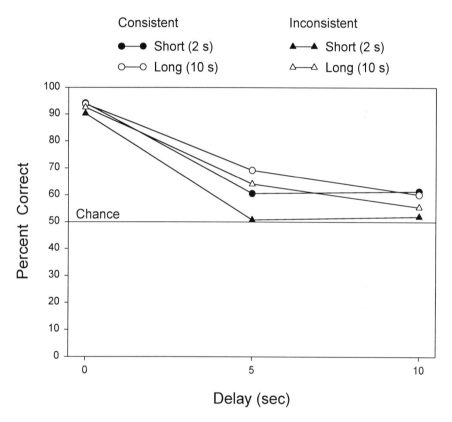

Figure 12. Mean percentage of correct responses on trials initiated by a short and long sample as a function of delay. In the consistent group, a 2–s presentation of either houselight or keylight was associated with one comparison, and a 10–s presentation of either houselight or keylight was associated with the alternative comparison. In the inconsistent group, a 2–s presentation of houselight and a 10–s presentation of keylight were associated with one comparison, and a 10–s presentation of houselight and a 2–s presentation of keylight were associated with the alternative comparison. The data were reported by Grant and Spetch (1993a, Exp. 3).

process. These results, viewed in conjunction with previous demonstrations of differential rates of forgetting on short– and long–sample trials in the choice–matching task in which a one–to–one (OTO) sample–to–comparison mapping was employed, suggest that naive pigeons are capable of coding samples of event duration either analogically or nonanalogically in the choice–matching task.

In further work concerning the effects of mapping arrangement on the coding of samples of duration in the choice–matching task, Grant and Spetch (1993a, Exp. 1) investigated whether the presence of nontemporal samples in an MTO mapping can alter the way in which temporal samples are coded. Two groups of pigeons were trained with two pairs of samples; line samples (a vertical and horizontal line, each 6 s in duration) and duration samples (2– and 10–s presentations of houselight). Color comparisons were always presented following the duration samples. In the MTO mapping, color comparisons were also presented following the vertical– and horizontal–line samples. In the OTO mapping, line–orientation comparisons were presented following the vertical– and horizontal–line samples. A third group served as a control and was trained only on the duration–matching task.

Following acquisition, a retention test involving delays of 0, 10, and 20 s was conducted. As shown in Fig. 13, a robust choose–short effect was obtained in both the Control and OTO group, as expected. Contrary to our expectations, however, the MTO group also demonstrated a choose–short effect, albeit the effect was not as strong as that in either the Control or OTO groups.

The results from the MTO group suggest that retrospective and analogical coding of event duration occurred in spite of the MTO mapping arrangement. This is somewhat surprising in that the work reviewed previously in this section revealed evidence of nonanalogical coding of event duration in two variations of an MTO mapping arrangement in which both pairs of samples were temporal. One possible explanation is that the introduction of nontemporal samples has little effect, if any, on the coding of temporal samples. Inspection of the acquisition data in the MTO group, however, suggested an alternative explanation. That inspection revealed that animals in the MTO group acquired accurate matching to temporal samples

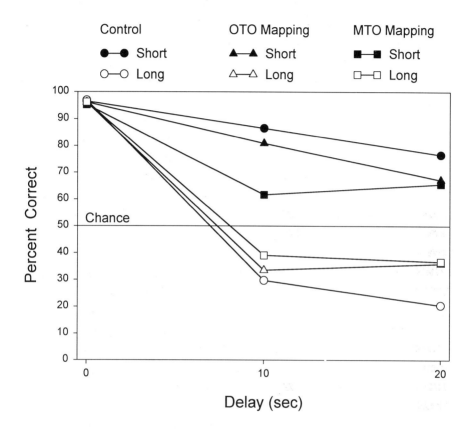

Figure 13. Mean percentage of correct responses on trials initiated by a short (2 s) and long (10 s) sample as a function of delay. In the MTO group, nontemporal samples (different line orientations) were followed by the same comparison stimuli as were employed on duration–matching trials. In the OTO group, the nontemporal samples were followed by comparison stimuli different from those employed on duration–matching trials. In both the MTO and OTO groups, matching to temporal and nontemporal samples was trained concurrently. In the control group, nontemporal samples were not employed. The data were reported by Grant and Spetch (1993a, Exp. 1).

considerably more quickly than they acquired accurate matching to nontemporal (line–orientation) samples. Thus, the failure of nontemporal samples to markedly influence the way in which temporal samples were coded may have been due to the animals learning to code the temporal samples before they learned to code the line–orientation samples.

To assess this possibility, Grant and Spetch (1993a, Exp. 2) conducted a second experiment in which animals in the OTO and MTO groups first acquired accurate matching with line–orientation samples (as previously, the comparisons were line orientations in the OTO group and colors in the MTO group). Following acquisition with the lines, half of the trials within each session now involved the temporal samples followed by color comparisons. As shown in Fig. 14, a robust choose–short effect was obtained in both the Control and OTO groups during retention testing, as was the case in the first experiment. However, the MTO group revealed no evidence of a choose–short effect. Instead, accuracy in this group declined at an approximately equivalent rate on trials initiated by short and long samples. Santi, Bridson, and Ducharme (1993) also found that an MTO mapping, involving 4–s red and green colors and 2– and 8–s durations (presentation of crossed diagonal lines on the center key) mapped onto line orientation comparisons, eliminated the choose–short effect. Although they trained the two problems concurrently, pigeons showed faster acquisition on color–sample than on temporal–sample trials. These findings suggest that, at least under some conditions, an MTO mapping incorporating nontemporal and temporal samples can result in nonanalogical coding of the temporal samples.

An MTO mapping may encourage a common–coding process in which samples associated with the same comparison activate the same code. For example, samples associated with the same comparison might activate a prospective representation of that comparison stimulus (e.g., "peck red"), or might activate the same intermediate code (e.g., "event A"). The finding that an MTO mapping eliminates the choose–short effect, at least under some conditions, is readily interpreted on the view that samples associated with the same comparison activated a single nonanalogical code (e.g., "peck red", "event A").

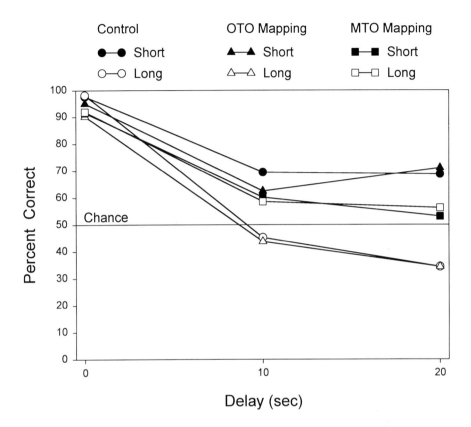

Figure 14. Mean percentage of correct responses on trials initiated by a short (2 s) and long (10 s) sample as a function of delay. In the MTO group, nontemporal samples (different line orientations) were followed by the same comparison stimuli as were employed on duration–matching trials. In the OTO group, the nontemporal samples were followed by comparison stimuli different from those employed on duration–matching trials. In both the MTO and OTO groups, matching to nontemporal samples was trained prior to training with temporal samples. In the control group, nontemporal samples were not employed. The data were reported by Grant and Spetch (1993a, Exp. 2).

To further test the notion that an MTO mapping involving nontemporal and temporal samples can result in common coding, Grant and Spetch (1994a) conducted a mediated–transfer test using birds from the OTO and MTO groups in their earlier experiment (Grant &

Spetch, 1993a, Exp. 2). If MTO training results in common coding, then subsequent training involving one set of samples and a new set of comparisons might transfer to the other set of samples (Urcuioli et al., 1989; Zentall, Steirn, Sherburne, & Urcuioli, 1991). To assess mediated transfer, birds in the OTO and MTO groups were given a phase of interim training in which one pair of samples from the original task (durations for half the birds and line orientations for the remaining birds) was associated with two new comparison stimuli (circle and dot shapes). In a subsequent transfer phase, the shape comparisons were presented on trials involving the alternative set of samples from the original task.

The contingencies during interim training and transfer testing with the shape comparisons were identical in both the OTO and MTO groups, and were arranged such that samples which were associated with the same color comparison in the original MTO mapping were now associated with *different* shape comparisons. To illustrate, suppose during MTO training the 2–s and vertical samples were associated with the red comparison, and the 10–s and horizontal samples were associated with the green comparison. Then, for example, during interim training vertical might be associated with the circle comparison, and horizontal with the dot comparison. Finally, during the transfer phase, circle and dot comparisons would follow duration samples; dot would be the correct comparison on 2–s sample trials and circle would be the correct comparison on 10–s sample trials. Thus, if the line sample and the duration sample that were associated with the same color comparison during original MTO training were represented by a single common code, then accuracy during the transfer test should be lower in the MTO group than in the OTO group (in which common coding, and hence mediated transfer, is precluded by the OTO mapping arrangement).

The transfer test did reveal evidence for common coding in the MTO group. Mean accuracy during the first 60–trial session of transfer testing was 61.8% correct in the OTO group and only 43.4% correct in the MTO group. A more molecular analysis of first session performance in the MTO group revealed even stronger evidence for common coding; the mean percentage of correct responses in each of the five 12–trial blocks that constituted the first session of transfer was 34.7, 39.3, 42.0, 52.5 and 48.3, respectively. Thus, when confronted with shape comparisons following samples which had not been explicitly trained with either of those comparisons, birds in the MTO group tended to select the shape comparison that had been correct on trials initiated by the sample that was associated, during original MTO training, with the same color comparison as the current sample.

This transfer result suggests that the MTO mapping resulted in a coding process in which samples associated with the same comparison activated a common code. The results of retention testing, described earlier in this section, further suggest that the content of the common codes is nonanalogical (e.g., "peck red", "event A").

Conclusion

A fundamental issue in the analysis of working memory is the nature of the representation, or code, that mediates accurate performance across a retention interval. In this chapter we addressed the nature of working memory representations in DMTS tasks in which the samples differ in duration. In the standard choice version of the DMTS task, in which each duration sample is mapped in a one-to-one relation to a comparison stimulus, naive pigeons code samples retrospectively and analogically. We maintain that, in such instances, pigeons represent different event durations as codes which specify values along a dimension that changes in a continuous and cumulative manner as a function of the passage of time. Such a coding process is analogical because the representative dimension shares with duration the property of changing in a continuous and cumulative manner. As one example of an analogical-coding process, pigeons might represent different durations as different intensities of a visual image. On this view, the intensity of the image increases as time spent in the presence of the sample increases, and decreases as time since termination of the sample increases. The phenomena considered in the first major section of this chapter encourage the view that durations are coded analogically, and discourage the view that durations are coded asymmetrically or categorically (either retrospectively or prospectively).

Research considered in the second major section of this chapter suggests, however, that duration samples are not always coded analogically. In the successive DMTS task, in which accuracy is assessed by a go/no-go response rather than by choice behavior, duration samples are coded nonanalogically, perhaps in an instructional, prospective manner. Evidence of nonanalogical coding of duration samples has also been obtained in two instances in the choice DMTS task. First, pigeons initially trained in successive DMTS demonstrate nonanalogical coding when subsequently transferred to a choice DMTS task in which the contingencies are consistent with those of prior training. Substantial positive transfer from successive to choice DMTS suggests that pigeons continue to use the nonanalogical coding process adopted in the successive task when transferred to the choice task. Second, pigeons trained with two sets of samples, either both temporal or one temporal and one nontemporal, mapped onto a single set of comparisons (an MTO mapping) also demonstrate nonanalogical coding of the temporal samples. The results of mediated transfer testing suggest that samples associated with the same comparison stimulus activate a common code (e.g., "peck red", "event A"). One caveat to this conclusion is that an MTO mapping is less likely to reduce the tendency to code temporal samples analogically if accurate matching with temporal samples is acquired prior to acquiring accurate matching with nontemporal samples.

In summary, substantial progress has been made over the past decade in identifying the coding processes used by pigeons in DMTS tasks with duration samples, and in determining some of the factors that induce analogical or nonanalogical forms of coding. As yet, less is known about the generality of these results across species (e.g., Santi, Weise & Kuiper, 1995) or the coding processes used in other types of tasks that may require retention of temporal information.

Author Notes

Preparation of this article was supported by Natural Sciences and Engineering Research Council of Canada Grants OGP 0443 to DSG and OGP 38861 to MLS. Correspondence concerning this article should be sent to either of the first two authors, Department of Psychology, University of Alberta, Edmonton, Alberta, Canada T6G 2E9. Electronic mail should be directed to dgrant@psych.ualberta.ca (DSG) or mspetch@psych.ualberta.ca (MLS).

References

Cabeza de Vaca, S., Brown, B. L., & Hemmes, N. S. (1994). Internal clock and memory processes in animal timing. *Journal of Experimental Psychology: Animal Behavior Processes, 20,* 184–198.

Colwill, R. M. (1984). Disruption of short–term memory for reinforcement by ambient illumination. *Quarterly Journal of Experimental Psychology, 36B,* 235–258.

DeLong, R. E., & Wasserman, E. A. (1985). Stimulus selection with duration as a relevant cue. *Learning and Motivation, 16,* 259–287.

Dougherty, D. H., & Wixted, J. T. (1996). Detecting a nonevent: Delayed presence–versus–absence discrimination in pigeons. *Journal of the Experimental Analysis of Behavior, 65,* 81–92.

Fetterman, J. G. (1995). The psychophysics of remembered duration. *Animal Learning & Behavior, 23,* 49–62.

Grant, D. S. (1981). Short–term memory in the pigeon. In N. E. Spear & R. R. Miller (Eds.), *Information processing in animals: Memory mechanisms* (pp. 227–256). Hillsdale, NJ: Erlbaum.

Grant, D. S. (1982). Prospective versus retrospective coding of samples of stimuli, responses, and reinforcers in delayed matching with pigeons. *Learning and Motivation, 13,* 265–280.

Grant, D. S. (1991). Symmetrical and asymmetrical coding of food and no–food samples in delayed matching in pigeons. *Journal of Experimental Psychology: Animal Behavior Processes, 17,* 186–193.

Grant, D. S. (1993). Coding processes in pigeons. In T. R. Zentall (Ed.), *Animal cognition: A tribute to Donald A. Riley* (pp. 193–216). Hillsdale, NJ: Erlbaum.

Grant, D. S. (in press). Directed forgetting in pigeons. In J. M. Golding & C. MacLeod (Eds.), *Intentional forgetting: Interdisciplinary approaches.* Hillsdale, NJ: Erlbaum.

Grant, D. S., & Kelly, R. (1996a). Choice matching to duration samples in pigeons: Effect of postsample cues to remember and forget. *Manuscript submitted for publication.*

Grant, D. S., & Kelly, R. (1996b). The effect of a variable training delay on the choose–short effect in pigeons. *Manuscript in preparation.*

Grant, D. S., & Kelly, R. (1996c). The role of sample discriminability and minimum wait time in the coding of event duration in pigeons. *Learning and Motivation, 27,* 243–259.

Grant, D. S., & Robinson, T. C. (1993). Cross–stimulus transfer of timing in pigeons. *Animal Learning & Behavior, 21,* 106–112.

Grant, D. S., & Spetch, M. L. (1991). Pigeons' memory for event duration: Differences between choice and successive matching tasks. *Learning and Motivation, 22,* 180–190.

Grant, D. S., & Spetch, M. L. (1993a). Analogical and nonanalogical coding of samples differing in duration in a choice–matching task in pigeons. *Journal of Experimental Psychology: Animal Behavior Processes, 19,* 15–25.

Grant, D. S., & Spetch, M. L. (1993b). Memory for duration in pigeons: Dissociation of choose–short and temporal–summation effects. *Animal Learning & Behavior, 21,* 384–390.

Grant, D. S., & Spetch, M. L. (1994a). Mediated transfer testing provides evidence for common coding of duration and line samples in many–to–one matching in pigeons. *Animal Learning & Behavior, 22,* 84–89.

Grant, D. S., & Spetch, M. L. (1994b). The role of asymmetrical coding of duration samples in producing the choose–short effect in pigeons. *Learning and Motivation, 25,* 413–430.

Honig, W. K. (1978). Studies of working memory in the pigeon. In S. H. Hulse, H. Fowler, & W. K. Honig (Eds.), *Cognitive processes in animal behavior* (pp. 211–248). Hillsdale, NJ: Erlbaum.

Honig, W. K., & Thompson, R. K. R. (1982). Retrospective and prospective processing in animal working memory. In G. H. Bower (Ed.), *The psychology of learning and motivation: Advances in research and theory* (Vol. 16, pp. 239–283). New York: Academic Press.

Kelly, R., & Spetch. M. L. (1996). [Retention functions following short and long samples in modified successive DMTS tasks]. Unpublished raw data.

Killeen, P. R., & Fetterman, J. G. (1988). A behavioral theory of timing. *Psychological Review, 95,* 274–295.

Kraemer, P. J., Mazmanian, D. S., & Roberts, W. A. (1985). The choose–short effect in pigeon memory for stimulus duration: Subjective shortening versus coding models. *Animal Learning & Behavior, 13,* 349–354.

Maki, W. S. (1981). Directed forgetting in animals. In N. E. Spear & R. R. Miller (Eds.), *Information processing in animals: Memory mechanisms* (pp. 199–225). Hillsdale, NJ: Erlbaum.

Maki, W. S., & Hegvik, D. K. (1980). Directed forgetting in pigeons. *Animal Learning & Behavior, 8,* 567–574.

Maki, W. S., Moe, J. C., & Bierley, C. M. (1977). Short–term memory for stimuli, responses, and reinforcers. *Journal of Experimental Psychology: Animal Behavior Processes, 3,* 156–177.

Maki, W. S., Olson, D., & Rego, S. (1981). Directed forgetting in pigeons: Analysis of cue functions. *Animal Learning & Behavior, 9,* 189–195.

Nelson, K. R., & Wasserman, E. A. (1978). Temporal factors influencing the pigeon's successive matching–to–sample performance: Sample duration, intertrial interval, and retention interval. *Journal of the Experimental Analysis of Behavior, 30,* 153–162.

Roitblat, H. L. (1980). Codes and coding processes in pigeon short–term memory. *Animal Learning & Behavior, 8,* 341–351.

Santi, A., Bridson, S., & Ducharme, M. J. (1993). Memory codes for temporal and nontemporal samples in many–to–one matching by pigeons. *Animal Learning & Behavior, 21,* 120–130.

Santi, A., Ducharme, M. J., & Bridson, S. (1992). Differential outcome expectancies and memory for temporal and nontemporal stimuli in pigeons. *Learning and Motivation, 23,* 156–169.

Santi, A., Weise, L., & Kuiper, D. (1995). Memory for event duration in rats. *Learning and Motivation, 26,* 83–100.

Sherburne, L. M., & Zentall, T. R. (1993a). Asymmetrical coding of food and no–food events by pigeons: Sample pecking versus food as the basis of the sample code. *Learning and Motivation, 24,* 141–155.

Sherburne, L. M., & Zentall, T. R. (1993b). Coding of feature and no–feature events by pigeons performing a delayed conditional discrimination. *Animal Learning & Behavior, 21,* 92–100.

Spetch, M. L. (1987). Systematic errors in pigeons' memory for event duration: Interaction between training and test delay. *Animal Learning & Behavior, 15,* 1–5.

Spetch, M. L. (1996). [Retention functions following short and long samples in a modified choice DMTS task]. Unpublished raw data.

Spetch, M. L., & Grant, D. S. (1993). Pigeons' memory for event duration in choice and successive matching–to–sample tasks. *Learning and Motivation, 24,* 156–174.

Spetch, M. L., Grant, D. S., & Kelly, R. (1996). Procedural determinants of coding processes in pigeons' memory for duration. *Learning and Motivation, 27,* 179–199.

Spetch, M. L., & Rusak, B. (1989). Pigeons' memory for event duration: Intertrial interval and delay effects. *Animal Learning & Behavior, 17,* 147–156.

Spetch, M. L., & Rusak, B. (1992a). Temporal context effects in pigeons' memory for event duration. *Learning and Motivation, 23*, 117–144.

Spetch, M. L., & Rusak, B. (1992b). "Time present and time past". In W. K. Honig & J. G. Fetterman (Eds.), *Cognitive aspects of stimulus control* (pp. 47–67). Hillsdale, NJ: Erlbaum.

Spetch, M. L., & Sinha, S. S. (1989). Proactive effects in pigeons' memory for event duration: Evidence for analogical retention. *Journal of Experimental Psychology: Animal Behavior Processes, 15*, 347–357.

Spetch, M. L., & Wilkie, D. M. (1982). A systematic bias in pigeons' memory for food and light durations. *Behavior Analysis Letters, 2*, 267–274.

Spetch, M. L., & Wilkie, D. M. (1983). Subjective shortening: A model of pigeons' memory for event duration. *Journal of Experimental Psychology: Animal Behavior Processes, 9*, 14–30.

Urcuioli, P. J., Zentall, T. R., Jackson–Smith, P., & Steirn, J. N. (1989). Evidence for common coding in many–to–one matching: Retention, intertrial interference, and transfer. *Journal of Experimental Psychology: Animal Behavior Processes, 15*, 264–273.

Wasserman, E. A., DeLong, R. E., & Larew, M. B. (1984). Temporal order and duration: Their discrimination and retention by pigeons. *Annals of the New York Academy of Sciences, 423*, 103–115.

Wearden, J. H., & Ferrara, A. (1993). Subjective shortening in humans' memory for stimulus duration. *Quarterly Journal of Experimental Psychology, 45B*, 163–186.

Wilkie, D. M. (1987). Stimulus intensity affects pigeons' timing behavior: Implications for an internal clock model. *Animal Learning & Behavior, 15*, 35–39.

Wilkie, D. M., & Willson, R. J. (1990). Discriminal distance analysis supports the hypothesis that pigeons retrospectively encode event duration. *Animal Learning & Behavior, 18*, 124–132.

Wilson, B., & Boakes, R. A. (1985). A comparison of the short–term memory performance of pigeons and jackdaws. *Animal Learning & Behavior, 13*, 285–290.

Wixted, J. T. (1993). A signal detection analysis of memory for nonoccurrence in pigeons. *Journal of Experimental Psychology: Animal Behavior Processes, 19*, 400–411.

Zentall, T. R., Steirn, J. N., Sherburne, L. M., & Urcuioli, P. J. (1991). Common coding in pigeons assessed through partial versus total reversals of many–to–one conditional and simple discriminations. *Journal of Experimental Psychology: Animal Behavior Processes, 17*, 194–201.

Zentall, T. R., Urcuioli, P. J., Jackson–Smith, P., & Steirn, J. N. (1991). Memory strategies in pigeons. In L. Dachowski & C. F. Flaherty (Eds.), *Current topics in animal learning: Brain, emotion and cognition* (pp. 119–139). Hillsdale, NJ: Erlbaum.

Time and Behaviour: Psychological and Neurobehavioural Analyses
C.M. Bradshaw and E. Szabadi (Editors)
265

CHAPTER 7

Ordinal, Phase, and Interval Timing

Jason A. R. Carr & Donald M. Wilkie

1. Overview

Learning in animals is thought to reflect the operation of multiple information–processing systems, each adapted by natural selection pressures to solve a *specific type of problem* that was repeatedly faced by the animals' ancestors. This idea, in a variety of guises, has been advocated by several authors including Cosmides and Tooby (1994), Gallistel (1989; 1990a), Gould and Marler (1987), Rozin and Kalat (1971), Sherry and Schacter (1987), Sutherland and Dyck (1984), Shettleworth (1993), and Timberlake and Silva (1994).

Obtaining food, water, and mates, and avoiding predation are perhaps the most ubiquitous, and fitness–determining, problems faced by animals. For many animals the location in *space* of these resources or predators changes predictably over *time*. To the extent that this has been true for a species, individuals that were able to anticipate and exploit *spatiotemporally–graded* resources or predators would have attained a potential fitness advantage over conspecifics who lacked this ability. Consequently, one might suspect that endogenous timing systems have evolved in animals that enable animals to anticipate and exploit biologically important spatiotemporal regularities. Aschoff (1989), Daan (1981), Enright (1970), Gallistel (1989; 1990a), and Mistlberger (1994) have made similar arguments.

In this chapter we will explore the notion that three broad types of timing systems have evolved in animals: ordinal, phase, and interval timers. We will suggest that each type of timing system is specifically adapted to provide animals with a capacity to process a *particular type* of temporal information. Gallistel (1989; 1990a; 1990b) has recently outlined a computational–representational approach to the study of animal cognition. Currently we are exploring the utility of this approach, specifically in the temporal information–processing domain. In this chapter, we will describe Gallistel's approach, and then characterize each type of timing system within a computational–representational framework.

To illustrate the types of temporal–problems that these three types of timing systems are specifically adapted to solve, we will review the field and laboratory evidence of time–place learning that has accumulated to–date. When reviewing this literature, we will argue that the type of timing system that an animal uses to anticipate and exploit a spatiotemporal regularity in a resource such as food availability depends on: (1) The temporal characteristics of the food source; and (2) the nature of the animal's interactions with any competitors for the food source.

We will then consider why a single, *general–purpose*, timing system has not evolved in animals. When addressing this question we will explore two lines of thought. First, we will outline Sherry and Schacter's (1987) notion of *functional incompatibility* and the implications that this principle may have had during the evolution of all the biological systems comprising an organism, including it's cognitive systems. Then we will suggest that multiple timing systems evolved because the operating characteristics of a cognitive system that enabled an animal to perform one type of timing were functionally incompatible with the efficient performance of other types of timing.

Secondly, we will briefly turn to the spatial navigation literature. We will suggest that animals posses multiple spatial navigation strategies, each of which is specifically adapted to solve a particular type of spatial problem. We will sketch out these spatial navigation strategies and the spatial problems that they are adapted to solve. Then using the spatial navigation literature as a case–in–point, we will suggest that a multiplicity of specifically adapted devices within a single problem domain is probably a common evolutionary outcome. Consequently, we should not be surprised that multiple timing systems have evolved in animals.

2.   The clockworks

To facilitate our discussion of ordinal, phase, and interval timing, we will characterize first the formal properties of biological oscillatory processes. In both invertebrates and vertebrates a wide variety of physiological and behavioural processes display oscillatory characteristics. The periodicity of these oscillations ranges from milliseconds to years (for reviews see Aschoff, 1989; Mistlberger & Rusak, 1994; Jacklet, 1985; Turek, 1985). Although the mechanisms envisioned vary, these endogenous oscillators, or their pacemakers, are widely thought to provide the physiological basis of biological timing systems (e.g., Aschoff, 1989; Church & Broadbent 1990; Gallistel, 1990a; Gibbon & Church, 1984; Mistlberger & Marchant, 1995).

In general, oscillators are self–sustaining and run continuously. Their periodicity can be adjusted, but only within set lower and upper limits.   For example, certain exogenous cues (Zietgebers) with a roughly circadian periodicity entrain endogenous oscillators within the circadian system so that each oscillator has the same periodicity as it's entraining cue (Mistlberger & Rusak, 1994). Two major Zietgebers for circadian rhythms are the daily light–dark cycle (LD cycle) and daily periods of food availability.   Each of these Zeitgebers appears to entrain a separate circadian pacemaker (Rosenwasser & Adler, 1986; Mistlberger & Rusak, 1994). The hypothalamic suprachiasmatic nuclei (SCN) appear to be the site of the light–entrained circadian pacemaker(s) (Klein, Moore, & Reppert, 1991; Welsh, Logothetis, Meister, & Reppert, 1995). The site of the food–entrained circadian pacemaker has not yet been determined (Mistlberger, 1994).

When a Zietgeber is removed [e.g., an animal is held in constant light (LL)], circadian endogenous oscillators persist but often attain a periodicity slightly different from 24 hr. When in this state an oscillator is said to be in free–run. This self–sustaining rhythmicity is thought to reflect the inherent periodicity of the circadian pacemaker(s) driving the oscillatory system. If a Zietgeber is phase advanced (occurs earlier each day), or phase delayed (occurs later each day), the oscillator(s) it controls drift in the direction of the phase shift and re-entrain to the Zietgeber.

3.   Temporal cognition in animals

Much laboratory work has examined timing in animals [for a review see Gallistel (1990a, chapters 7–9)]. Taken together, this work suggests that animals posses *interval timers* and *phase timers* (Gallistel, 1990a; Gibbon, 1984). We have recently argued, on the basis of empirical evidence, that animals also posses *ordinal timers* (Carr & Wilkie, MS). These three types of timing systems all solve the same basic problem as they all enable animals to anticipate *when* something will occur (i.e., they all solve problems in the temporal domain). What does differ across ordinal, phase, and interval timing is the *nature* of the temporal information that these timing systems provide to an animal. Each type of timing system is specifically adapted to provide animals with a unique type of temporal information (Carr & Wilkie, MS).

3.1.   A Representational–Computational Approach

Borrowing the usage of the concept of representation from mathematics, Gallistel (1989; 1990a; 1990b) suggested that a cognitive system represents some aspect of the environment because a *formal correspondence* exists between the operations of the cognitive system and the operations of the environmental system that it is specifically adapted to represent. Therefore, a representation is thought to exist in the form of a *relationship* between the represented (external) and representing (cognitive) systems. Because of this formal correspondence between the operations of cognitive systems and represented environmental systems, or *functional isomorphism*, an animal is able to perform relational and computational operations within cognitive systems to generate *predictions* about the state of represented environmental systems. The relational and combinatorial operations that may be employed validly in a cognitive system (and the resultant conclusions that can be made about the represented environmental system), depends on the nature of the functional isomorphism that exists between the two systems. If the functional isomorphism is weak, only weak computational operations may be employed [e.g., the equals operation (=), or greater than (>) or less than (<)], whereas if the functional isomorphism is rich, strong computational operations may be employed (e.g., multiplication and division).

Gallistel's formal criteria for the existence of a representation are outlined in Table 1.

Table 1.   Gallistel's (1990b) Criteria for the Existence of a Representation

1.   A *mapping* exists from entities or events in the represented (environmental) system to representatives in the representing (cognitive) system.

2.   A *formal correspondence* exists between relational and combinatorial operations involving the cognitive representatives and relational and combinatorial operations involving the represented environmental entities or events.

3.   Because of the mapping process, and the formal correspondence between the cognitive and environmental systems, relational and combinatorial operations in the cognitive system generate *valid predictions* about the environmental system.

4.   Animals utilize the predictive ability of their cognitive systems to produce behaviour *adapted to represented environmental systems*.

Gallistel's approach to representation is very similar to that proposed by S. S. Stevens (1951) in the context of Measurement Theory. Stevens suggested that valid measurement requires the use of a procedure that maps observations or reports of some psychological variable (e.g., response rate, perceived loudness, or perceived sweetness) to numerical representatives. The mapping procedure inherent in the measurement process creates an isomorphism between the measured system and the measurement scale, and because of this isomorphism the measurement scale can be used as a *model* of the psychological system of interest. Additionally, Stevens argued that not all isomorphisms are equal: The nature of the formal correspondence between a psychological variable and a measurement scale determines the relational and combinatorial operations that may be validly employed within a measurement scale. This in turn determines what statements can be validly made about the psychological variable of interest. Four types of scales may be derived: nominal, ordinal, interval, and ratio. These four types of scales are described in greater detail in Table 2.

We have recently applied Gallistel's computational–representational approach to animal timing systems (Carr & Wilkie, MS). Within a computational–representational framework, an endogenous timing system is thought to represent the temporal characteristics of a biologically–important environmental system because there is a formal correspondence between the operations of the timing system and the operations of the represented environmental system. Because of this functional isomorphism, animals are able to make predictions about a represented environmental system (i.e., determine *when* something will happen) by performing relational and combinatorial operations within the appropriate timing system. Furthermore, the nature of the functional isomorphism that exists between a timing system and the environmental system(s) that it represents, determines the relational and combinatorial operations which may be validly employed within the timing system, and therefore the type of predictions that can be made within the timing system.

Table 2. **Scales of Measurement (Stevens, 1951)**

Scale	Mathematical Structure	Valid Operations	Typical Examples
Nominal	Permutation Group $x'=f(x)$, where $f(x)$ means any one–to–one substitution	=	"Numbering" of football players, nationality, sex
Ordinal	Isotonic Group $x'=f(x)$, where $f(x)$ means any increasing monotonic function	=, <, >	Pleasantness of odors, position in class, percentile norms
Interval	General Linear Group $x' = ax + b$	=, <, >, −, +	Temperature (^{0}F or ^{0}C), Year (AD)
Ratio	Similarity Group $x' = ax$	=, <, >, +, −, *, /	Distance, age, weight, temperature (^{0}K)

3.2.   Three types of timing systems

3.2.1. Ordinal timing

Ordinal timing allows animals to anticipate events that reliably occur in a certain *order within a period of time*. In essence, this type of timing system provides an animal with a representation of a temporal sequence (e.g., food is available at location 1 *then* at location 2, *each day*). Therefore, with an ordinal timer an animal can anticipate, and exploit, what will happen next in time by maintaining a representation of its current position in a temporal sequence (Collet, Fry, & Wehner, 1993; Terrace, 1993).

Two lines of evidence support the notion that animals have evolved ordinal timing systems. First, many animals follow fixed routes from site to site in a foraging circuit. As this strategy is reminiscent of fur trappers checking their widely scattered traps, it is called traplining (Gill, 1995). For example, tropical euglossine honeybees, *Eulaema cingulata,* and *Euglosia surinamensis* (Janzen, 1970), and pied wagtails, *Motacilla alba,* (Davies & Houston, 1981) make foraging trips on which they circuit about feeding sites, often in the same order on every trip.

Related evidence comes from the serial learning literature. For example, Terrace and his colleagues and D'Amato and Columbo have shown that pigeons, *Columba livia,* and monkeys, *Cebus apella,* can learn to peck or press a set of test stimuli in a certain *order each trial* when trained on a simultaneous–chaining paradigm (D'Amato & Colombo, 1988; Straub & Terrace, 1981; Terrace, 1993). In the simultaneous–chaining paradigm all the elements in the sequence are present throughout each trial and reinforcement is only provided at the end of a correctly executed trial. This design rules out the possibility that responding to one element in the sequence serves as a discriminative stimulus for responding to the next element (a non-temporally based strategy) and forces the subject to develop, and use, a representation of the sequence to respond to the test stimuli in the correct order (see Terrace, 1993).

There appear to be some interesting phylogenetic differences in the richness of the representation of a temporal sequence utilized by animals. For example, pigeons trained on a simultaneous–chaining task (Straub & Terrace, 1981), and bees trained to fly through three successive choice points within a Plexiglas chamber (Collet, Fry, & Wehner, 1993), are able to learn which elements are *first* and *last* in the sequence, but have difficulty learning the ordinal position of the middle element(s). In contrast, monkeys trained on a simultaneous–chaining task appear to learn the ordinal position of all sequence elements (D'Amato & Colombo, 1988; Swartz, Chen, & Terrace, 1990).

The field evidence that some animals make foraging trips on which they visit sites in a fixed order, and the laboratory evidence that animals can learn to respond to a set of stimuli in a certain order each trial, suggest that animals can represent temporal sequences, monitor their current position in a sequence, and adapt their behaviour to the next event in the sequence. It is this representation of the order of a set of events in time that we have suggested can be thought of as ordinal timing (Carr & Wilkie, MS). Later in the chapter we will describe the empirical support for this concept in greater detail.

A computational–representational characterization of ordinal timing

As the correspondence between an ordinal timer and the represented external system only preserves information about the order of a set of events within a period of time (and not for example, the time that elapses between the events, or the time–of–day at which they occur), the only valid relational and computational operations within an ordinal timer are same as ($=$), and earlier than ($<$) or later ($>$) than. Consequently, ordinal timing provides an animal with a representation of time that is at the ordinal level of measurement. A schematic representation of ordinal timing is presented in Figure 1A.

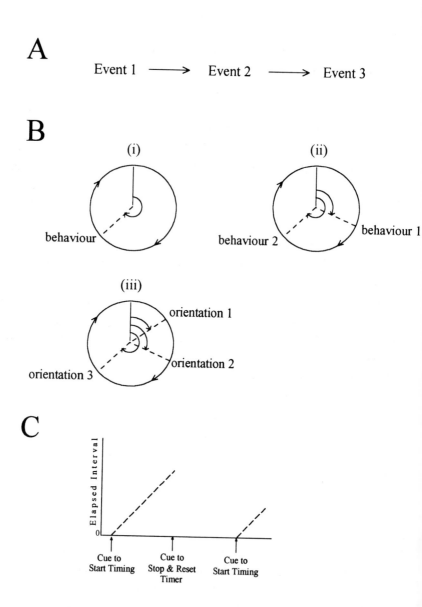

Figure 1. Schematic representations of the three types of timing system. A) Ordinal timing. B) (i) Go/no–go phase discrimination, (ii) Multiple–choice phase discrimination, and (iii) Conditional phase discrimination. (C) Interval timing. The start, stop, and reset functions of interval timers are illustrated. See text for details.

3.2.2. Phase Timing

Phase timing allows animals to anticipate events that are cyclic, or reliably occur with a fixed periodicity (e.g., every 5 min, every 12.4 hr, or every day at 14:00). Circadian phase timing is by far the most extensively studied type of phase timing. The anticipation of daily meals (Mistlberger, 1994), and the daily sleep–awake cycle displayed by many animals (Binkley, 1990; Klein et al. 1991), appear to be controlled by independent circadian clocks. Other work suggests that bees (Gould, 1980), desert ants (Wehner & Lanfranconi, 1981), and homing pigeons (Keeton, 1969), use a circadian clock to perform sun–compass navigation.

There is also some evidence that animals use phase timers to track cyclically available resources of non–circadian periodicities. For example, sparrows, canaries, and finches appear to use a phase timer to anticipate food availability when on a fixed time (FT) schedules in the range of 4 to 10 min [Stein (1951) as reported in Gallistel, (1990a)]. Additionally, oystercatchers, *Haematopus ostralegus*, (Daan & Koene, 1981), terns, *Sterna hirundo*, (Becker, Frank, & Sudmann, 1993), and iguanas, *Amblyrhynchus cristatus*, (Wikelski & Hau, 1995), appear to use an endogenous oscillator to track the tidal cycle. Also, the migratory restlessness displayed by many birds (Gwinner, 1986), the reproductive state of female lizards, *Cnemidophorus uniparens* (Cuellar & Cuellar, 1977), and the cycles of fattening and hibernation displayed by ground squirrels, *Citellus lateralis*, appear to be controlled by circannual phase timers (Pengelley & Asmundson, 1974).

Although the temporal information inherent in all forms of phase timing is thought to be provided by the phase angle of an endogenous oscillator, the type of temporal information provided by phase timing (and the underlying computational/neurobiological processes), may depend on the type of phase timing. For example, the anticipation of daily meals demonstrated by rats is thought to be produced automatically when a food–entrained circadian phase timer attains some fixed phase angle each day (Mistlberger & Marchant, 1995). Consequently, for a rat to anticipate two daily meals, the food–entrained circadian system must decompose into two independent sub–

oscillators (Stephan, 1989). Crystal (unpublished manuscript) called this type of phase timing "go/no–go phase discrimination". A schematic representation of go/no–go phase discrimination is presented in Figure 1B(i).

In another type of phase timing, the phase angle of an oscillator determines which of a set of behaviours is appropriate. For example, the orientation (north or south) of the migratory restlessness (Zugunruhe) displayed by nocturnal migratory passerine birds depends on the phase angle of a circannual oscillator (Gwinner, 1986). Crystal called this type of phase timing "multiple–choice phase discrimination". A schematic representation of multiple–choice phase discrimination is presented in Figure 1B(ii).

In a third type of phase timing some aspect of an animal's behaviour necessary to attain a goal is conditional on the phase angle of an oscillator. For example, when a circadian clock is used in sun-compass navigation by bees, ants, and birds, these animals must be able to discriminate the current phase angle of a circadian oscillator at any time–of–day, and use that information in conjunction with information about the position of the sun to set a course (Dyer, in press). In other words, these animals must be able to use a circadian oscillator as a continuously consulted clock. Crystal called this type of phase timing "conditional phase discrimination". A schematic representation of conditional phase discrimination is presented in Figure 1B(iii).

Because all forms of phase timing are based on the phase angle of an oscillator, phase timers display the properties of oscillators. Therefore, they run continuously; they are self–sustaining; and they are entrainable and therefore can be phase–shifted and adjusted to cycle at different periodicities (but generally only within a narrow range).

A computational–representational characterization of phase timing

With a phase timer an animal can learn the phase angle of events from an arbitrary zero–point of an oscillator, or use the current phase angle of an oscillator in some other computation (e.g., sun–compass navigation). An oscillator does not have a true zero–point as it runs continuously. Therefore an arbitrary zero–point is necessary to represent the start of an oscillation. Consequently, equal differences between the phase angles of events correspond to equal differences between their time–of–occurrence. However, it is impossible to use phase information to determine that some event is some proportion later or earlier *in time* than another event. For example, consider three events A, B, and C, that occur at phase angles of 60^0, 120^0, and 180^0 respectively from an arbitrary zero–point in an oscillator. With this information we know that the amount of time between A and B is equal to the amount of time between B and C. However, as the phase angles of A, B, and C are based on an arbitrary zero–point, is incorrect to say that event C occurs three times later in time than event A. Therefore, equals (=), earlier than (<) and later than (>), and addition (+) and subtraction (–), are all valid operations within a phase timer. The use of multiplication (*) and division (/) to generate proportions or ratios is clearly invalid. As a result, phase timing provides an animal with a representation of time which is at the interval level of measurement.

3.2.3. *Interval timing*

Interval timing allows animals to anticipate events that reliably occur a fixed amount of time after some external event (e.g., event 2 occurs 10 s *after* event 1). The classic example of interval timing is the temporal control ("scalloping") exhibited by animals responding on fixed interval (FI) schedules ranging in length from a few seconds to many hours (Ferster & Skinner, 1957). Interval timing is also thought to be a fundamental component of classical conditioning (Gallistel, 1990a; Gibbon & Balsam, 1981).

Information processing and connectionist models of interval timing have been developed (Church & Broadbent, 1990). In the information processing model, the passage of time is captured by the accumulation of pulses emitted by a pacemaker. In the connectionist model, the time of the beginning and end of an interval is represented by recording the status of a bank of endogenous oscillators of ranging periodicities. The length of the elapsed interval is then calculated by subtracting the time at the end of the interval from the time at which the interval began.

Biologically relevant external cues control the operation of interval timers as they stop, reset, and restart timing (Roberts & Holder, 1984; Holder & Roberts, 1985). Because interval timers can be stopped, reset, and restarted, they behave much like a stopwatch.

A computational−representational characterization of interval timing

Timing from any point in time is possible with an interval timer because it has a resetable zero−point. Once timing has begun, subjective time runs roughly linear with real time, and response decisions are made on the basis of the ratio of the animal's remembered length of the interval to the animal's measure of the interval that has thus−far−elapsed (Gibbon & Church, 1981; S. Roberts, 1981; Cheng & W. Roberts, 1991). Because interval timers have a true zero point and their representation of the elapsed interval increases roughly linearly with real time, equals (=), earlier than (<) and later than (>), addition (+) and subtraction (−), multiplication (*), and division (/) are all valid operations within interval timers. Interval timing is therefore at the ratio level of measurement. A schematic representation of interval timing is presented in Figure 1C.

3.2.4. Why these very different behaviours all constitute timing

The operating characteristics and temporal information provided by ordinal, phase, and interval timers are outlined in Table 3. Inspection of Table 3 reveals that each type of timing system has a unique set of operating characteristics and provides animals with a unique type of temporal information. However, despite their differences ordinal, phase, and interval timing all enable animals to anticipate *when* something will occur. Because these systems all solve problems in the temporal domain, we would argue that they are all

Table 3. The Operating Characteristics and Temporal Information provided by Ordinal, Phase, and Interval Timing

Timing System	Operating Characteristics	Type of Temporal Information	Valid Operations	Level of Measurement	Typical Field and Laboratory Examples
Ordinal	discrete, counts up, resetable.	current position, p, in a temporal sequence of n elements	$=, <, >$	ordinal	foraging circuits, serial pattern learning.
Phase	continuous, self-sustaining, entrainable.	current phase angle, α, in a cyclic process	$=, <, >, -, +$	interval	meal anticipation, tidal anticipation, sun-compass navigation.
Interval	discrete, times up, started, stopped & reset by events	temporal interval, i, that has elapsed since some event	$=, <, >, +, -, *, /$	ratio	FI scallop, pattern of flower re-visits by nectavorous birds.

types of timing systems, irrespective of differences in their operating characteristics, or qualitative differences in the nature of the temporal information that they provide to an animal.

This is not an unusual proposition. Problem–based classification is also evident in other domains in animal cognition. Consider two animals locomoting in space, one using a map–based strategy and a second using path–integration. Most people would agree that both animals are using spatial navigation mechanisms, despite dramatic differences in the mechanisms that are thought to underlie these strategies and differences in the situations when the use of each strategy is appropriate (Dyer, in press). We say they are both spatial navigation strategies because they both solve spatial problems. We will elaborate on this idea later in the chapter.

4.   **Field evidence that animals use ordinal, phase, and interval timers to track spatiotemporal regularities in food availability**

4.1.   Field Evidence of Ordinal Timer Use

Ordinal timers allow animals to anticipate events that tend to occur in a certain order within a period of time. Central to this form of timing is an animal's ability represent temporal sequences, and to maintain a representation of it's current position in a temporal sequence as it successively encounters each element in the sequence. Some trapline foragers likely use ordinal timers as they visit a set of feeding sites in the same *order* on each foraging trip.

Euglossine bees provide an interesting example of trapline foraging. Female euglossine bees serve a vital function in tropical rainforests as they outcross many widely scattered species of wooded plants (Janzen, 1971; 1974). The plants on a single bee's trapline are often spaced more than several hundred yards apart, and Janzen estimated that a typical bee's trapline could be as long as 24 km. A typical euglossine bee's trapline is presented in Figure 2A. On this

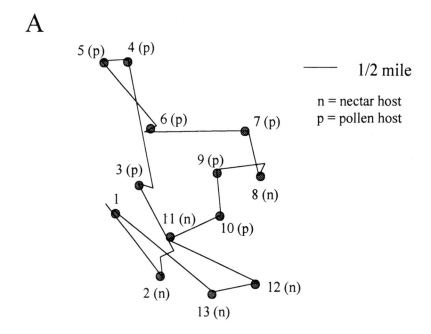

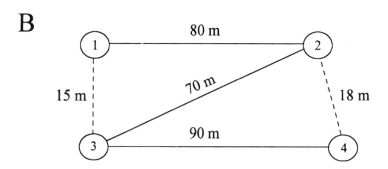

Figure 2 A) Typical foraging route of a tropical bee. Numbers refer to successive sites visited. The data are redrawn from Janzen (1974) with permission. B) The route taken by a tropical bee between four successive sites on it's trapline. See the text for details. The data are redrawn from Janzen (1971) with permission.

route, a single bee visited 13 widely scattered sites in the same order each day. These sites could include ground plants, shrubs, understory trees, vines, and epiphytic shrubs. These bees return to sites on their trapline using some form of spatial memory, and not by beacon homing (using perhaps olfactory or visual cues) as they return to sites even after all the flowers have been removed (Janzen 1974).

Interestingly, these bees often visit sites in their traplines in a far from optimum order. A sample flight–path of one bee is presented in Figure 2B. The bee's optimum route would be to fly from site 1, to site 3, to site 2, and then to site 4. This route is approximately 103 m long. Instead the bee regularly flew from site 1, to site 2, to site 3, then to site 4 each day. This route is approximately 240 m long, a 133 % increase in length over the optimum route. Janzen (1971) and Collet et al. (1993) suggested that bees develop these less–than–optimum temporal routes as they visit sites in their traplines in the ordinal position in which they were encountered on their first foraging flight.

Pied wagtails, *Motacilla alba*, also use foraging circuits. These birds eat dead insects that have washed up onto muddy riverbanks (Davies & Houston, 1981). In the long–run, riverbanks are re–stocked with insects in a monotonic, time–dependent manner. Therefore, when a wagtail depletes a stretch of riverbank of insects, time must pass before enough insects have accumulated to warrant feeding in that area again.

During the winter some wagtails defend stretches of riverbank and chase off potential poachers. Territory owners systematically circulate about their territories, walking up one bank, flying over the river, and then walking down the other bank. As they perform their foraging circuits, territory owners leave recently depleted riverbank behind them and move on to forage in areas of their territory that been restocking with insects for the longest period of time. However, things are not quite this simple for territory owners as they must frequently break their foraging routes and chase off potential poachers from other areas in their territory. Often a single intruder bird is chased by many territory owners as it flies through their adjacent territories.

Wagtails' foraging circuits could reflect the use of a rigid motor sequence (e.g., the territory owners could walk at a constant speed and turn when they reach a boundary of their territory). However, we believe this is unlikely. When territory owners resume foraging after chasing an intruder, they must remember where they were in their circuit when they were interrupted and resume foraging where they stopped. Otherwise, they risk resuming foraging in an area that they have recently depleted or leaving un-depleted portions of riverbank behind them. Efficient execution of this foraging technique seems to require that wagtails constantly maintain a representation of their current position in their foraging route, a process likely based on ordinal timing [for a similar argument see Biebach, Gordijn, & Krebs (1989)].

These foraging strategies employed by euglossine bees and wagtails entail visiting locations in space in a certain *order* over a period of time. Although these animals may simultaneously use many strategies to execute their foraging routes (see Collet et al., 1993), these field examples suggest that animals routinely represent sequences, monitor their current position in a sequence, and adapt their behaviour to the next event in the sequence. Some hummingbirds also employ trapline foraging, but with a slight twist due to a difference in how they defend their renewable food source. Consequently, we will discuss their foraging strategy in a later section.

4.2.   Field Evidence Of Phase Timer Use

Phase timing allows animals to anticipate events that are cyclic, or reliably occur with a fixed periodicity. There is a great deal of field evidence that a wide variety of animals use phase timers to anticipate cyclic spatiotemporal-regularities in the behaviour of their prey. The prevalence of this time-place learning ability may reflect the fact that the behaviour of most animals is adapted to some cyclic environmental system, be it the daily light-dark cycle, the tidal cycle, or the lunar cycle (Binkely, 1990; Daan, 1981; Enright, 1970).

Rijnsdorp, Daan, and Dijkstra (1981) studied kestrels', *Falco tinnunculus*, daily activity in a large flat area of reclaimed land in the Netherlands. These raptors prey mainly on common voles, *Microtus arvalis*. At Rijnsdorp et al.'s study site, kestrels and voles display strikingly intermeshed spatiotemporal-regularities in their foraging behaviour, suggesting that animals routinely represent, and anticipate, spatiotemporal regularities in the behaviour of their prey and predators.

Voles are herbivores, and because of the low quality of their food, and the requirements of their digestive process, they must feed every few hours (Hansson, 1971; Daan & Slopsema, 1978). Therefore voles must leave their burrows during daylight to make foraging trips on the surface. During these daylight foraging trips voles are at a high risk for predation by kestrels. Interestingly, a population of voles do not venture onto the surface randomly throughout the day. Instead their surface foraging trips (as measured by trapping counts and kestrel strikes) display a pronounced, roughly 2–3 hr, ultradian rhythm (Daan & Slopsema, 1978). The peaks in vole trappings throughout the day at various times–of–the–year are presented in Figure 3. The horizontal lines indicate the 1–hr periods with the highest vole activity.

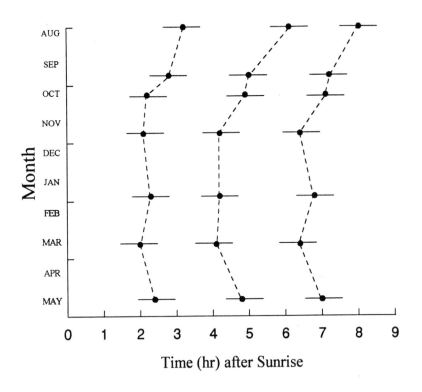

Figure 3 The times–of–day of the 1–hr periods when vole surface activity peaked (horizontal lines). On the x–axis, 0 corresponds to sunrise. The y–axis illustrates the seasonal variation in this activity rhythm. The data are redrawn from Daan and Slopsema (1978) with permission.

Inspection of Figure 3 reveals that vole trappings peaked every 2–3 hr. Additionally, despite seasonal changes in the time of sunrise, the first peak in daylight vole trappings consistently occurred about 2 hr after sunrise. This observation suggests that this ultradian activity rhythm is phase–locked to dawn.

Daan and Slopsema suggested that this ultradian rhythm in vole feeding behaviour is the product of two pressures: Due to digestive requirements the voles must eat several times during the day, but foraging on the surface during daylight exposes voles to an increased risk of predation by kestrels. Daan and Slopsema further suggested that an individual vole's probability of predation during daylight foraging is lower if it forages on the surface at the same time–of–day as many of its conspecifics. Foraging in–phase with population peaks might decrease an individual vole's probability of predation as any kestrels present in the area may be distracted by other voles, satiated, or otherwise constrained. Consequently kestrels themselves may have selected for this ultradian rhythm in vole surface activity by disproportionately preying on voles whose diurnal surface activity was out–of–phase with the rest of the vole population (Daan & Slopsema, 1978).

What about the kestrels? We know that their main prey displays a strong ultradian rhythm in daytime surface activity. Apparently kestrels are able to represent, and exploit, this cyclic pattern of vole availability as Rijnsdorp et al. found that kestrels restricted their flight–hunting to the times–of–day when vole surface activity was at its highest (see Figure 4). Additionally, Rijnsdorp et al. found that individual kestrels displayed idiosyncratic daily habits. For example, individual birds tended to flight–hunt in one area at one time–of–day, and to flight–hunt in another area at another time–of–day. However, field observations of flight–hunting activity, and field experiments that employed experimenter–released prey, suggested that these flight–hunting habits were plastic: When a kestrel was successful hunting in an area, it tended to return and hunt in that area 24 hr later. Rijnsdorp et al. (1981) estimated that by caching prey caught early in the day, and by restricting her flight–hunting to the times–of–the–day when a peak yield could be expected, one female kestrel reduced her flight–hunting time by 10–22%.

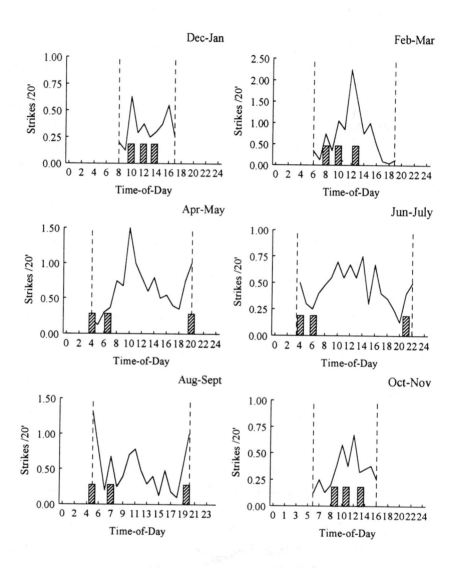

Figure 4. Hunting strikes by kestrels (solid lines) as a function of the time–of–day (x–axis). The crosshatched blocks indicate the times–of–day of the 1–hr periods when vole surface activity peaked during each observation period. The data are redrawn with permission from Rijnsdorp, Daan, and Dijkstra (1981).

Kestrels are not the only birds that track the location of food availability over the course of the day. Wilkie, Carr, Siegenthaler,

Lenger, Liu, and Kwok (in press) studied the flocking behaviour of a variety of scavenging birds including pigeons, *Columba liva*, starlings, *Sturnus vulgaris*, gulls, *Larus sp.,* and crows, *Corvus caurinus* at two outdoor locations. One location was a busy outdoor market, and the other was a concrete plaza adjacent to a university student center. People tend to congregate and eat food at these locations, and the number of people present peaks, on average, just after mid–day. At both of these locations the number of birds present peaked just before the number of people peaked. A multiple regression analysis found that the number of birds present at these locations was best predicted by the time–of–day, and not by the number of people present. At a control location where the number of people present declined throughout the day and little or no food is consumed, none of these relationships were observed. Taken together, these findings suggest that these scavenging birds used an endogenous timer to represent, and anticipate, this daily peak in food availability.

Daan and Koene (1981) studied oystercatchers', *Haematopus ostralegus*, foraging flights from their inland roosts to nearby tidal mudflats. During low tide the oystercatchers foraged on the mudflats for mussels. During high tide the oystercatchers congregated in inland fields. Here they occasionally supplemented their diet of mussels with earthworms. The inland fields and nearby tidal mudflats were separated by a high dike. This dike provided a visual barrier between these two sites. Observers sat on the dike and recorded the number of oystercatchers in the fields, on the mudflats, and in–flight between these two sites. The timing of oystercatchers' foraging flights is crucial: If they arrive at the mudflats too early the mussels are unobtainable as they are still covered by deep water, and if they arrive at the mudflats too late the mussels are unobtainable as they have closed their valves to prevent desiccation and predation. The amount of exposed mudflat, the amount of exposed musselbed, the number of oystercatchers on the musselbeds, and the proportion of the oystercatchers on the musselbeds that were foraging, as a function the tidal cycle is presented in Figure 5. Although the oystercatchers could not see the flats from their roosts, they clearly flew to the flats just as the mussels were exposed by low tide. This suggests that these oystercatchers were using an endogenous timing system to track mussel availability. Alternatively, these oystercatchers could have used some stimulus (e.g., the smell of exposed musselbeds) to control their foraging flights. However, it is unlikely that these oystercatchers employed this strategy as the time

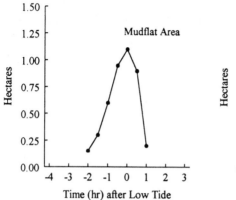

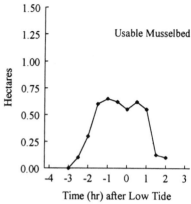

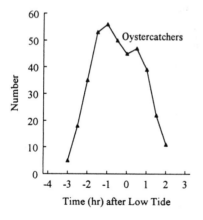

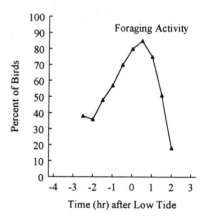

Figure 5. Foraging behaviour of oystercatchers, redrawn with permission from Daan and Koene (1981). On the x–axis, 0 represents the time of low tide. The top panels show the amount of exposed mudflat (left) and amount of usable musselbed (right) as a function of the tidal cycle. The bottom panels show the number of oystercatchers present (left) and their foraging frequency (right) as a function of the tidal cycle.

of their foraging flights was best predicted by the tabulated time of low tide, and not by the estimated amount of exposed musselbed.

Apparently oystercatchers are not the only seabirds that track the tidal cycle. Becker, Frank, and Sudmann (1993) studied the foraging activities of common terns, *Sterna hirundo*, in the Wadden Sea. They studied these birds when they were either incubating eggs or raising chicks. During the incubation period parent terns must repeatedly relieve their incubating partner so they can forage for food, and during the chick–raising period both birds must provide their offspring with a steady supply of food. Clearly for terns foraging efficiency is at a premium during this time.

Becker et al. tagged these birds (only one bird per breeding pair) with light radiotransmitters, and tracked their daily movements from sunrise to sunset over an average of 10 days. The terns' foraging trips lasted an average of 115 min (range 10 to 563 min), and during this time the terns traveled an average of 30 km (range 6 to 70 km). Although Becker et al. found individual differences in the terns' foraging activity and site preferences, some overall patterns emerged. First, there was no diurnal pattern to the birds' site visits (i.e., they did not tend to visit certain sites at certain times–of–the–day), however the terns did forage more overall in the morning than in the afternoon. Secondly, on long foraging trips terns tended to visit two or three sites, and they visited each site at a site–specific point in the tidal cycle. The number of birds that visited three locations in the Wadden Sea as a function of the tidal cycle is presented in Figure 6. Inspection of Figure 6 reveals that the terns tended to visit Strandplate during flood, Wattengebiete during the first four hours of the outgoing tide, and Mellum during ebb. This tidal pattern in site visits is mirrored by a tidal pattern in the type of food fed to nestlings (Frank, 1992).

Becker et al. (1993) suggested that terns may acquire this detailed knowledge of the spatiotemporal characteristics of food availability in the Wadden Sea during their pre–laying and incubation periods –– a time of the year when it is possible for parent terns to make long exploratory flights. This knowledge is then used during the chick–feeding period when terns must provide their chicks with a steady supply of food. Becker et al. also speculated that the need for detailed knowledge of the spatiotemporal characteristics of food availability may have contributed to the development of the site tenacity that is common in terns and other long-lived seabirds.

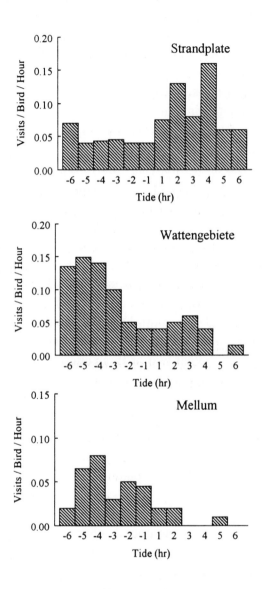

Figure 6. The spatiotemporal pattern of visits by terns at three different foraging areas in the Wadden Sea. The data are redrawn with permission from Becker, Frank, and Sudmann (1993).

Nectar production in some flowers (e.g., buckwheat) clearly depends on the time–of–day. Not surprisingly then, other field work suggests that honeybees use circadian phase timers to organize their daily foraging activities. In one experiment Wahl (1932, as cited in Gallistel, 1990a) provided bees sugar water in a beaker at one feeding station from 09:00 to 10:30 and in a beaker at a second feeding station from 15:30 to 17:00. Then on a test day the beakers were present at both stations but no sugar water was available. Wahl's results are presented in Figure 7. There are two main points of interest in Wahl's data. First, the total number of visits to both feeding stations clearly increased just before the two usual feeding times. Evidently Wahl's bees anticipated sugar water availability at the times–of–day of the usual feeding periods. Secondly, while the honeybees did frequently visit the incorrect stations during the morning and afternoon of the test day, they did not treat the correct and incorrect stations the same. The majority of their visits to the correct station were land–and–enter–visits, whereas the majority of their visits to the incorrect feeding station were fly–bys. This suggests that the bees expected sugar water to be available at the correct feeding station during each of the daily feeding periods. The fly–bys at the alternate station at each feeding time presumably reflect checks for bee activity (indicating unexpected food availability) as the test day was conducted under extinction.

As the temporal characteristics of resource availability in these field studies, and in the bees natural foraging context, is well described by a cyclic (circadian) process, one might suspect that honeybees represent, and anticipate, daily food availability with a circadian phase timer. This notion is supported by two lines of evidence. First, Beling (1929) as reported in Gallistel (1990a), fed bees sugar water every 19 hr for 10 days. On the following day no sugar water was available, and the bees' food–seeking behaviour was recorded. If honeybees anticipate daily resource availability with an interval timer, their food–seeking behaviour would be expected to peak 19 hr after the last feeding. Alternatively, if bees anticipate daily resource availability with a circadian phase timer, their food seeking behaviour would be expected to peak closer to 24 hr after the last feeding. Evidently bees track daily food availability with a circadian phase timer as their food seeking behaviour peaked 24 hr after their last meal.

SITE A

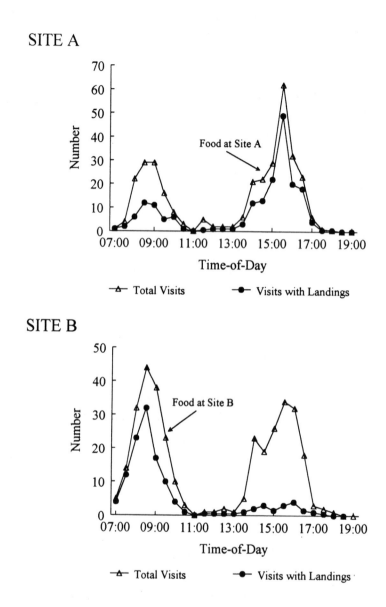

Figure 7. Foraging flights by honeybees, fed at two sites (A and B) at different times–of–the–day. Open triangles represent the total number of visits to a site (fly-bys and land–and–enters), and the filled circles represent only land–and–enter visits. The data, originally collected by Wahl (1932), and reproduced in Gallistel 1990a, were redrawn with permission.

If bees use the phase angle of a circadian phase timer to anticipate daily regularities in nectar availability, their time–place behaviour should exhibit other properties of biological oscillators. One other property of biological oscillators displayed by their time–place behaviour is phase–shifting. Renner (1960) worked with bees in an open field in Long Island. He trained his bees to obtain sugar water from a feeding site northwest of their hive at the same time each day. He then flew the hive overnight to a field in California. In California he set up eight feeding stations at 45^0 intervals about the hive. Each of the eight feeding stations was the same distance from the hive as the original feeding station was in Long Island. In order to obtain an uncontaminated measure of where and when the bees expected sugar water, no sugar water was available at the feeding stations in California.

The overnight flight to California constituted a 3–hr clock shift for the bees. This means that the bees' circadian clocks would have been 3 hr ahead on their first morning in California. This time difference also means that at any point–in–time during the bees' first day in California, the azimuthal angle of the sun would have been, on average, 45^0 behind it's current angle in Long Island. During their first day in California the bees' visits to the feeding sites peaked roughly 24 hr after their last feeding in Long Island. This occurred even though the sun was not in the same position as it was during their training flights in Long Island. Over the next 3 days the peak in the bees' visits to the feeding stations shifted towards the correct local time. This strongly suggests that the bees' circadian phase timers were re–entraining to the local sunrise/sunset schedule, resulting in a gradual phase shift in their time–place behaviour. Additionally, as the bees' circadian clocks were initially 3 hr hours ahead, their sun–compass navigation was approximately 45^0 incorrect on their first day in California.

It appears that kestrels, oystercatchers, terns, and bees are able to abstract, and exploit, spatiotemporal regularities in their primary food source. Their prey is well characterized by a cyclic (ultradian, circadian, circatidal, or circalunar) temporal pattern of availability. Consequently, these putative examples of time–place learning probably reflect the use of ultradian, circadian, circalunar, or circatidal phase–timers.

4.3.   Field Evidence Of Interval Timer Use

Interval timers allow animals to anticipate events that reliably occur a fixed interval of time after another event. There is not a great deal of field evidence of animals using interval timers to track spatiotemporally–graded resources. This lack of field evidence stands in marked contrast to the vast experimental work that has examined interval timing in laboratory animals (e.g., Church, 1984; Church & Broadbent, 1990; Gibbon, 1991; Killeen & Fetterman, 1988). Clearly much more field work is needed in this area.

In one relevant study, Gill (1988) studied the spatiotemporal characteristics of male long–tailed hermit hummingbird, *Phaethornis superciliosus*, foraging activity in a tropical rainforest in southwestern Costa Rica. These hummingbirds congregate at leks and try to attract females with their brilliant displays. However, every few minutes they must leave the lek to obtain nectar from a site on their traplines.

These traplines commonly contain 10 or more feeding sites. These hummingbirds do not appear to visit all the sites in their traplines in a fixed order on each foraging bout. Instead they may visit a subset of their sites on each foraging bout (Gill, 1988). Flowers produce nectar at a constant rate (that depends, in many cases, on the time–of–day). Consequently, once a lek male long–tailed hummingbird depletes a site on it's trapline, the amount of nectar available on a return visit depends on the temporal interval that has elapsed since the bird last depleted the flower. The longer the elapsed interval, the larger the amount of accumulated nectar. That is, unless the accumulating nectar was poached by another hummingbird. When this happens, the bird may revisit his feeding site and find it recently depleted. This is surely not a trivial matter for lek males given the potentially large distances between their feeding sites and the lek, and their incredibly high metabolic rate.

Wagtails prevent poaching from their foraging circuit using interference competition −− they actively chase intruder birds out of their territory. Long–tailed Hermit hummingbirds do not do this.

Instead, they defend their feeding sites using a form of exploitative competition called *defense by depletion* (Gill, 1988). Field experiments conducted by Gill using artificial feeders that provided nectar on FI reinforcement schedules suggest that these hummingbirds adjust the timing of their site visits to match the timing of nectar replenishment. Consequently, when these hummingbirds have near exclusive control of a feeding site, they distribute their revisits roughly evenly in time. However, when these hummingbirds visit a site on their trapline and find that it has been depleted recently by another hummingbird, they return to that site repeatedly afterwards. Often these repeat visits are only a few minutes apart. Although the trapliner obtains little nectar on these frequent defensive revisits, it is thought he minimizes future losses to the intruder, thereby discouraging further poaching (Gill, 1988; Paton & Carpenter, 1984).

The site–defense strategies employed by wagtails and long–tailed hermit hummingbirds clearly necessitate the use of different timing systems. Wagtails require a timing system that enables them to represent their current position in their foraging route, and enables them to resume foraging at the correct position in their route once they have chased off potential poachers. On the other hand, lek male long–tailed hummingbirds require a timing system that enables them to measure the amount of time that has elapsed since they last depleted a flower. This information can then be used to schedule site re–visits, both productive and defensive.

5. **Laboratory evidence that animals use ordinal, phase, and interval timers to track spatiotemporal regularities in food availability**

Laboratory time–place learning tasks examine if, and how, animals exploit food availability that moves in space over time. Different time–place learning tasks have been developed, and in each task food availability displays different spatiotemporal characteristics. We distinguish between *daily* and *sequential* time–place learning tasks.

5.1   Daily Time–Place Learning Tasks

In *daily time–place learning tasks* the location in space of food availability depends on the time–of–day. For example, an animal may be required to exploit a food source that is available at one point in space at one time–of–day, and at another point in space at another time–of–day. This time–place learning task is explicitly designed to model the type of foraging problems faced by the bees and kestrels that we described earlier.

5.1.1.   Laboratory Demonstrations of Daily Time–Place Learning in Insects

The nectivorous tropical ant, *Paraponera clavata,* can anticipate daily meals that occur at one location in space every 24 hr (Harrison & Breed, 1987). Additionally, Schatz, Beugnon, and Lachaud, (1994) demonstrated daily time–place learning in another nectivorous ant, *Ectatomma ruidum.* Schatz et al. placed these ants in a plaster nest. The nest was connected to a central area. From the central area the ants could walk along tubes to forage in three outside arenas. Honey was available in each of the outside arenas for a 1–hr period each day: One arena always contained honey between 09:00 and 10:00, the second arena always contained honey between 11:00 and 12:00, and the third arena always contained honey between 15:30 to 16:30. Light onset was at 08:00 and the ants were held in a 12:12 LD cycle.

After 15 days of training the ants learnt to visit preferentially the correct arenas throughout the course of each day. On the 22nd day of training no honey was available in any of the arenas at any time–of–day. This test was done to determine whether the ants were using an endogenous timing system or some stimulus associated with a baited arena to track food availability. If the ants were using an endogenous timing system to track the location of honey, their pattern of arena visits would be largely unaffected when honey was not available in any arena at any time–of–the–day. On the other hand, if the ants' foraging behaviour was controlled by some stimulus associated with a baited arena, the spatiotemporal pattern of their arena visits would be markedly disrupted as soon as honey was withheld. Evidently the ants used an endogenous timing system to exploit the daily spatiotemporal pattern of honey availability as on the test day they continued to visit each arena at the correct time–of–day.

In the paradigm employed by Schatz et al. honey was available at each arena at a specific time–of day, and light onset was at 08:00 every day. Consequently, the time–place behaviour displayed by these ants could reflect the operation of a circadian phase timer, or an interval timer started each day by light onset. Unfortunately Schatz et al. do not report the results of any further manipulations (e.g., housing the ants in LL), that might allow us to chose between these possible mechanisms.

5.1.2. *Laboratory Attempts to Demonstrate Daily Time–Place Learning in Fish*

Some work indicates that fish can anticipate daily feedings at one location in space. For a review of this work see Mistlberger (1994). Additionally, daily time–place learning has been studied in fish. Reebs (1993) tested individuals and groups of convict cichlids, *Cichlasoma nigrofasciatum*, in large aquariums. In a series of experiments these fish were held in a constant 14:10 LD cycle, and received food at two locations, each at different times–of–the–day, and at four locations, each at different times–of–the–day. Meals were preceded by a reliable signal (the air supply to the filter was turned off every 5–s during the preceding 60–s period). The convict cichlids displayed little ability to learn the time–of–day when food would be available at each feeding location. Instead, the fish simply patrolled all their feeding locations when the feeding signal occurred.

However, after switching to golden shiners, *Notemigonous crysoleucas*, and modifying the training protocol, Reebs (1996) found evidence of time–place learning ability in a fish. The golden shiners were tested in groups of eight, and they were housed in a 12:12 LD cycle. (It should be noted that one of the problems created by testing groups of subjects is that it is impossible to know how many animals actually learned all the contingencies.) In the revised paradigm meals were not preceded by a reliable exogenous signal. Reebs found that these fish could anticipate two meals per day at two different locations, and three meals per day at two different locations. Within the training given, these fish were not able to anticipate three daily meals at three different locations.

These fish received their feedings at the same times each day, and they were housed in a stable LD cycle. Therefore, they could have used an interval timer started by light onset each day or a phase timer to track the location of food availability. To test between these alternative mechanisms Reebs phase advanced the fishes' LD cycle by 6 hr. If the golden shiners were using an interval timer started by light onset, their time–place behaviour would immediately advance by 6 hr. Alternatively, if the golden shiners were using the phase angle of a light–entrained phase timer to track the location of food availability, their time–place behaviour would be largely undisturbed when they were first switched to LL. Over time, however, the spatiotemporal pattern of their foraging behaviour would advance by 6 hr. Evidently, the golden shiners' time–place behaviour was driven by a light–entrained phase timer as their time–place behaviour gradually advanced by 6 hr over the 3–day period that followed the 6–hr phase shift.

5.1.3. Laboratory Demonstrations of Daily Time –Place Learning in Birds

Daily time–place learning has been extensively studied in birds. Biebach, Gordijn, and Krebs (1989) studied daily time–place learning in garden warblers, *Sylvia borin*. Their birds lived in a large apparatus that consisted of a central area and four surrounding rooms. Lighting was provided on a 12:12 LD cycle, and light onset was at 06:00 each day. Food was successively available in each of the four surrounding rooms for one 3–hr interval during the 12–hr light period: From 06:00 to 09:00 food was available in Room 1, from 09:00 to 12:00 food was available in Room 2, from 12:00 to 15:00 food was available in Room 3, and finally from 15:00 to 18:00 food was available in Room 4. Every room entry (reinforced and non–reinforced) was followed by a 280–s time–out period.

Figure 8A presents the mean number of entries the warblers made to Rooms 1, 2, 3 and 4 as a function of the time–of–day. These data were collected during the first 5 days after each bird attained asymptotic performance. There are two main points of interest in

Biebach et al.'s results. First, the garden warblers clearly learned to enter each room during the correct (i.e., reinforced) 3–hr period of each day. Secondly, the warblers began to enter each room just before entries to each room were reinforced. This suggests that the warblers represented, and anticipated, the transition points at 09:00, 12:00, and 15:00 each day when food availability changed from one room to the next in the daily sequence. This anticipation suggests that the warblers tracked the location of food availability using an endogenous timer.

To test whether the warblers' were using an endogenous timer to track the location of food availability, the warblers received a test day during which all four rooms provided food throughout the entire day. Inspection of Figure 9 reveals that even when all room entries were reinforced the warblers continued to visit each room during the correct (i.e., previously reinforced) time–of–day, and the birds continued to enter each room just before the time–of–day when they usually provided food. This outcome strongly suggests that the warblers' room preferences depended on the status of an endogenous timing mechanism.

As food was available in each room for a certain period of time each day, and the birds were housed in a stable LD cycle, the warblers could have used a circadian phase timer, or an interval timer started each day by light onset, to anticipate the transition times–of–the–day when food availability moved from one room to the next in the daily sequence. To test between these two possible mechanisms, Biebach, Falk, and Krebs (1991) trained warblers on the task employed by Biebach et al. (1989) and then switched the birds to a LL lighting schedule. To obtain an uncontaminated measure of where the warblers expected food throughout the course of a day, food was continuously made available in all four rooms. If the warblers were tracking the location of food availability using an interval timer started each day by light onset, the spatiotemporal pattern in their room choices would be greatly disrupted during the first day they were held in LL.

On the first day of the LL test the warblers' pattern of room entries was largely unaffected. Over the next few days the periodicity of their daily pattern of room entries dropped from 24 hr to ca. 23 hr. This caused the warblers to enter each room progressively earlier each day. This result strongly suggests that the warblers' time–place behaviour was driven by the phase angle of a light–entrained circadian phase timer that was going into free–run.

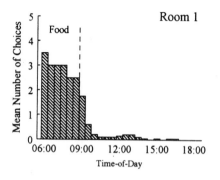
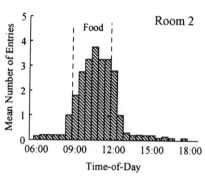

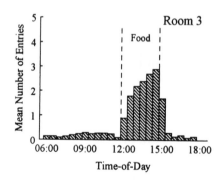
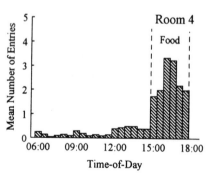

Figure 8   Foraging behaviour of garden warblers that could obtain food in four different rooms, each during a different 3–hr period of the day. The data are redrawn with permission from Biebach, Gordijn, and Krebs (1989).

In another experiment Biebach et al. (1991) phase advanced light onset by 6 hr. Again food was continuously available in all four rooms. If the warblers were using an interval timer started each day by

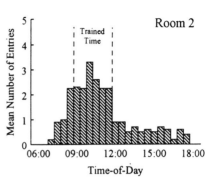

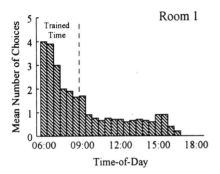

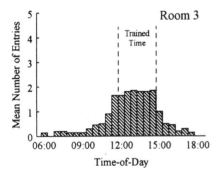

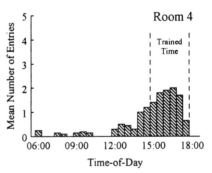

Figure 9. Foraging behaviour of garden warblers during the open–hopper test session when food was available in all four rooms throughout the day. The data are redrawn with permission from Biebach, Gordijn, and Krebs (1989).

light onset, they would enter and leave each room 6 hr early on the first day of the phase shift. On the other hand, if the warblers were using the phase angle of a light–entrained phase timer, they would enter and leave each room early on the first day of the phase shift, but the change in the pattern of their room entries would be much less than 6 hr. However, every day thereafter the warblers would enter and leave each room earlier than the day before until a full 6-hr phase shift was realized.

The 6-hr phase shift in their LD cycle caused a partial shift (mean of 2.6 hr) in the warblers' pattern of room entries on the first day. Over the next few days their pattern of room entries gradually advanced by 6 hr. This result also strongly suggests that the warblers' time–place behaviour was driven by the phase angle of a light–entrained phase timer. In this case the light–entrained phase timer was re–entraining to the new LD cycle.

However, the warblers were not solely using a light–entrained phase timer to track the location of food availability over the course of each day. The results of another experimental manipulation suggest that the warblers' time–place behaviour was also weakly controlled by an ordinal timer. In an experiment reported by Krebs and Biebach (1989) the warblers were prevented from entering any rooms during the first 3-hr period of a test day. During this period Room 1 usually provided food. After the locked period, all four rooms were unlocked and the warblers could then receive food in any of the four rooms.

If the warblers' time–place behaviour was driven solely by the phase angle of a phase timer they would prefer Room 2 when all four rooms were unlocked. Instead, five of their nine birds preferred Room 1. This result suggests that the warblers' time–place behaviour was partially determined by a representation of the *order* in which they should visit the four surrounding rooms each day. We have argued that this type of temporal information is provided by an ordinal timer. In another experiment, Krebs et al. blocked entries to all four rooms during the third 3-hr period of a test day (i.e., when Room 3 usually provided food). Krebs et al. unlocked all the rooms in the fourth 3-hr period (i.e., when Room 4 usually provided food), and the warblers entered Rooms 3 and 4, but overall they preferred Room 4. This suggests that warblers, like bees and pigeons, may have trouble learning the middle elements of a sequence.

Saksida and Wilkie (1994) investigated daily time–place learning in pigeons. Saksida and Wilkie adopted a different training and testing procedure than Biebach et al. Their pigeons received test sessions in a large, clear Plexiglas, square testing chamber twice per day; once between 09:00 and 10:00 and at a second time between 15:30 and 16:30. A number of salient, distal, spatial cues were visible from within the test chamber that was located on a table top in a well lit room. A pecking key was mounted on each of the four chamber walls. All four keys were illuminated during test sessions.

Each bird could peck one key for grain in morning sessions, and could peck a second key for grain during afternoon sessions. Key assignment was counterbalanced across the pigeons. Each test session began with a brief period of variable length during which pecks to all four keys were recorded, but no reinforcement was provided. Anticipatory key pecks (i.e., premature and non–reinforced) during this period were used to infer where the pigeons expected food to be available during each test session.

The pigeons clearly learned to peck the correct key in morning and afternoon sessions. Saksida and Wilkie's pigeons could have anticipated the location of food availability by simply alternating between the two keys which provided them food (i.e., Key 1 → Key 2 → Key 1, and so on). However, this was not the case: When either morning or afternoon sessions were skipped, their pigeons still anticipated the location of food during the following test session. This result suggests that their pigeons were using an interval or phase timer to discriminate between morning and afternoon sessions.

Saksida and Wilkie performed various LD manipulations similar to those employed by Biebach et al. (1991). The results of these manipulations suggested that their pigeons also used the phase angle of a light–entrained circadian phase timer to discriminate between morning and afternoon test sessions. For example, placing the birds in LL initially had no effect on their time–place behaviour. However, over the next few days their ability to anticipate the location of food availability in morning and afternoon sessions gradually declined. Similarly, a 6–hr phase advance of the pigeons' LD cycle caused a small decline in their performance on the first day. Accuracy then declined over the next six sessions.

However, Saksida and Wilkie's pigeons did not exclusively e a phase timer to track the location of food availability over the course of a day. When their pigeons did not receive a morning session, they did prefer their afternoon keys during the following afternoon session. However, in these sessions their pigeons displayed a reduced preference for their afternoon keys and an increased preference for their morning keys. Taking the work with garden warblers and pigeons together, it appears that in a daily time–place learning task these birds primarily track the location of food by consulting the phase angle of a light–entrained phase timer. However, their time–place behaviour was also partially driven by a representation of the *order* in which food was available at their daily feeding sites. We have argued that this type of temporal information is provided by an ordinal timer. When the warblers' and pigeons' ordinal and phase timers predicted that food would be available at different locations (during the blocked Room 1 and skipped morning session probes), the result was a sort of behavioural compromise. The pigeon and warbler data suggesting that animals can concurrently use two different timing systems implies that timing modes can be selected in parallel. It does not seem to be the case that timing modes are selected in a serial, or all or nothing manner (e.g., if ordinal timing doesn't work then next try phase timing, if that doesn't work then next try interval timing). Animals may use all three timing systems to track spatiotemporal regularities, and the relative behavioural control gained by any single timing system may depend on the characteristics of the temporal problem.

5.1.4. *Laboratory Demonstrations of Daily Time–Place Learning in Rodents*

In a recent abstract Daan, Leiwakabessy, Overkamp, and Gerkema (1994) outlined an experiment investigating daily time–place learning in house mice (C57B1). Their mice lived in a central cage connected to four outer rooms. Each of the outer cages contained a running wheel. Wheel revolutions in each outer room produced food during one 3–hr interval of the light period: Wheel running produced food in Room 1 from 06:00 to 09:00, in Room 2 from 09:00 to 12:00, in Room 3 from 12:00 to 15:00, and in Room 4 from 15:00 to 18:00. Light onset was at 06:00.

Daan et al. report that their mice learned to wheel run in each outer room at the correct time–of–day. When their lighting was switched from LD to LL, and food was produced by wheel running in all four rooms throughout the entire day, their mice's pattern of wheel running initially conformed to the prior spatiotemporal pattern of food availability. Their pattern of wheel running then gradually fell out of phase over the next 5 days. This result strongly suggests that their wheel running was controlled by the phase angle of a light–entrained circadian phase timer that was going into free–run.

Boulos and Logothetis (1990) examined time–place learning in the laboratory rat. Their rats lived in circular light–tight chambers. Food was provided at one lever at one time–of–day and at a second lever at a second time–of–day. Intact rats and rats with ablated suprachiasmatic nuclei (SCNX) were tested in LL and LD. Most rats displayed increases in lever pressing in anticipation of both daily meals, but fewer than half of their rats anticipated the location of both their morning and afternoon meals. Some of their SCNX rats acquired the task, but their intact rats housed in a stable LD cycle performed the task best. As the SCN is the site of the LD–entrained pacemaker in the rat, these data suggest that daily time–place learning in the rat may be primarily controlled by a food–entrained circadian phase timer. However, as Boulos and Logothetis did not include any tests to prove that their rats were not simply alternating between the two response levers that provided them food, it is impossible to conclude that their rats' ability to anticipate the location of food reflected the use of a circadian phase timer.

Mistlberger, de Groot, Bossert, and Marchant (in press) recently reported strong evidence that rats can acquire a daily time–place learning task. Their rats were held in a 12:12 LD cycle and each day they received two test sessions on a T–maze. Morning and afternoon test sessions were 7 hr apart. Intact rats and rats bearing lesions of the SCN were tested. A response lever and hopper was mounted on two arms of the T–maze. For all the rats, responses on one lever produced food in morning sessions (2–3 hr after light onset), and responses on a second lever produced food in afternoon sessions (1–2 hr before light offset). In an attempt to remove any confounding odor trails (Galef &

Buckley, 1996), the T–maze arms were wiped down with disinfectant after each rat.

Both their intact and SCN ablated rats learned to press the correct lever during morning and afternoon sessions. Mistlberger et al. then skipped two morning sessions, two afternoon sessions, and one block of three consecutive test sessions. In the sessions following the majority of the skipped sessions their rats clearly anticipated which lever would provide food. This rules out the possibility that their rats used an alternation strategy to anticipate the location of food (i.e., Arm 1 → Arm 2 → Arm 1, and so on), and suggests instead that their rats used the status of an endogenous timer to discriminate between morning and afternoon test sessions.

Mistlberger et al.'s rats received their morning and afternoon test sessions at the same time every day and they were housed in a stable LD cycle. Consequently, their rats' time–place behaviour could reflect the operation of at least two different types of timers. First, they could have discriminated between morning and afternoon sessions by consulting the phase angle of a food–entrained phase timer. We can rule out the use of a light–entrained phase timer as Mistlberger et al.'s SCN ablated rats acquired their task. Alternatively, their rats could have used an interval timer to discriminate between morning and afternoon test sessions: They could have started an interval timer at light onset each day, and pressed the morning lever when the timer had accumulated to one value and pressed the afternoon lever when it accumulated to a second, higher, value. Because their rats were unaffected by skipped test sessions, we can rule out the use of an interval timer strategy that capitalized on the asymmetrical intersession intervals (7 hr versus 16 hr) inherent in Mistlberger et al.'s procedure.

To test whether their rats used a food–entrained phase timer or an interval timer started each day by light onset, Mistlberger et al. inverted their rats' LD cycle twice and held their rats in constant dark (DD) once. Their rats' time–place behaviour was largely unaffected by these lighting manipulations, suggesting that they used a food–entrained phase timer to discriminate between morning and afternoon sessions.

However, we recently examined daily time–place learning in rats using a procedure very similar to that employed by Mistlberger et al. and we obtained different results (Carr & Wilkie, MS). Rats received two daily test sessions in a large, square, Plexiglas test chamber located on a table top in a well lit room. A response lever was mounted on each of the four chamber walls. The rats were transported in a quasi–random order to the testing chamber twice per day; once between 09:30 and 10:30 and at second time between 15:30 and 16:30. A number of salient distal spatial cues were clearly visible from within the test chamber. Each rat could press one lever for food during 09:30 sessions, and could press a second, different, lever for food during 15:30 sessions. Lever assignment in 09:30 and 15:30 sessions was counterbalanced across the rats.

As in Saksida and Wilkie (1995), anticipatory responses recorded at the beginning of each test session were used to infer where the rats expected food to be available during each session. This dependent variable was expressed in the form of a mean percent of all responses score for each of the three types of levers for each rat: (1) the correct lever in that session (MPCorr), (2) the correct lever in the alternate daily session (MPAlt), or (3) the incorrect levers that never provided food (MPInc). Figure 10 presents the rats' weekly overall MPCorr scores in the 09:30 and 15:30 test sessions over the Acquisition and First Baseline Periods. The rat's MPCorr increased from near 25% (chance in a four lever box) to approximately 90% in 09:30 sessions and 80% in 15:30 sessions. Clearly our rats learned which levers would provide them food during 09:30 and 15:30 test sessions.

We first determined the effect skipping 09:30 and 15:30 test sessions had on our rats' time–place behaviour. Over a 40 day period our rats skipped four 09:30 sessions and four 15:30 sessions. Each skipped session was separated by at least 3 days when the rats received both 09:30 and 15:30 sessions. The rats' overall MPCorr, MPAlt, and MPInc scores in the 09:30 and 15:30 sessions immediately before a skipped session, in the sessions immediately after a skipped session,and in the second session after a skipped session are presented in Figure 11. In the 09:30 sessions that followed a skipped 15:30 session the rats continued to expect food at their correct (i.e., 09:30) levers. However, in the 15:30 sessions that followed a skipped 09:30 session the rats incorrectly expected food at their alternate (i.e., 09:30) levers. The effect of skipping a 09:30 session was not permanent as the rats' MPCorr returned to 80–90% during the second 15:30 session after a skipped 09:30 session.

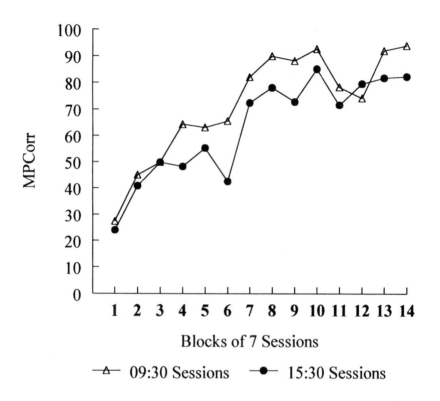

Figure 10.   The rats' weekly mean 09:30 and 15:30 MPCorr scores over the 14 week Acquisition and First Baseline Period. Data are redrawn from Carr and Wilkie (MS).

These results suggest that receiving a 09:30 test session, *and not the passage of time*, was necessary for our rats to anticipate the location of food in 15:30 sessions, and receiving a 15:30 session was not necessary for our rats to anticipate the location of food during 09:30 sessions. This outcome suggests that our rats did not use the phase angle of a phase timer, the accumulated value of an interval timer, or an alternation strategy to discriminate between 09:30 and 15:30 test sessions. Instead, these results suggest that our rats learned to press

09:30 SESSIONS

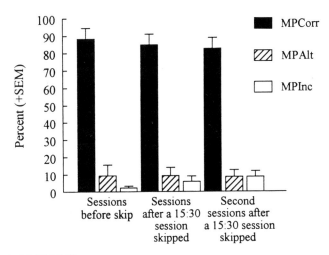

15:30 SESSIONS

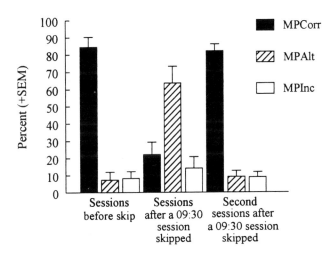

Figure 11. The rats' mean 09:30 and 15:30 MPCorr, MPAlt, and MPInc scores (+ SEM) during the sessions immediately before skipped sessions, during the sessions immediately after a skipped session, and during the second sessions after skipped sessions. Data are redrawn from Carr and Wilkie (MS).

one lever during their *first* session of each day and to press a second lever during their *second* session of each day. It therefore appears that our rats' time–place behaviour was driven primarily by an ordinal timer –– a representation of the *order* in which food was available at their two daily feeding sites (i.e., lever press at Location 1 *then* at Location 2, *each day*). We called this strategy a *daily route*. As our rats' time–place behaviour was not disrupted when 15:30 sessions were skipped, some event other than 15:30 sessions must have been capable of resetting their routes each day.

The baseline 09:30 sessions and the 15:30 sessions that followed a skipped 09:30 session were both our rats' first session of the day. Therefore if our rats were solely using a daily route to track the location of food, they should have treated these two types of test sessions identically. This was clearly not the case. In the 15:30 sessions that followed a skipped 09:30 session our rats did not favor their 09:30 levers as strongly as they preferred their 09:30 levers in baseline 09:30 sessions. Or from the other point of view, during the 15:30 sessions that followed a skipped 09:30 session, our rats pressed their 15:30 levers relatively more than they usually pressed their 15:30 levers in baseline 09:30 sessions. The ordinal position of these two types of sessions was the same (first in the day), but they occurred at different times–of–the–day (09:30 versus 15:30). We therefore suspected that our rats' time–place behaviour was driven by two types of timers: (1) An ordinal timer that had primary control over the rats' time–place behaviour; and (2) a phase timer that had relatively weak control over the rats' time–place behaviour.

To look for further evidence of partial behavioural control by a phase timer we measured where our rats' expected food 1.15 hr, 2.30 hr, and 3.45 hr after the end of a 09:30 session. The 09:30 sessions typically ended at 10:30. If our rats' time–place behaviour was driven solely by a representation of their current position in their daily route, they should prefer their 15:30 levers equally 1.15, 2.30, 3.45, and 5 hr after a 09:30 session. Alternatively, if our rats' time–place behaviour was at least partially driven by the phase angle of a phase timer, a temporal gradient should be evident in their preference for their 15:30 levers during the interpolated probe sessions: They should prefer their 15:30 levers least at 11:45 (when their ordinal and phase timers predict food at their 15:30 and 09:30 levers respectively) and their preference for their 15:30 levers should then increase the closer the probe session is in time to 15:30 (when their ordinal and phase timers both predict food at their 15:30 levers).

The mean proportion of our rats' responses that were on their 09:30, 15:30, and Incorrect levers as a function of the time–of–day is presented in Figure 12. At 11:45 our rats mainly pressed their 15:30 levers. This is consistent with primary behavioural control by an ordinal timer. However, at 11:45 the rats pressed their 15:30 levers relatively less, and they pressed their incorrect levers relatively more, than they did during baseline 15:30 sessions. This effect was also evident in the probe sessions at 13:00 and 14:15, but the magnitude of the effect decreased the closer the probe session was in time to 15:30. This outcome is clearly consistent with our suggestion that our rats' time–place behaviour was primarily driven by an ordinal timer, but a phase timer had weak, but detectable, behavioural control. Our rats' increased preference for their Incorrect levers at 11:45 and 13:00 may reflect exploratory sampling as they already received food at their first location of–the–day and these probes were too early in–the–day for food to be available at the second location in their daily routes.

As our rats anticipated the location of food in the 09:30 sessions that followed a skipped 15:30 session, we knew that some event other than 15:30 test sessions could reset the rats' routes each day. To test whether the rats reset their routes at light onset each day, we switched the rats' colony lighting to LL. If placing the rats in LL immediately caused them to expect food at their 15:30 levers during the first 09:30 session after the switch to LL, we would know that the daily colony LD cycle served as the necessary daily reset stimulus. We would also know that receiving a 15:30 session was not sufficient for the rats to reset their daily routes.

The rats were largely unaffected by the change to a LL lighting schedule. This result suggests two conclusions. First, the daily transitions of the colony LD cycle was not necessary for the rats to reset their routes each day. As we explicitly varied the time–of–occurrence of many other potential cues, it appears that the rats' routes were reset endogenously each day. Second, we suggested that the rats' time–place behaviour was partially controlled by a phase timer. If the rats' time–

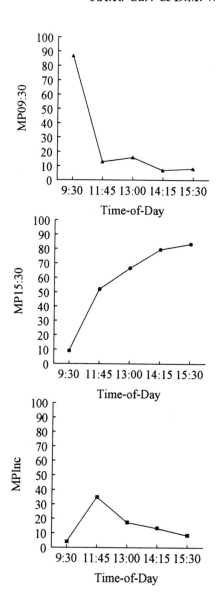

Figure 12. The rats' mean MP09:30, MP15:30, and MPInc scores during the baseline sessions at 09:30 and 15:30, and during the interpolated probe sessions at 11:45, 13:00, and 14:15. Data are redrawn from Carr and Wilkie (MS).

place behaviour was partially controlled by the phase angle of alight–entrained phase–timer, housing the rats in LL for 9 days should have caused a detectable disruption in the rats' time–place behaviour that increased as the LL period progressed. This was clearly not the case. It appears then, that our rats' time–place behaviour was weakly controlled by the phase angle of a food–entrained phase timer.

Mistlberger et al. found that rats use the phase angle of a food–entrained phase timer to discriminate between morning and afternoon sessions, while we found that rats primarily use a daily route to discriminate between morning and afternoon sessions. Our training protocols were virtually identical, but there were some potentially important differences. First, the use of a T–maze by Mistlberger et al. may have made the spatial component of their task easier. However, it is not clear to us how a difference in the spatial components of our tasks could have resulted in our rats using different types of timers to track the location of food. A second, perhaps more important difference, is in the length of training. Mistlberger's animals received their first skip–session probe after 25 days of training, whereas our animals received their first skip–session probe after 98 days of training. Perhaps when rats are trained on this type of time–place learning task, the relative behavioural control held by ordinal and phase timers changes over training.

We are currently examining the behaviour of rats in a modified daily time–place task. In this task, rats receive three different types of training days; morning session only days, afternoon session only days, and morning and afternoon session days. Rats receive these three types of training days in a random order. As these rats do not experience a reliable temporal sequence of food availability each day, they can not track the location of food availability with an ordinal–timer (e.g., a daily route). However, this task is solvable with a phase timer. Therefore, if rats are able to use a phase timer to track daily spatiotemporal regularities in food availability, they should be able to track the location of food in this modified task.

To explore in greater detail the characteristics of ordinal timing, we will examine the behaviour of rats in a daily time–place learning task with more than two daily locations. If rats are able to acquire a task with three or four daily locations, we will use skip–session probes to determine the nature of their representation of the daily pattern of food availability. This line of research would enable us to compare the characteristics of the ordinal timer rats use in daily time–place learning tasks with the results of previous investigations of serial learning.

5.2.   Sequential Time–Place Learning Tasks

In *sequential time–place learning tasks* the location of food availability depends on the amount of time that has elapsed since the beginning of the test session, and subjects receive test sessions at various times–of–the–day. At the beginning of each session food is available at one location (pecking key or lever) for a certain period of time. Food is then available at a second location for another period of time and so on throughout the course of the test session. Typically in sequential time–place learning experiments three or four levers (or keys) have provided food in succession. These experiments have been conducted in large Plexiglas test chambers housed in experimental rooms that contained a variety of distal landmarks.

5.2.1.   Laboratory Demonstrations of Sequential Time–Place Learning in Birds

Wilkie and Willson (1992) published the first laboratory demonstration of sequential time–place learning. In their first experiment, pecks to each of three keys were reinforced during one third of each 90 s trial: Pecks to Key 1 produced grain for the first 30 s, pecks to Key 2 produced grain for the next 30 s, and pecks to Key 3 produced grain during the last 30 s of each trial. Grain access was provided on a random ratio 20 schedule. Pigeons received 20 trials per day, and test sessions took place at various times–of–the–day. Their pigeons clearly learned to peck each key during the correct (i.e., reinforced) point in each session. Additionally, their response rate on each key increased just before each key was scheduled to provide food. This suggests that the pigeons represented, and anticipated, the transition points 30 and 60 s into each trial when food availability moved from one key to the next in the sequence.

In a second experiment, Wilkie and Willson employed a similar design, but made two changes. First, four keys provided food in succession. Second, each key provided grain for 15 min during each 60- min session. Again their pigeons learned to peck each key during the correct point in each session and they anticipated the transitions points 15, 30 and 45 min into each session when the location of food availability changed. Two lines of evidence further support the notion that Wilkie and Willson's pigeons used an endogenous timer to track the location of food over each test session. First, a correlational analysis suggested that the pigeons were not moving from key to key after they obtained a certain amount of reinforcer deliveries. This rules out the use of a counting–based strategy. The second line of evidence was provided by probe tests during which no food was provided during the 5 min before and after the three transition points in each session when food availability moved from one key to the next in the sequence. During these probe sessions their pigeons mainly pecked the correct key in the first 5 min of each unrewarded period, and then switched to the next key in the sequence during the last 5 min of each unrewarded period. As the pigeons continued to change keys at the correct times–into–the session, the use of an endogenous timer was strongly supported.

Wilkie, Saksida, Sampson, and Lee (1994) continued the investigation of pigeons in successive time–place learning tasks. Again, each of four response keys provided food for one 15–min interval during each 60–min test session. Figure 13 presents one pigeon's average response rate on each of the four keys as a function of the time–into–the–session. Inspection of Figure 13 reveals that this pigeon clearly learned to peak each key during the correct period in each test session, and he anticipated the three transition points in each session when food availability moved from one key to the next key in the sequence.

Wilkie et al. (1994) then conducted a series of experiments designed to characterize the timer pigeons use to track the location of food availability in a successive time–place learning experiment. These experiments capitalized on a fundamental difference in the operating characteristics of interval and phase timers (see Table 3). Phase timers are self–sustaining and run continuously –– they can not be stopped. On the other hand, interval timers can be started, stopped, reset, and restarted by external cues. Wilkie et al. reasoned that if pigeons performing a sequential time–place learning task can stop, restart, and reset their time–place behaviour, they are likely using an interval timer to track the location of food availability.

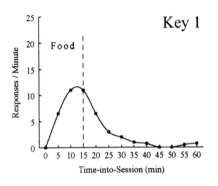
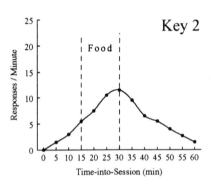

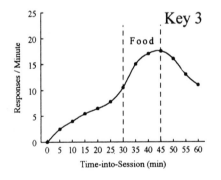
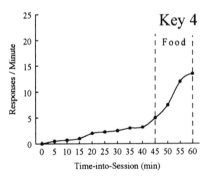

Figure 13. Average response rate on the four keys that provided food successively during each 60 min session in a successive time–place learning task. Data are for subject Clinton. Data are redrawn from Wilkie et al. (1994) with permission.

In one experiment Wilkie et al. tested whether their pigeons could stop and restart their time–place behaviour. Pigeons received baseline sessions and intermixed probe sessions. During probe sessions, their pigeons could peck Key 1 for food for the first 15 min of the session as per usual. Then all four response keys were darkened, and no key pecks were reinforced for a 20–min time–out period. After the time–out period had elapsed, all four response keys were turned on for a 5–min test period. During the 5–min test period all key pecks were recorded but none were reinforced. If the pigeons used an interval timer they would stop timing during the time–out period and then restart timing when the keys were re–illuminated. Consequently, they would prefer Key 2 during the 5–min test period. On the other hand, if their pigeons used the phase angle of a phase timer to track food availability, they would not stop timing during the time–out period, and during the 5–min test period they would prefer Key 3. Wilkie et al.'s pigeons appeared to use an interval timer to track the location of food as they clearly preferred Key 2 during the 5–min test period.

In another experiment, Wilkie et al. tested whether their pigeons could reset their time–place behaviour in a successive time–place learning experiment. Their pigeons again received baseline sessions and intermixed probe sessions. Probe sessions were the same as those used to detect a stop/restart function except during the 20–min time–out period their pigeons were transported from the testing chamber back to their home cages, and then from their home cages back to the testing chamber. Again, the time–out period was followed by a 5–min test period during which all four response keys were illuminated but no key pecks were reinforced. If their pigeons were using an interval timer to track food availability, they would reset their time–place behaviour when they were returned to the testing chamber. Consequently, they would prefer Key 1 during the 5–min test period. On the other hand, if the pigeons were using a phase timer, they would not reset their time–place behaviour when returned to the test chamber. Consequently, they would prefer Key 3 during the 5–min test period. Again their pigeons' appeared to use an interval timer to track the location of food as they preferred Key 1 during the 5–min test period.

Overall the results of Wilkie and Willson (1992) and Wilkie et al. (1994) suggest that within the temporal and spatial parameters investigated, pigeons use an interval timer to represent and anticipate food availability in sequential time–place learning tasks. As described by Wilkie et al. (1994) an interval timer can be used in at least two ways in a sequential time–place learning task. First, an interval timer could be started at the beginning of each session and run continuously throughout each session. The animal would then move from location to location as the interval timer accumulated to three, successively larger, criterion values. Alternatively, the animal could start an interval timer when it began to forage at each location. Then when its interval timer accumulated to some critical value, the animal would move on to the next location in the sequence, then reset and restart its interval timer.

6.   Why have these multiple timing systems evolved ?

We have proposed that animals posses ordinal, phase, and interval timers and we have reviewed both field and laboratory evidence of these processes. Each of these timing systems provides animals with a different type of temporal information. Ordinal timers enable animals to anticipate events that occur in a certain order within a period of time, phase timers enable animals to anticipate events that occur with a certain periodicity, and interval timers allow animals to anticipate events that occur a certain amount of time after another event. On occasions it seems that animals may use two or more of these systems in parallel but differentially weight the output from the different systems. In the last section of this chapter we will explore some ideas as to why multiple timing systems have evolved in animals, and suggest that multiple systems within a single problem domain may be a common evolutionary outcome.

6.1.   *Functional Incompatibility*

Sherry and Schacter (1987) considered whether multiple, specialized, memory systems or a single, general–purpose, memory system was likely on evolutionary and information–processing grounds. Sherry and Schacter argued that multiple memory systems is the more likely evolutionary outcome because a memory system that solves one memorial problem well, cannot, because of it's specialized nature, adequately solve other fundamentally different types of memorial problems. Here we briefly describe their argument, and extend their argument into the domain of timing systems.

First, Sherry and Schacter suggested that memory systems can be characterized by certain *rules–of–operation*. These might include the timing of learning, restrictions on what is learnt, and the interval between acquisition and performance. They then suggested that during evolutionary history animals likely evolved multiple memory systems when there was selective pressure for a species to solve a new type of memorial problem, and the rules–of–operation of existing memory systems made those existing memory systems unsuitable to solve the new problem. They called this mismatch between the rules–of–operation of existing memory systems and the requirements of the new problem *functional incompatibility*.

Of course, not all new memorial problems require a new, dedicated, cognitive solution as some new problems can be solved by existing memory systems. Sherry and Schacter called this chance match between the requirements of a new problem and the rules–of–operation of an existing memory system *exaptive generalization*. Sherry and Schacter also differentiated between strong and weak views of multiple memory systems. They suggested that a strong view of multiple memory systems holds that each memory system has fundamentally different rules–of–operation and completely independent component processes. In contrast, a weak view of multiple memory systems holds that each memory system has fundamentally different rules–of–operation, but memory systems share some component processes.

To illustrate and support their notion of functional incompatibility, Sherry and Schacter described other examples of multiple, dedicated, problem–solving systems within a single problem domain. One of their examples was drawn from the insect vision literature. Some insects (e.g., the bumblebee) have two visual systems. One – the frontal oscli system – is a primitive, but fast, system that basically detects light from dark. The second system is a compound lateral eye system that is capable of detailed resolution, but is complex and slow in its processing speed. Insects are thought to use the frontal system to detect the horizon and thereby keep a level flight path. In contrast, insects are thought to use their lateral system to detect their hive, flowers, landmarks, etc. It appears that insects have evolved multiple visual systems, each with a unique set of operating characteristics, and each specifically adapted to provide them with a specific type of visual information.

Insects are not the only animals that have evolved multiple visual systems to solve different visual problems. Pigeons also posses two broad visual systems (see Goodale & Graves, 1980; Martinoya & Block, 1981; Roberts, Phelps, Macuda, Brodbeck, & Russ, 1996). The binocular myopic frontal system processes items immediately in front of the pigeon, about 25 degrees below the horizon, and about 10 degrees on either side of the beak. The hypermyopic monocular lateral system processes panoramic information. The frontal system seems to be used to detect and peck at food items on the ground and the lateral system seems to be used to avoid obstructions, predators, and the like when walking and flying.

Throughout this chapter we have suggested that animals have evolved three types of timing systems. The operating characteristics, or in Sherry and Shacter's nomenclature, the rules–of–operation, of ordinal, phase, and interval timing clearly differ (see Table 3). Ordinal timers have as many states as there are elements in the represented sequence, they count up, and they are resetable. Phase timers are self-sustaining, run continuously, and are entrainable. Finally, interval timers time up and can be started, stopped, reset, and restarted. The rules–of–operation of these timing systems are not arbitrary: They allow each timing system to provide animals with a particular type of temporal information. Therefore, if there has been selective pressure for animals to represent temporal sequences, periodic events, and elapsed intervals, animals may have evolved dedicated timing systems because the rules–of–operation of any individual timing system precluded that system from efficiently computing other types of temporal information.

6.2. Multiple Systems and a Computational–Representational Approach

A computational–representation approach to timing also predicts multiple, dedicated timing systems. If cognitive systems represent external systems because a functional isomorphism exists between the two systems, then represented external systems that have fundamentally different characteristics must be represented by different cognitive systems. If reliable sequences, periodic events, and elapsed intervals are fundamental, and unique, types of temporal information, they must then be represented by separate, dedicated, cognitive systems.

6.3 . Multiple Systems as a Common Evolutionary Outcome

We have argued that animals have evolved three different types of timing systems, each dedicated to provide animals with a specific type of temporal information. This multiplicity of systems has parallels in other problem domains and therefore should not be considered unusual. Consider the literature on spatial navigation (for a recent review see Dyer, in press; see also the many papers in the January 1996 issue of the *Journal of Experimental Biology*). Most mobile animals will at times find themselves a distance from a goal such as a nest, a mate, offspring, a food source, a source of water, and so forth. A vast literature suggests that animals do not use a single navigational strategy to get from their current position to their goal. Instead, animals posses many spatial navigation systems, each specifically adapted to solve a particular type of navigational problem. One strategy, called piloting, entails following a series of landmarks, or beacons, to arrive at the goal. Another strategy is variously called path integration, dead reckoning, or route reversal. The basic notion is that a foraging animal keeps a record of its angle (provided by an endogenous compass or the vestibular system) and distance traveled on the outward bound journey, and then uses this record to computes it's homeward trajectory (Dyer, in press). In this navigational system landmarks are not used, except possibly to recalibrate the system. A third strategy entails the use of a map and compass: Animals use an endogenous compass to orient themselves and a spatial map to compute a heading to be followed for a certain distance.

Animals likely evolved these different navigation systems because a single spatial navigation strategy did not provide a useful solution for all navigation problems. Piloting, for example, works only within an animal's familiar world. Such a strategy would not enable a homing pigeon to find its way home when released from a distant, unfamiliar site. Additionally, while piloting is very accurate, this strategy is useless when landmarks are disrupted (e.g., moved, covered by snow). Dead reckoning does work in environments such as the open sea and desert that have no visual landmarks to serve as beacons. Dead reckoning has problems, however, as this spatial navigation system tends to accumulate error when used over large distances. Map and compass navigation is the most flexible and powerful navigation system. However, it requires multiple landmarks and complex representational and computational processes in the spatial and temporal domains that likely entail a high cognitive overhead. Considering the multiple spatial navigation systems animals clearly have evolved, it seems likely that multiple, dedicated, problem solving devices within a single problem domain is a common evolutionary outcome.

Acknowledgements

Preparation of this chapter was made possible by a grant from by the Natural Sciences and Engineering Research Council of Canada (NSERC). J. A. R. C. is supported by a Sir Dudley Spurling Postgraduate Scholarship. Maria Phelps, Christine Stager, and Jennifer Galloway provided helpful suggestions on the manuscript.

References

Aschoff, J. (1989). Temporal orientation: Circadian clocks in animals and humans. *Animal Behaviour, 37*, 881–896.

Becker, P. H., Frank, D., & Sudmann, S. R. (1993). Temporal and spatial pattern of common tern (*Sterna hirundo*) foraging in the Wadden Sea. *Oecologia, 93,* 389–393.

Biebach, H., Gordijn, M., & Krebs, J. R. (1989). Time–and–place learning by garden warblers, *Sylvia borin*. *Animal Behaviour, 37,* 353–360.

Biebach, H., Falk, H., & Krebs, J. R. (1991). The effect of constant light and phase shifts on a learned time–place association in garden warblers (*Sylvia borin*): Hourglass or circadian clock? *Journal of Biological Rhythms, 6,* 353–365.

Binkley, S. (1990). *The clockwork sparrow: Time, clocks, and calendars in biological organisms.* New Jersey: Prentice Hall.

Boulos, Z., & Logothetis, D. E., (1990). Rats anticipate and discriminate between two daily feeding times. *Physiology & Behavior, 48,* 523–529.

Carr, J. A. R., & Wilkie, D. M. (MS). Rats use an ordinal timer in a daily time–place learning task. *Journal of Experimental Psychology: Animal Behavior Processes.*

Cheng, K., & Roberts, W. A. (1991). Three psychophysical properties of timing in pigeons. *Learning & Motivation, 22,* 112–128.

Church, R. M., & Broadbent, H. A. (1990). Alternate representations of time, number, and rate. *Cognition, 37,* 55–81.

Collet, T. S., Fry, S. N., & Wehner, R. (1993). Sequence learning by honeybees. *Journal of Comparative Physiology A, 172,* 693–706.

Cosmides, L., & Tooby, J. (1994). Beyond intuition and instinct blindness: Toward an evolutionary rigorous cognitive science. *Cognition, 50,* 41–77.

Crystal, J. D. (unpublished manuscript). A function of biological rhythms: Discrimination of phase.

Cuellar, H. S., & Cuellar, O. (1977). Refractoriness in female lizard reproduction: A probable cirannual clock. *Science, 197,* 495–497.

Daan, S. (1981). Adaptive strategies in behavior. In J. Aschoff (Ed.), *Handbook of behavioral neurobiology* (pp. 275–298) Vol. 4: Biological Rhythms. New York: Plenum Press.

Daan, S., & Koene, P. (1981). On the timing of foraging flights by oystercatchers, *Haematopus ostralegus*, on tidal mudflats. *Netherlands Journal of Sea Research, 15,* 1–22.

Daan, S., Leiwakabessy, W., Overkamp, G., & Gerkema, M. P. (1994). *Time –place– association in mice: a circadian function.* Abstract, Fourth Meeting of the Society for Research on Biological Rhythms, Jacksonville, Fl.

Daan, S., & Slopsema, S. (1978). Short–term rhythms in foraging behaviour of the common vole, *Microtus arvalis. Journal of Comparative Physiology, 127,* 215–227.

D'Amato, M. R., & Colombo, M. (1988). Representation of serial order in monkeys (*Cebus apella*). *Journal of Experimental Psychology: Animal Behavior Process, 14,* 131–139.

Davies, N. B., & Houston, A. I. (1981). Owners and satellites: The economics of territory defense in the pied wagtail, *Motacilla alba. Journal of Animal Ecology, 50,* 157–80.

Dyer, F. C. (in press). Cognitive ecology of navigation. In R. Dukas (Ed.), *Cognitive ecology*. Chicago: University of Chicago Press

Enright, J. T. (1970). Ecological aspects of endogenous rhythmicity. *Annual Review of Ecology and Systematics*, *1*, 221–238.

Ferster, C. B., & Skinner, B. F. (1957). *Schedules of reinforcement*. New York: Appleton–Century–Crofts.

Frank, D. (1992). The influence of feeding conditions on food provisioning of chicks in Common Terns *Sterna hirundo* nesting in the German Wadden Sea. *Ardea*, *80*, 45–55.

Galef, B. G. Jr., & Buckley, L. L. (1996). Use of foraging trails by Norway rats. *Animal Behaviour*, *51*, 765–771.

Gallistel, C. R. (1989). Animal cognition: The representation of space, time, and number. *Annual Review of Psychology*, *40*, 155–189.

Gallistel, C. R. (1990a). *The organization of learning*. Cambridge, Massachusetts: MIT Press:

Gallistel, C. R. (1990b). Representation in animal cognition: An introduction. *Cognition*, *37*, 1–22.

Gibbon, J. (1991). Origins of scalar timing. *Learning & Motivation*, *22*, 3–38.

Gibbon, J. (1984). Introduction. In J. Gibbon & L. Allan (Eds.). *Timing and time perception* (pp. 469). New York: Annals of the New York Academy of Sciences.

Gibbon, J., & Balsam, P. (1981). Spreading associations in time. In C. M. Locurto, H. S. Terrace, & J Gibbon (Eds.), *Autoshaping and conditioning theory* (pp. 219–253). New York: Academic.

Gibbon, J., & Church, R. M. (1981). Time left: Linear versus logarithmic subjective time. *Journal of Experimental Psychology: Animal Behavior Process*, *7*, 87–107.

Gibbon, J., & Church, R. M. (1984). Sources of variance in an information processing theory of timing. In H. L. Roitblat, T. G. Bever, & H. S. Terrace (Eds.), *Animal cognition* (pp. 465–488). Hillsdale, NJ: Erlbaum.

Gill, F. B. (1988). Trapline foraging by hermit hummingbirds: Competition for an undefended renewable resource. *Ecology*, *69*, 1933–1942.

Gill, F. B. (1995). *Ornithology*. New York: W. H. Freeman & Co.

Goodale, M., & Graves, J. A. (1980). The relationship between scanning patterns and monocular discrimination learning in the pigeon. *Physiology & Behavior*, *25*, 39–43.

Gould, J. L. (1980). Sun compensation by bees. *Science*, *207*, 545–547.

Gould, J. L., & Marler, P. (1987). Learning by instinct. *Scientific American*, *255*, 72–85.

Gwinner, E. (1986). Circannual rhythms in the control of avian migrations. *Advances in the study of behavior*, *16*, 191–228.

Hansson, L. (1971). Small rodent food, feeding, and population dynamics. *Oikos*, *22*, 183–198.

Harrison, J. M., & Breed, M. D. (1987). Temporal learning in the giant tropical ant, *Paraponera clavata*. *Physiological Entomology*, *12*, 317–320.

Holder, M. D., & Roberts, S. (1985). A comparison of timing and classical conditioning. *Journal of Experimental Psychology: Animal Behavior Processes, 11*, 172–193.

Jacklet, J. W. (1985). Neurobiology of circadian rhythms generators. *Trends in Neurosciences, 8*, 69–73.

Janzen, D. H. (1970). Euglossine bees as long–distance pollinators of tropical plants. *Science, 171*, 203–205.

Janzen, D. H. (1974). The deflowering of central america. *Natural History, 83*, 48–53.

Keeton, W. T. (1969). Orientation by pigeons: Is the sun necessary? *Science, 165*, 922–928.

Killeen, P. R., & Fetterman, J. G. (1988). A behavioral theory of timing. *Psychological Review, 95*, 274–295.

Klein, D. C., Moore, R. Y., & Reppert, S. M. (1991). *Suprachiasmatic nucleus: The mind's clock*. New York: Oxford University Press.

Krebs, J. R., & Biebach, H. (1989). Time–place learning by garden warblers (*Sylvia borin*): Route or map? *Ethology, 83*, 248–256.

Martinoya, C., & Bloch, S. (1981). Depth perception in the pigeon: Looking for the participation of binocular cues. *Advances in Physiological Science: Sensory Functions, 16*, 477–482.

Mistlberger, R. E. (1994). Circadian food–anticipatory activity: Formal models and physiological mechanisms. *Neuroscience and Biobehavioral Reviews, 18*, 171–195.

Mistlberger, R. E., de Groot, M. H. M., Bossert, J. M., & Marchant, E. G. (in press). Discrimination of circadian phase in intact and suprachiasmatic nuclei-ablated rats. *Brain Research*.

Mistlberger, R. E., & Marchant, E. (1995). Computational and entrainment models of circadian food–anticipatory activity: Evidence from non–24–hr feeding schedules. *Behavioral Neuroscience, 109*, 790–798.

Mistlberger, R. E., & Rusak, B. (1994). Circadian rhythms in mammals: Formal properties and environmental influences. In M. H. Kryger, T. Roth, & W. C. Dement (Eds.) *Principles and practice of sleep medicine*. (Second Edition) Philadelphia: W.B.: Saunders Co.

Paton, C. C., & Carpenter, F. L. (1984). Peripheral foraging by territorial rufous hummingbirds: defense by exploitation. *Ecology, 65*, 1808–1819.

Pengelley, E., & Asmundson, S. (1974). Circannual rhythmicity in hibernating mammals. In E. Pengelley (Ed.), *Circannual clocks* (pp. 95–160). New York: Academic Press Inc.

Reebs, S. G. (1993). A test of time–place learning in a cichlid fish. *Behavioural Processes, 30*, 273–282.

Reebs, S. G. (1996). Time–place learning in golden shiners (Pisces: Cyprinidae). *Behavioural Processes, 36*, 253–262.

Renner, M. (1960). Contributions of the honey bee to the study of time sense and astronomical orientation. *Cold Spring Harbour Symposium on Quantitative Biology, 25*, 361–367.

Rijnsdorp, A., Daan, S., & Dijkstra, C. (1981). Hunting in the kestrel, *Falco tinnunculus*, and the adaptive significance of daily habits. *Oecologia, 50*, 391–406.

Roberts, S. (1981). Isolation of an internal clock. *Journal of Experimental Psychology: Animal Behavior Processes, 7*, 242–268.

Roberts, S., & Holder, M. D. (1984). What starts an internal clock? *Journal of Experimental Psychology: Animal Behavior Processes, 10*, 273–296.

Roberts, W., Phelps, M. T., Macuda, T., Brodbeck, D. R., & Russ, T. (1996). Intraocular transfer and simultaneous processing of stimuli presented in different visual fields of the pigeon. *Behavioral Neuroscience, 110*, 290–299.

Rosenwasser, A. M., & Adler, N. T. (1986). Structure and function in circadian timing systems: Evidence for multiple coupled circadian oscillators. *Neuroscience and Biobehavioral Reviews, 10*, 431–448.

Rozin, P., & Kalat, J. W. (1971). Specific hungers and poison avoidance as adaptive specializations of learning. *Psychological Review, 78*, 459–486.

Saksida, L., & Wilkie, D. M. (1994). Time-of-day discrimination by pigeons, *Columbia livia*. *Animal Learning & Behavior, 22*, 143–154.

Schatz, B., Beugnon, G., & Lachaud, J. P. (1994). Time–place learning by an invertebrate, the ant *Ectatomma ruidum* Roger. *Animal Behaviour, 48*, 236–238.

Sherry, D. F., & Schacter, D. L. (1987). The evolution of multiple memory systems. *Psychological Review, 94*, 1–16.

Shettleworth, S. J. (1993). Varieties of learning and memory in animals. *Journal of Experimental Psychology: Animal Behavior Processes, 19*, 5–14.

Straub, R. O., & Terrace, H. S. (1981). Generalization of serial learning in the pigeon. *Animal Learning & Behavior, 9*, 454–468.

Stephan, F. K. (1989). Entrainment of activity to multiple feeding times in rats with suprachiasmatic lesions. *Physiology & Behaviour, 46*, 489–497.

Stevens, S. S. (1951). Mathematics, measurement, and psychophysics. In S. S. Stevens (Ed.), *Handbook of experimental psychology* (pp. 1–49). New York: John Wiley & Sons, Inc.

Sutherland, R. J., & Dyck, R. H. (1984). Place navigation by rats in a swimming pool. *Canadian Journal of Psychology, 38*, 322–347.

Swartz, K., Chen, S., & Terrace, H. S. (1990). Multiple list acquisition by rhesus monkeys. *Journal of Experimental Psychology: Animal Behavior Processes. 17*, 396–410.

Terrace, H. S. (1993). The phylogeny and ontogeny of serial memory: List learning by pigeons and monkeys. *Psychological Science, 4*, 162–169.

Timberlake, W., & Silva, F. J. (1994). Observation of behavior, inference of function, and the study of learning. *Psychonomic Bulletin & Review, 1*, 73–88.

Turek, F. W. (1985). Circadian neural rhythms in mammals. *Annual Review of Physiology, 47*, 49–64.

Wehner, R., & Lanfranconi, B. (1981). What do ants know about the rotation of the sky? *Nature, 293*, 731–733.

Welsh, D. K., Logothetis, D. E., Meister, M., & Reppert, S. M. (1995). Individual neurons dissociated from rat suprachiasmatic nucleus express independently phased circadian firing rhythms. *Neuron, 14*, 697–706.

Wikelski, M., & Hau, M. (1995). Is there an endogenous tidal rhythm in marine iguanas? *Journal of Biological Rhythms, 10*, 335–350.

Wilkie, D. M., Carr, J. A. R., Siegenthaler, A., Lenger, B., Liu, M., & Kwok, M. (in press). Field observations of time–place behaviour in scavenging birds. *Behavioural Processes*.

Wilkie, D. M., Saksida, L. M., Samson, P., & Lee, A. (1994). Properties of time–place learning by pigeons, *Columba livia*. *Behavioural Processes, 31*, 39–56.

Wilkie, D. M., & Willson, R. J. (1992). Time–place learning by pigeons, *Columba livia*. *Journal of the Experimental Analysis of Behavior, 57*, 145–158.

Time and Behaviour: Psychological and Neurobehavioural Analyses
C.M. Bradshaw and E. Szabadi (Editors)
© 1997 Elsevier Science B.V. All rights reserved.

CHAPTER 8

Cooperation, conflict and compromise between circadian and interval clocks in pigeons

John Gibbon, Stephen Fairhurst & Beverly Goldberg

In the past decade enormous strides have been made in our understanding of the circadian time sense, the endogenous oscillatory system that coordinates our activity rest cycles, temperature cycles and many other endogenous rhythms. Very recently, genes and gene products have been isolated with recombinant DNA techniques establishing definitively the biological bases of this kind of timing. In contrast, over the last one or two decades, considerably less work has been done on the neurobiological systems which measure intervals of time. These "stopwatch" (or "hourglass") interval timing systems permit a wide variety of species to gauge short intervals of time which are critical to their survival. A foraging animal, for example, must be able to estimate rates of return of a patchily distributed resource so that it may make decisions about "where the grass is greener" on succeeding foraging bouts. Less is known about the neurological substrate of this kind of timing, although some recent work has developed evidence for a central role of dopaminergic systems in the basal ganglia (Meck, 1986, 1996a, 1996b). Interval timing has a venerable history of study in the psychology laboratory, however, and several contrasts between the behavioral processes underlying interval and circadian timing are now becoming available.

The present chapter will first summarize some of the properties of both systems, and then describe an experiment studying interval timing in the hours range embedded in circadian food anticipation timing. The two systems may be put in cooperative or conflicting roles by manipulation of time of day, and/or cue onsets for interval timing, and the results of these manipulations will be seen to involve a compromise between the two timing systems.

Circadian / Interval Timing Properties

Table 1 describes basic properties of the two kinds of timing systems. A hallmark of the circadian system is its endogenous oscillatory character: when animal subjects are released from a strict Light–Dark (LD) cycle into continuous dim light, their activity–rest rhythm continues to cycle (free run) at a period close to 24 hours.

Figure 1 shows bar–pressing activity for a rat in constant dim light, with all barpresses rewarded by food. Successive days are indicated by successive rows downward, and two days are plotted horizontally together (raster plot). Activity shows a period of slightly longer than 24 hours. This free–run character is not seen for interval timing systems which generally require considerable training before a clear oscillation in the absence of reinforcement at successive intervals may be seen. An example is shown in Figure 2 below (unpublished data from our laboratory). The data presented are key peck responding by pigeons (N=5) on a fixed interval schedule in which 3 s access to food is primed for the first response after 16 s. The schedule runs continuously, and in the condition shown here, feedings are omitted with probability $\frac{1}{2}$, so that food occurs at time zero on the ordinate, and then on $\frac{1}{2}$ the occasions at 16 s, on $\frac{1}{4}$th of the occasions not until 32 s, and on $\frac{1}{8}$th of the occasions, not until 48 s. Anticipation of feeding, the initial rise to 16 s, shows a typical fixed interval scallop reflecting anticipation of food. The decline afterward is taken only from trials which did not receive food at 16 s. It reflects what is now a common index of timing often studied under the "peak" procedure (Catania, 1970, Roberts, 1981). On some trials food is omitted, and subjects indicate their anticipation of the possible feeding by a peak in responding at the appropriate time. A small rise to the next possible feeding time (32 s) is seen in these data, but it is noteworthy that extensive training was given to these birds before this small, second peak appears.

Table 1: Properties of circadian and interval timing systems

Circadian	Interval
Timing Properties	
Endogenous oscillation: = free run	Requires reset: = one shot
Entrainment range: Limited. Approximately 8 hr maximum	Training range: Broad. Approx 3–4 orders of magnitude – seconds to hours
Phase shift adjustment: Slow. Several cycles usually required.	Phase shift immediate. Arbitrary onset phase.
Variance Properties	
High level of precision: $\sigma/\mu=.01-.05$	Low level of precision: $\sigma/\mu=.10-.35$
Relationship to entrainment period (?)	Scalar property: Superposition in relative time. $f_r(rt)=(1/r)f(t)$, $\sigma/\mu=\gamma$.

Figure 1. Raster plot for a rat barpressing for food in constant dim light.

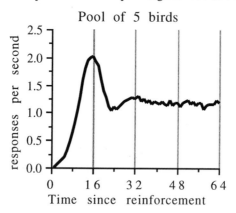

Figure 2. Response rate versus time since reinforcement for a pool of 5 birds. Reinforcement occurs at time 0, at 16 s on half the occasions, at 32 s on one quarter of occasions, at 48 s on one eighth of occasions.

The second row in Table 1 describes another fundamental difference between the two kinds of timing. The entrainment range for circadian timing, that is the range over which LD cycles differing from 24 hours may nevertheless entrain the activity–rest cycle so that coherent activity–rest periods are maintained, is limited to approximately eight hours. Subjects may be entrained with LD cycles of approximately 19 hours up to approximately 27 hours (Aschoff, 1980). But beyond this range entrainment fails, and behavior becomes arhythmic.

Interval timing, in contrast, has been generally studied in the seconds to minutes range. That is, over a span of about 2 orders of magnitude. We will present below some evidence that this range may be extended to include hours, or approximately 3+ orders of magnitude. Thus, these timing systems appear to be designed for a rather specialized range of time periods in the circadian case, but in the interval case a very broad range, undoubtedly reflecting the environmental constraints these systems encounter. Light/Dark (LD) cycles, even in our prehistoric past did not encompass changes over more than a few hours, while interval timing requirements for animals foraging for food may frequently entail bouts lasting from seconds to many minutes, and perhaps hours.

The third property in Table 1 describes the ease of adjustment of the two systems to changes in phase. In the circadian system, phase shifting, particularly phase advances (earlier Dawn/Dusk), are well known to require several cycles before synchronization with the new phase is complete. Americans traveling east as opposed to west have a considerably harder time adjusting to the European LD schedule than, say, to the Hawaiian LD schedule. The theoretical reasons underlying this difference in ease of adjustment of phase are beyond the scope of this chapter, but they reflect, essentially, the fact that our endogenous circadian timer free-runs at a period somewhat longer than twenty four hours, and adjustment to a later dawn and dusk is generally smoother than the reverse.

Phase shifting of interval timing has not been extensively studied previously. We report below some unpublished results from the experiment described above (Figure 2) showing that such phase shifting is almost instantaneous, requiring but one new onset time for virtually complete adjustment. In Figure 3 below pigeons were subjected to a phase shift from their standard 16 s cycle either early (8 s) or late (24 s), or maintained at the standard 16 s FI, when the next two cycles went unreinforced.

The left panel shows four cycles of the procedure. The left-most cycle, between 0 and 16, is data in which reinforcement was primed at 16 s. The three curves are identical. After reinforcement at 16 s, all three functions begin to rise in a standard fixed interval scallop. The early shift function (thin line) stops after an additional 8 s, when an unexpected reinforcement is delivered, thereafter rising again to peak about 16 s later. The non–shift function (heavy line) continues upward until reinforcement is delivered at 32 s, then resets to zero and rises to a peak about 16 s later. The late shift function (dotted line) reaches a peak near 16 s, then begins to decline until an unexpected reinforcement is delivered at 40 s. It then resets to zero, and rises to a peak about 16 s later. Thus, when reinforcement is delivered earlier than expected (8 s after the last reinforcement) or later than expected (24 s after the last reinforcement) the interval timer resets virtually perfectly and begins to time 16 s from that point. This instantaneous phase reset is shown in the right hand panel of Figure 3, where the functions for expected (standard non–shift condition), or unusually early, or unusually late reinforcers are plotted from reinforcement delivery at time zero. All three functions peak at about 16 s after the last reinforcement and a decline thereafter, just as in Figure 2.

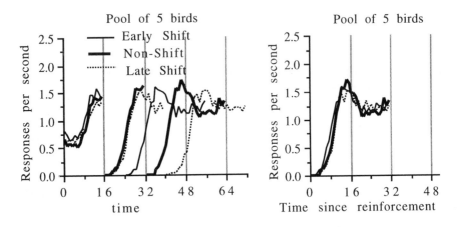

Figure 3. *Left:* Response rates for a pool of five birds when reinforcement is delivered at expected and unexpected times. Reinforcement always occurs at 16 s, and, on half of occasions, at an expected time of 32 s (non–shift). On the remaining occasions, reinforcement occurs at an unexpected time, either early (at 24 s), or late (at 40 s). *Right:* Responding after the delivery of reinforcement at the expected or unexpected time. The time of reinforcement is set at zero.

It is clear that reinforcement at an unexpected time nonetheless starts the interval clock timing the next 16 s interval quite precisely, no matter whether the unexpected reinforcement occurs early or late. Evidently the two timing systems are designed very differently. One reflects a heavy weight on regularity of period (circadian timing) but which is nevertheless susceptible to small phase adjustments such as occur in day lengths over the course of each year. The other (interval timing) undoubtedly reflects evolutionary pressure to permit timing from arbitrary epochs matching the variable (Poisson) temporal nature of the world that foragers face. Thus, this timing system requires immediate adjustment to arbitrary onset phases. The interval timer, like a stopwatch, can be begun, and re-begun, at any point either during or after a timed cycle. The manner in which phase shifts occur for the circadian clock has been studied extensively (phase response curve) and of course there is no parallel in interval timing since this system produces a virtually instant adjustment to arbitrary phase-shifted onsets.

The two systems differ in their variability and the sources of this variability as well. The circadian system has an extremely high level of precision. Onsets of activity-rest cycles may differ in some species (e.g. the golden hamster) by as little as 1% of the 24 hour period. The level of precision in interval timing is considerably less. Common estimates of the coefficient of variation (standard deviation/ mean) in behavior anticipating feeding in the peak procedure, for example, may range from about 10% to 35% of the interval being timed.

The last row describes a fundamental difference between the two systems. Variability in the circadian system has not been extensively studied with respect to its relation to the period. In contrast, interval timing systems are now well known to reflect what is known as the scalar property, that is, superposition of time estimate errors at proportional distances from the target time. This is a Weber Law-like property for time, and it is reflected in a very wide range of procedures (cf. Gibbon, 1991, 1992). Figure 4 shows data for 3 pigeons trained in the peak procedure at a 30 s interval and a 50 s interval.

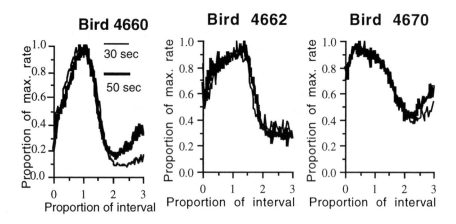

Figure 4. Responding by three birds in a peak procedure with a reinforced interval of 30 s (thin line) and 50 s (thick line). Responding is plotted as proportion of maximum rate versus proportion of reinforced interval.

Despite idiosyncratic differences in the form of the peak function for different subjects (they do not necessarily peak at the reinforcement delivery time), the functions for each subject show rough superposition when relative rate is plotted at proportions of the target interval. We illustrate this property later as a hallmark of interval timing in the study reported below, extending interval timing to the hours range.

Circadian and interval timing interaction

We report next an experimental study of two pigeons, living in isolation chambers for several years. The experiments were designed to illuminate several problems:

1.   Is interval timing seen in the hours range, as in situations approximating foraging problems for birds, that is, approximating those that animals actually experience in traveling to and from patchily distributed resources in the field? In particular, are properties such as the scalar property seen in the longer time ranges?

2.   Under common circumstances in the field, animals must choose between several resource sites upon beginning their daily foraging travels. Does this choice involve an assessment of the time to reach successive patches, and if so, is this choice an interval timer function or a circadian timer function? By the latter, we mean, is the time of day at which successful foraging occurs somehow marked and stored subjectively, so that subjects may decide on the basis of time of day how to organize foraging choices? There is considerable evidence that time of day may be indeed a controlling cue for foraging decisions (e.g., Biebach et al, 1994).

3.   Alternatively, interval timing initiated by dawn might control foragers' estimates of delays to different feeding bouts. If so, differences between the properties of the two timing systems ought to appear subsequent to changes in phase. To answer questions of this sort, we have investigated:

4.   Conflict between circadian and interval timing by phase shifting one or the other of the two systems in ways to be described below. These manipulations are illuminating for several reasons. For one, the question of which system might dominate in the field is addressed by such a manipulation, and for another, the flexibility of subjects to adapt to whichever timing system seems momentarily more advantageous is also addressed.

General Method

Two pigeons were housed continuously in large, sound attenuating chambers. The outer chamber was 38" long, 23" high, 19.5" deep. The living quarters inside were 15.5" long, 15.25" high, 14.75" deep. They worked for all of the food that they received by pecking an illuminated key. They were initially trained on a baseline condition with a 14:10 LD regimen in which the house light was illuminated at 6 AM and extinguished at 8 PM. The *Baseline* condition programmed three cue episodes: at 7 am (breakfast), at 12 noon (lunch), and again at 5 PM (dinner). The cue episodes began with illumination of the key light (red) and were of two types: feeds and omissions, each occurring on a random 50% of the occasions, except that no more than two feedings were allowed per calendar day, and no more than six omissions of feeding in a row (e.g. the equivalent of a 48 hr span) were permitted. During each fed cue episode, feeding was possible for responding on a variable ratio, three–response reinforcement schedule beginning 30 minutes after the key light was turned on. That is, no reinforcement was available until 30 minutes after the key light began, and reinforcement – 3.5 s access to a grain hopper – was available on average, for every third response for an additional 30 minutes, after which the key light was extinguished. When feeds were not programmed, on omission trials, the key light simply remained on for 150 minutes. Responses were collected in five minute time bins.

Experiment 1. Redundant Interval and Circadian Timing.

1.1.   Baseline

The baseline procedure diagrammed at the top of Figure 5 describes the LD cycle (dark or light) and possible cued feeding periods or omission periods at three positions during the circadian day.The hatched rectangle within the cue–on periods represents potential feeding times. On occasions when responding was fed in a cue, the remaining cue–on time was deleted, since the key did not remain lit past the 30 minute feeding period. Below the diagram, raster plots show responding. Outside of cue–on periods responding is represented by horizontal stripes – there was very little such responding. Responding during the cue is indicated in gray, and during fed portions of a given cue in black. Successive days are indicated by successive rows downward, and two days are plotted together (common in raster plots of this type to indicate periodicity visually).

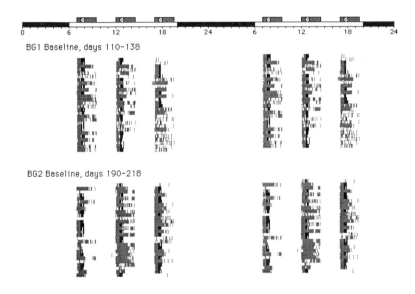

Figure 5. Procedure and raster plots for two birds in the Baseline condition. Black and white areas in the procedure diagram correspond to periods when the houselight is off or on. Gray areas represent cuelight on periods. Feeding, if available, occurs in the hatched area, and the cue goes off at the end of feeding. If feeding is not available, the cuelight continues on through the rest of the gray area. In the *raster plot*, responding during feeding is shown in black, responding during the remainder of cue–on time in gray. Plotted are 5–minute bins in which responding is greater than the average for the day. Shown are days 110–138 of the condition for BG1 and days 190–218 for BG2.

One subject (BG1) was trained for 23 days on an initial Baseline period, and the other (BG2) was trained for 32 days on the initial baseline period with all meals fed. The data shown in Figure 5 are for the subsequent training, when one half of the possible feeding periods went unfed (omission trials).

In Figure 6, summary data from omission periods are presented for birds BG1 and BG2 in the right and left column respectively, plotted both in terms of absolute rate of responding (top row) and relative rate of responding (bottom row).

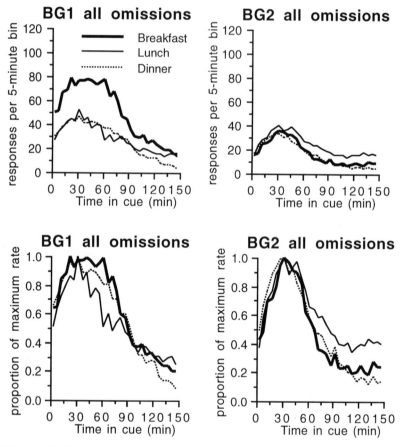

Figure 6. Responding during meal omissions in the Baseline condition for both birds. Absolute response rate is shown in the upper row, responding as proportion of maximum rate in the lower row.

The absolute rates of responding for the two birds do not differ systematically between time of day episodes (the breakfast meal for BG1, when deprivation is the greatest is an exception), but differ substantially between the two subjects. In the bottom row, these data are transformed as the proportion of the maximum rate, and it can be seen that there is rough agreement between all anticipation functions. Subjects respond maximally at about the time when the feeding period would begin. During the latter portion of the 30 minute potential feeding period, responding declines steadily when food has not been obtained earlier.

1.2. The Scalar Property

For this phase of training, the feeding opportunity times were shifted from 30 minutes to 90 minutes after cue onset. The procedure is diagrammed in Figure 7, in the same format as figure 5. Data from the last 28 days of the 6.5 months (BG1) and 5.5 months (BG2) of this condition are shown in the raster plot.

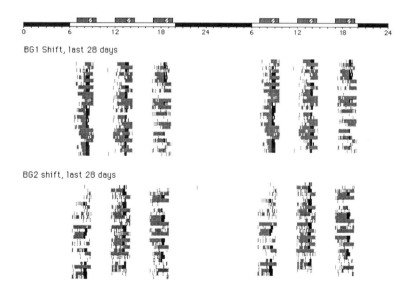

Figure 7. Procedure and raster plots for the shift (90−min) condition. The format of the plot is as in Fig. 5. The last 28 days of the condition are shown for each bird.

These data are re–plotted in Figure 8, as proportions of the maximum response rate in each of the cue episodes.

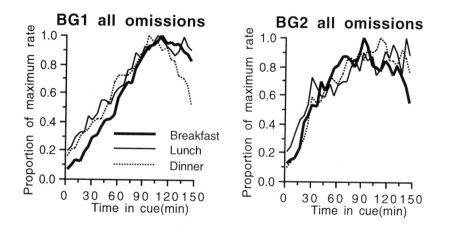

Figure 8. Responding as proportion of maximum rate during omission trials in the shift (90–minute) condition.

Rough superposition, with a peak at 90 minutes, is seen for both birds. Note that the decline beyond the peak is not as readily observed in all cases here, as the cue was extinguished after a total of two and a half hours for both the 30 minute and 90 minute conditions.

Both subjects were then returned to the baseline (30 minute) condition. Data pooled over 3.5 months for each are shown in Figure 9, as proportions of the maximum response rate in each cue episode. Again, rough superposition is seen for both birds.

In Figure 10 pooled data from the three meals in the 90 minute condition and in the second baseline (30 minute) condition are plotted relative to the pre–feeding interval (30 minutes or 90 minutes). Responding in the two conditions shows rough superposition at equal proportions of the target interval, similar to what was seen for intervals in the seconds range in Figure 4. Individual meals (e.g. breakfast in the two conditions) show similar superposition.

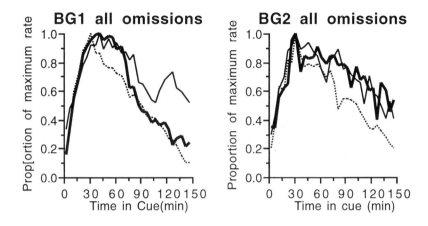

Figure 9. Responding as proportion of maximum rate during omission trials in the second Baseline condition.

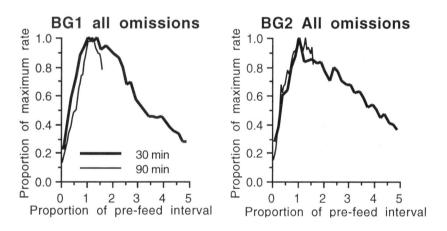

Figure 10. Responding during the second Baseline condition (pool of all meals, thick line) and during the shift condition (pool of all meals, thin line) plotted against proportion of the prefeeding interval.

Discussion

These results replicate and extend the scalar property to interval timing in the hours range. However, the degree of superposition is perhaps less compelling here, and there are differences between these data and those commonly seen in the seconds to minutes range. First, the initial rise of these functions is approximately linear, whereas peak functions in the seconds to minutes range typically appear bell shaped. Second, the right wing, that is the decline after reinforcement time, does not always show superposition. Because of the proximity of the next upcoming feeding, the key light signals could not be maintained beyond their original 2.5 h period in the 90 minute condition, and it is well documented that considerable training with long probes is necessary to see a clear decline post food time. Thus, it is not demonstrated, although it is suggested, that the decline beyond the expected food time, as well as the rise toward feeding time, is scalar in the .5 – 1.5 h range.

Therefore it appears that behavior in the hours range cues exhibit the hallmark property of interval timing, the scalar superposition finding in relative time.

In addition to the differences noted above between hours range timing and seconds to minutes range timing, timing in the hours range raises the issue of redundancy with circadian, time of day (TOD) mechanisms for anticipating feedings. Such mechanisms have been documented in the hours range. However, the precision of such timing would make it unlikely that TOD mechanisms would operate in the seconds and minutes range, since these would not be expected to generate reliably different time of day positions for food anticipation. In the hours range however they may well do so. Thus in some sense a finding of the scalar property in hours range timing is unusual. It means that the contribution of TOD mechanisms may be minimal when the range being studied spans but one hour as in the present case. Within this span, it appears that interval timing from the cue dominates and hence the scalar property holds.

TOD mechanisms may well be operative however but simply redundant with cue interval timing. Moreover the most salient event of the day for these subjects is probably dawn, and thus a dawn initiated interval timing mechanism may also participate. The next two experiments will examine three alternative food anticipation mechanisms which might operate redundantly in our baseline condition:

1. An interval timer, started by the onset of each cue, might readily demonstrate the scalar property, as it does in the short interval range.

2. An interval timer started at the onset of the house light, that is, started by dawn, might also demonstrate the scalar variance property.

3. Finally, a circadian clock, marking the time of day for food anticipation, might itself exhibit the scalar property, although current thinking (in the absence of much data) suggests that this is probably not a property of the circadian time–of–day mechanism (cf. Aschoff, 1980, 1984).

In both conditions (30 and 90 m delays), all three timing mechanisms might well be operative, since they are completely redundant with respect to the time at which feeding is to be expected. However, the scalar property under hypothesis two, that is, time since dawn timed by an interval clock, should show broader variance at lunch, and broader still at dinner, as these times are respectively one, six and eleven hours from dawn. Clearly, the three timing functions look rather similar, and, in particular, the dinner cue does not result in a vastly increased variance relative to breakfast. Thus, it seems unlikely that hypothesis two, an interval timer based on timing all three meals from dawn, is operative. It is important to note, however, that while it is unlikely that lunch and dinner are timed by an interval timer initiated at dawn, it remains possible that breakfast might be timed by such a clock. Subsequent experiments will examine that question.

These considerations suggest that either interval timing from cue–light onset, interval timing from dawn, at least for breakfast, or circadian timing of three different times of day, are all feasible interpretations of these data, although the third hypothesis is doubtful in light of the scalar property. The next experiment examined the role of circadian time of day timing by successively eliminating the putative interval timers associated with cue onset and dawn.

Experiment 2. Elimination of Circadian or Interval timing

2.1.   Time of Day Anticipation Only – Constant Cue Light

In this experiment, the cue light and house light were made redundant. As shown in Figure 11, the cue light came on at 0600, and went off at 2000 with the house light.

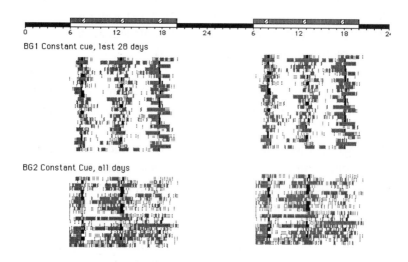

Figure 11. Procedure and raster plots in the constant cuelight condition. The last 28 days for BG1 and all 21 days for BG2 are shown.

Feedings remained available on the 50% schedule at their previous times of day, that is, 0730, 1230, and 1730, 30 minutes after the former cue onset times. Under this condition, hypothesis 1 is no longer a feasible timing mechanism since cue onset and offset are now identical to the houselight, or dawn and dusk times. It remains possible here that interval timing beginning with dawn is a mechanism for any timing observed during the day. More specifically, timing of lunch and dinner is no longer expected to be as precise as seen in the baseline conditions. Here, the only available cues are either circadian cues, for the time of day at which feeding sometimes occurs, or interval timing from dawn, which should introduce about than ten times as much variability for dinner as that observed under the dawn–breakfast interval.

Raster plots for the last 28 of 43 days for BG1, and all 21 days for BG2 are shown in Figure 11. Responding is more variable than in the baseline conditions, but is still concentrated near meal times.

Figure 12 shows responding during the houselight and cuelight on period (thin line) with the previous cued baseline condition superimposed (thick line). For the constant cue data, all days are included up to the start of breakfast feeding time; from there to the start of lunch, only days on which breakfast was omitted are included; from there to the start of dinner, only days on which both breakfast and lunch were omitted, and after the start of dinner, only days on which all meals were omitted. For the (second) baseline data, all omissions are included.

Timing of the breakfast meal is quite sharp. as might be expected from a combination of cue– and dawn–initiated clocks. But the variability in the peaks around the three meals is clearly less than what would be expected from an interval clock alone. This suggests that at least part of the timing of meals is due to a circadian clock. Were a single interval clock initiated with dawn timing the dinner interval, for example, responding should be about 11 times as variable around the dinner availability time as it is around the breakfast availability time. Timing of dinner is clearly less variable than that.

2.2. Circadian timing attenuated

In the next conditions, an attempt was made to attenuate or eliminate the circadian clock. Only BG1 experienced these conditions; they proved to be quite stressful and could not be maintained for long.

BG1 constant cue and 2nd baseline

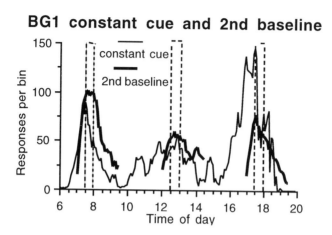

BG2 constant cue and 2nd baseline

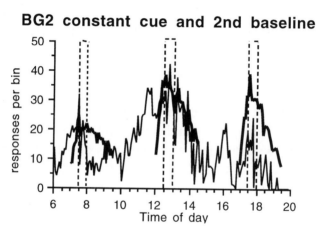

Figure 12. Responding during cue–on time in the constant cue condition (thin line). Data for all days is included up to the start of breakfast (first dotted rectangle). Thereafter data for days on which breakfast was omitted are included up to the start of lunch. After that data for days on which both breakfast and lunch were omitted is included up to the start of dinner. After the start of dinner, only data for days in which all meals were omitted is included. Superimposed (thick line) is data for all omissions in the second Baseline condition.

First, the houselight was left on throughout the 24–hour day (Light Light, LL), while the cuelight continued to go on at 0600 and off at 2000. The procedure and raster plot for the 17 days of this condition are shown in Figure 13.

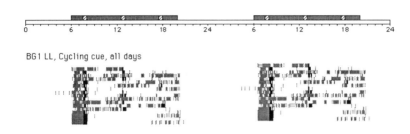

BG1 LL, Cycling cue, all days

Figure 13. Procedure and raster plot for the constant houselight, cycling cue condition. Only BG1 entered this condition. All 17 days of data are shown.

Figure 14a shows average responding throughout the day (cue on period). Like Figure 12, it includes only data up to the first feeding of the day. While responding appears to be higher near meal times than away from meal times, it is considerably more scattered than was the case when the houselight and cuelight came on together.

An interesting feature of these data is the location of the peak near breakfast (Figure 14b). In the prior condition, both the houselight and cuelight could be used to predict breakfast, but the prediction based on houselight onset was the same (90 minutes) as it had been in the baseline conditions, whereas the cuelight prediction (also now 90 minutes) was different from what it had been for the baseline (30 minutes). Under those circumstances breakfast was accurately timed at 90 minutes after houselight and cuelight onset. With the houselight removed as a cue in the current condition, the breakfast peak reverted to 30 minutes after cue onset. This suggests that the prior 90–minute timing was based primarily on houselight onset.

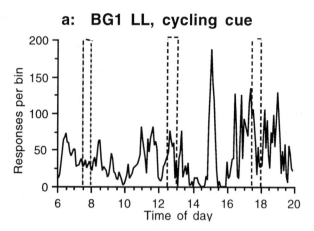

a: BG1 LL, cycling cue

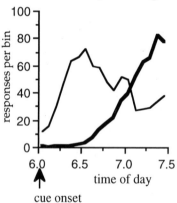

b: BG1 responding in cue

Figure 14. a. Responding during the cue–on time in the constant houselight, cycling cue condition. Data are plotted in the same format as in Figure 12. **b.** Responding during the interval between cue light onset and the start of breakfast feeding time when the cuelight and houselight cycled together (thick line), and when only the cuelight cycled (thin line).

2.3.   No cue onsets – constant light

Finally, both the houselight and cuelight were left on throughout the day, so that neither the houselight nor the cuelight could serve as a cue for interval timing. The procedure and raster plot for the 7 days of this condition are shown in Figure 15.

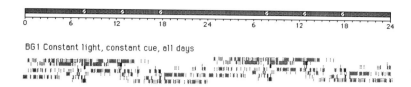

Figure 15. Procedure and raster plot for BG1 in the constant houselight and cue condition. All data are shown.

Figure 16 shows average responding throughout the day, with (as in Figure 14) each day's data showing responding up to the start of the first scheduled feeding.

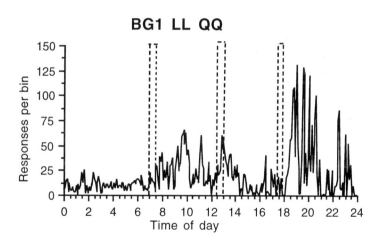

Figure 16. Responding in the constant houselight and cue condition. The format is similar to figs 12 and 14, except that data starts at midnight and continues (when all meals are omitted) to midnight.

Responding is now scattered throughout the day and night, with no clear concentration near meal times. Scheduled meals were often missed, and the subject showed clear signs of stress. Hence this condition was terminated after one week.

The constant, bright houselight evidently disrupted the circadian timing system so that no timed behavior was possible (Aschoff, 1984). In the following phases we attempted to make the houselight irrelevant by maintaining constant dark (DD).

First however, both subjects were returned to the baseline condition for 4.5 months to reinstate their standard timing behavior.

2.4.   Time of day anticipation – constant dark

Constant dark proved considerably less stressful than constant light, so that considerable data could be collected for these conditions. The first condition was like the Baseline, except that the houselight was off at all times. This condition was maintained for approximately 4.5 months. Figure 17 shows the procedure and the performance for the last 4 weeks for each bird.

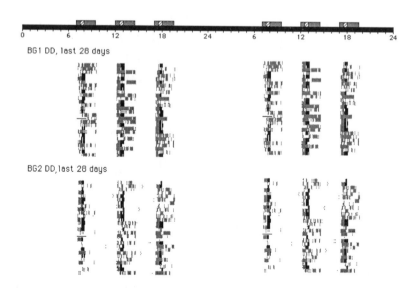

Figure 17. Procedure and raster plot for the constant dark condition. The raster plots show that last 28 days of the condition for each bird.

The data in the raster plot closely resemble the baseline, as does the average data for each meal (Figure 18), except in the case of breakfast for BG1.

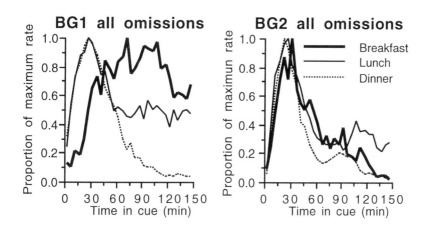

Figure 18. Responding during omission trials in the constant dark condition.

Responding rises later and more slowly here than for the other meals. The level of responding in this meal was much lower than in other meals, contrary to the preceding baseline. These features, along with examination of data from individual days, suggest that the bird, perhaps still roosting, did not always notice the onset of the relatively dim cuelight, so that responding began sometimes later in the interval, leading to a broader, later, and lower peak than was the case when a clear day/night signal was available.

The major finding here is that the meal anticipation cycle is well maintained in the absence of a dawn–initiated interval clock, and in the absence of an externally signaled circadian light/dark cycle. There are, however, events that can serve as circadian zeitgebers – keylight onset at fixed times of day, and the (probabilistic) occurrence of meals at fixed times of day.

Phase 2.4 then demonstrated that without the disruption of constant, intense light and with the cue lights cycling as in the baseline condition, responding in the cue looks very similar to that observed under the baseline with the houselight cycle. Evidently, the zeitgeber of cue–light–plus–food–availability was sufficient to maintain the circadian and interval timing systems.

In the final phase of this experiment we eliminated the cue clock as well, so that the only available cues for timing meals were their actual delivery at the scheduled times.

2.5.   Constant Dark/Constant Cue

The cue light remained on and the houselight off continuously. This condition was maintained for approximately 3 months. Figure 19 shows the procedure and a raster for the last 4 weeks for both subjects for this condition.

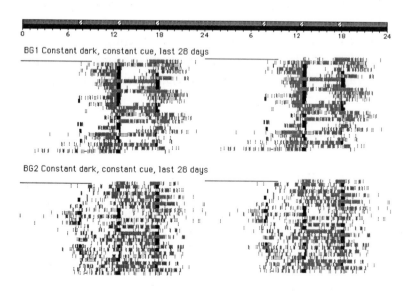

Figure 19. Procedure and raster plot for the constant dark, constant cue condition. The raster plots show the last 28 days of the condition for each bird.

Responding is broadly distributed throughout the time when meals are available. For both birds activity was concentrated within the 8–12 hour period surrounding possible feeding times, and virtually absent during the "subjective night" between the last possible feeding period of the day (dinner) and the first possible feeding period (breakfast).

For BG1, responding often did not begin until after the time for the breakfast meal had passed, a tendency this subject also displayed when meals were cued in constant darkness. BG2 shows a similar tendency, but not as extreme. For both birds responding appears to begin earlier in the "day" when the previous dinner meal was omitted, than when it was fed.

Figure 20 shows average responding throughout the 24–hour period for both birds, with (as in Figures 14 and 16) each day's data shown only up to the start of the day's first fed meal. BG1 shows a broad rise and fall during the "day", with some peaking near the lunch and dinner mealtimes. BG2 shows a similar pattern, but with sharper peaks. This contrasts with the data obtained under constant light and constant cue, where responding was scattered throughout the 24 hours. Evidently constant light disrupts the circadian clock, while constant dark allows it to continue to function, with meals serving as zeitgebers. There is no clear tendency for the circadian cycle to free–run at a period other than 24 hours. Thus the presence of the meals alone may serve to keep subjects entrained to the 24–hour cycle.

Discussion

Experiment 2 explored several phases in which the circadian and interval cues were degraded, by maintaining either the house light or the cue on or off continuously. When the houselight was off continuously, animals could use either the cue light plus the feeding schedules (2.4) or just the feeding schedules alone (2.5) to maintain entrainment to the 24 hour food anticipation schedules. Time–of–day food anticipation, then, clearly is operative in cases in which interval timing, induced by the cues, and time of day timing induced by periodic potential feeding times are both present. While the precision of the anticipation is attenuated when the interval timer initiated by cue light onset is no longer present (2.5), is also clear that the anticipation of food at dinner with a circadian LD schedule (Figure 12) and without (Figure 20) shows clear peaks induced by the feeding schedule alone.

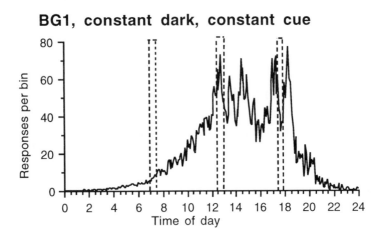

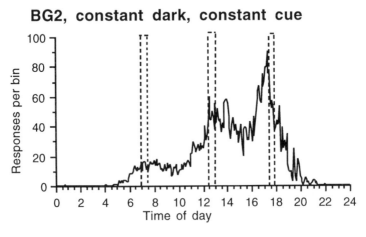

Figure 20. Responding in the constant dark, constant cue condition. The format is the same as that in figures 12, 14, and 16.

In sum, then, Experiment 2 demonstrated (1) circadian anticipation of potential feeding times when no interval cues were present. This interval timing was about as precise at lunch as it was at dinner and thus is unlikely to show the scalar variance property. (2) Cue initiated interval timing was present in the absence of an entrainment schedule for TOD timing except to the extent that meals may serve as zeitgebers, and in the absence of the possibility of a dawn initiated interval timer. (3) It is likely that a dawn initiated interval clock is contributing to precise timing at breakfast, when it is available and when the keylight cue is made redundant (Figure 12). This inference is suggested though not demonstrated as the LD schedule clearly entrained circadian timing as well since anticipation of dinner is clearly not an order of magnitude more variable than anticipation of breakfast.

In the final experiment we introduced conflicts between the three potential time anticipation mechanisms by probing unreinforced delay or advance phase shifts from the baseline condition, so that one of the three putative clocks "dissents" from the timing prediction of the other two.

Experiment 3: Conflict and compromise between clocks

The three hypothesized clocks are three means of predicting the time of the same event, the occurrence of a meal. In timing breakfast for example, a dawn–initiated interval clock would predict a meal after 1.5 hours, a cue–initiated clock a meal after 30 minutes, and a circadian clock a meal at 0730 hours. In the baseline condition, breakfast occurs at 7:30, 30 minutes after cue onset and 1.5 hours after houselight onset, so that all three clocks agree. But if, for example, the time of houselight onset is changed, then its prediction of feeding 1.5 hours after onset will be in conflict the predictions of the cue–clock, and the circadian, time–of–day clock. Similarly, changing the time of cuelight onset sets its prediction in conflict with the dawn and circadian clocks, and changing the time of both houselight and cuelight onset by the same amount would put a dawn clock and a cue clock in agreement, but both in conflict with the circadian clock. Since the circadian clock requires several cycles to change phase, changing the times of houselight and cuelight onset on a single, unreinforced probe day should not alter the circadian clock's prediction substantially.

Changes in the pattern of responding under different situations of conflict between clocks should reveal their relative strength, and the method of resolving such conflict: compromise at a middle position, some responding at both, or disregarding a "dissenting" clock.

The next series of conditions presented just such timing conflicts. On most days the standard baseline condition was presented. On those (rare) days when all meals were to be omitted the times of houselight onset, cuelight onset, or both were shifted earlier (phase advanced) or later (phase delayed) by an hour. Thus the feeding–time predictions of two of the clocks were in agreement, while the third predicted a feeding time an hour earlier or later.

For each condition there were 15 omission days in about 6 months of data.

3.1.   Dawn and Cuelight Advanced

In the first condition, the houselight and cuelight onset times were advanced by one hour on omission days (Figure 21). The third diagram indicates the feeding times predicted by each clock, by hatched bars. The houselight and cue light initiated interval timers agree in their predictions that food may occur 30 minutes after cue onset (top bar) or 1.5 hours after houselight onset (middle bar at 0630 hours), while the circadian clock would predict feeding at the previous feeding time, 0730 hours (bottom bar).

Thus, the two hypothesized interval clocks, initiated by houselight and cuelight, agree in their predictions, while the circadian clock predicts that feeding should begin an hour later. If the two interval clocks predominate, responding should peak at the same time after dawn and after cue onset as it does on standard baseline days, but an hour earlier in time–of–day. If the circadian clock predominates, peak responding should be at the same time of day as the standard, and an hour later after dawn or time since cue initiation. If a compromise is made, the peak should be somewhat later after dawn or cue onset, and somewhat earlier in time of day.

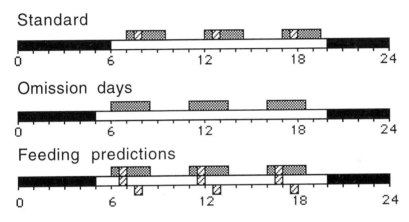

Figure 21. Procedure for the dawn and cuelight advance condition. Standard days are shown at the top, omission days (on which food is never available) in the center. In the bottom diagram, the hatched area shows the time of feeding as predicted by each of the clocks.

Figure 22 shows responding during standard and early–shifted breakfast omissions. Columns correspond to subjects. The top row presents the data for each bird in terms of time since cue onset. The bottom row presents the data in terms of time of day. In each plot the dashed rectangle represents the potential breakfast feeding periods during the standard days. The lunch and dinner meals are not presented here as they look very similar to the breakfast meal. A summary of all meals is presented later (Figure 25).

The shifted data (heavy line) peaks later than the standard when measured in terms of time in the cue (top row), and considerably earlier than the standard when measured in time of day. The degree to which the peak is delayed in the cue and advanced in time of day varies somewhat from bird to bird, but in all cases the interval clock predictions, equal to the standard in time in cue, but 1 hour early in time of day, control the location of these peaks more strongly than time–of–day. Subjects are delayed on average in the time in cue by about 15 minutes, and advanced in time of day by about 45 minutes. This is particularly true for the breakfast omissions, when a dawn initiated interval clock is more precise than such a clock would be at the later meals.

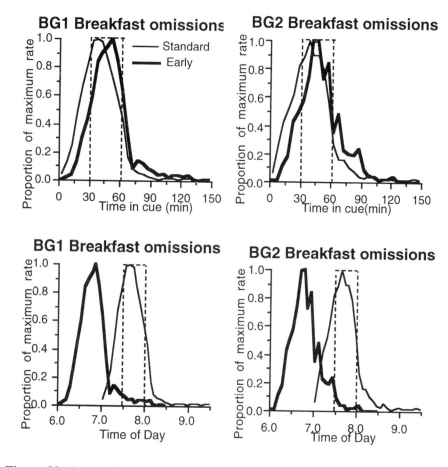

Figure 22. Responding during breakfast omissions in the standard (thin line) and advanced dawn and cuelight conditions (heavy line). In the upper row functions are plotted against time since cue onset, in the lower row against time of day. Data for other meals closely resembles breakfast.

It is important to note that the peaks obtained in the shifted condition are virtually uni-modal. That is, the compromise between the interval and circadian clock predictions for feeding time appears to "settle" on a single value largely controlled by the cue/dawn advanced onset, and less controlled by the previous circadian time of day. There were occasional exceptions to a strict compromise between the three clock predictions. One subject (BG2) showed an occasional subsidiary lower peak in the advance shift data which coincided with the onset of feeding predicted by time of day. However here the major peak occurs at the compromise time according to dawn or cue light clocks. Thus in this case perhaps the subject responded at both the compromise time and the circadian time rather than only compromising . However the other two meals for this subject showed a strict compromise, just as is seen for the breakfast meal in Figure 22.

Thus the advance shifts seem to be dominated by the interval timing systems rather than the circadian timing systems. It is as though the three clocks contributed roughly equally to the compromise prediction with a weight 2/3 for the interval clocks and 1/3 for the prior circadian prediction.

While none of these meals are fed, it is still true that the interval clock predictions occur earlier in time of day than the circadian prediction. Since feeding is not obtained there, perhaps the subsequent circadian time of day prediction has less force. On the other hand, it is clear that it has *some* force since a compromise, rather only interval clock predictions control the peak. In the next phase, delay shifts of dawn and cue are examined. Under these conditions the earliest prediction (in terms of time of day) now is that afforded by the circadian clock, while the interval predicted feeding times are later.

The above inferences depend heavily on the assumption of a slow phase adjustment for the circadian clock. Two caveats must be noted however: (1) While the circadian system takes several cycles for full adjustment, partial adjustment is often seen on the first phase advance or delay to a new phase position. In the present case an hour phase advance of the houselight might initiate the slow phase adjustment for the circadian clock resulting in a small phase advance, say, perhaps 15 minutes (although that is large) and hence the adjusted peak that is seen might indeed be dominated by the circadian system with little contribution from the interval clocks if the circadian system was phase advanced by a small amount upon encountering an early dawn. Moreover avian species occasionally show shorter than 24 hour rather than longer than 24 hour natural periodicity under free run, and hence a phase adjustment to an advance might be more rapid for these species than for mammals or humans with a natural periodicity longer than 24 hours. It is thus of critical importance that the same regime be observed under phase delays of the houselight and cuelight. Under these conditions if the circadian clock adjustment was rapid for advances it would be slow for delays (and *vice versa*). Thus if we see the same sort of compromise between the interval and circadian time predictions, it is most likely that the circadian system has adjusted very little to a one hour advance or, in the experiment below, delay.

3.2.   Dawn and cuelight delayed

In the next condition, the times of houselight and keylight onsets were shifted an hour later than normal on omission days (Figure 23). The bottom diagram again shows the feeding predictions for the three clocks. The houselight and cue light clocks are in agreement and predict food 1 hour later than the circadian clock which predicts food at the previous time of day.

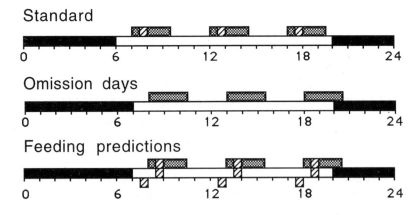

Figure 23. Procedure for the dawn and cuelight delay condition.

If a compromise like the one seen with phase advances is made, we would expect responding to peak slightly earlier than the standard (perhaps 15 minutes earlier) when measured in time since cue onset, and later than the standard (perhaps 45 minutes later) when measured in time of day. Since the peak in responding is usually near the start of the feeding interval, we cannot expect to see a peak at the same time of day as the standard, in as much as the shifted cuelight does not come on until the standard feeding time, in time of day, has ended.

The expected pattern was seen at all meals. Responding in the breakfast meal (Figure 24) is typical. The peaks in the shifted data again show a dominance by the cue/dawn delay shift by about the same amount. That is the peak in the cue is about 15 minutes or so earlier than in the standard, and the peak in terms of time of day about 45 minutes later. The data from the shifted meals also show narrower peaks than those from the standard condition.

In sum, the dawn/cue advance and delay conditions show a compromise between the two predictions with little evidence of simultaneous expectations. In both conditions responding peaks at a point intermediate between the anticipated feeding according to time in the cue and that according to time–of–day.

It is striking that in both cases the compromise position for the peak is at nearly exactly the same absolute value, that is about 15 minutes shift from the standard peak in terms of time in the cue and about 45 minutes shift from the standard peak in terms of time of day.

In Figure 25, the advance and delay conditions are shown averaged over all meal periods for both animals. The match to the same relative weight– about 2/3 to 1/3 in favor of dawn and cue clocks is striking. And it is important to note that asymmetry is *not* observed here between advances and delays. But if the circadian system were dominating the advance, it should have less weight for delays, and *vice versa*. Evidently these clocks contribute equally to a compromise prediction no matter in which direction the phase shift.

The uni–modal nature of the peaks is seen more clearly here than in individual meals. This feature of the peaks suggests that, given a compromise prediction, a single mechanism is used to time that prediction. Also clearly seen is the progressive broadening of the peak functions with increasing time–in–cue of the peak. This increase in variability with increasing peak time is the hallmark of a scalar interval timer. This suggests that the prediction is a compromise between those of the three clocks, but that the prediction is timed by an interval timer starting at cue onset.

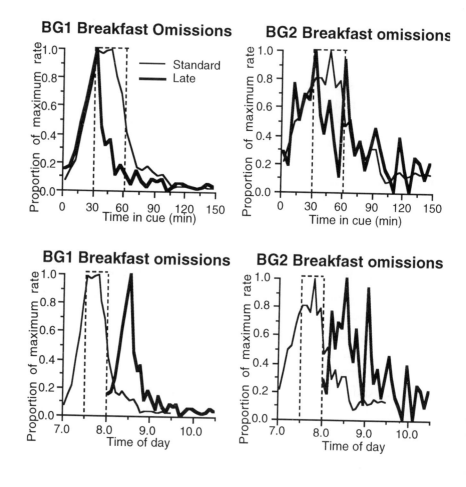

Figure 24. Responding during breakfast omissions in the standard and delayed houselight and cuelight condition.

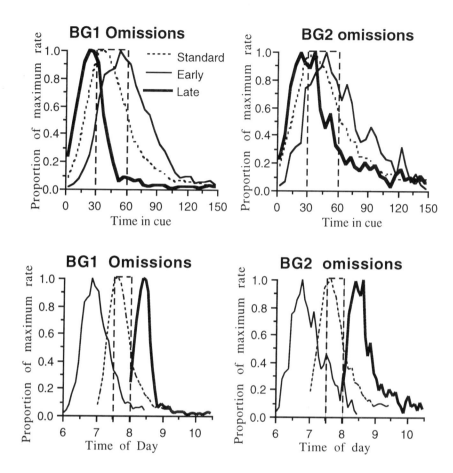

Figure 25. Responding in the standard (dotted line, pool of all standard omissions) advanced houselight and cuelight (thin line, pool of all advance omissions), and delayed houselight and cuelight (thick line, pool of all delayed omissions). Responding is plotted against time in cue in the upper row and against time of day in the lower row.

3.3. Dawn delayed

In the next conditions, the time of houselight onset was shifted, while the cuelight continued to come on at its normal time. The circadian clock and cuelight–initiated clock are thus in agreement, while the dawn–initiated clock gives a dissenting prediction. In the first of these conditions, the houselight, on omission days, came on an hour later than usual (Figure 26), coincident with the breakfast cuelight.

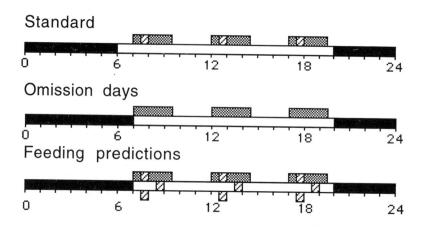

Figure 26. Procedure for the delayed houselight condition.

The strongest conflict is at breakfast, when the dawn–initiated interval clock would be most reliable, and the cuelight comes on simultaneously with the houselight rather than some time later. The circadian and cue clocks both predict feeding after a half hour, while the dawn initiated clock predicts it in one and a half hours.

Figure 27 shows that again a compromise is reached which leads to a peak nearly midway between the two predictions: somewhat more than an hour into the cue for BG1 (top row), somewhat less for BG2 (bottom row). All three meals are shown in Figure 27 in terms of time in the cue only. The reason is that in this condition when dawn is shifted but the other two clocks agree, we may expect stronger effects at breakfast than at lunch and dinner due to less reliability of an interval clock initiated with dawn. Breakfast as measured in time since dawn is slightly more than an hour after dawn (BG1) or slightly less (BG2). The conflict between a dawn initiated clock and the other two appears less pronounced for lunch and dinner as the reliability of a dawn interval clock should be degraded in these longer delays. While it is hard to read a clear peak for BG2 at lunch it is clear that by dinner the peaks obtained for both subjects occur close to the standard (thin line functions) showing stronger dominance by the cue and circadian clocks. Thus in this phase shift, delaying the dawn clock puts it at variance with predictions of the cue initiated clock and that afforded by time of day. The result is a compromise between the two predictions for breakfast, but domination by the cue and TOD clocks for lunch and dinner.

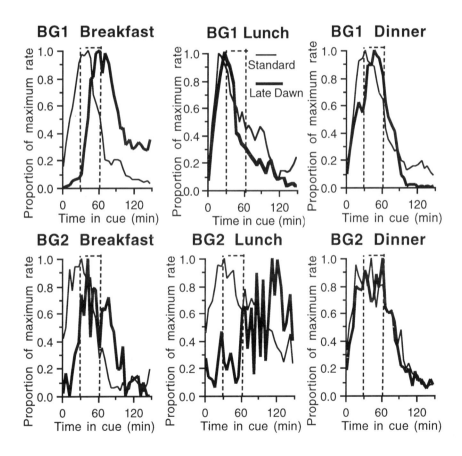

Figure 27. Responding in standard and delayed dawn omissions shown separately for each meal, BG1 in the upper row, and BG2 in the lower row.

Again for the dawn delay condition, one might expect perhaps a small phase adjustment of the circadian clock to later times. However there is no reason that such an adjustment should not operate equally at all meal positions, yet clearly the compromise shows a strong effect of a delayed dawn at breakfast and very little effect of the delayed dawn at lunch and dinner.

However, here as in the earlier phase shift data it is important to study the symmetric dawn advance phase shift so that any dissimilarity between the two contributed by differential ease of phase adjustment of the circadian system might be observed.

3.4.   Dawn advanced

In the next condition, the houselight was turned on an hour early on omission days (Figure 28).

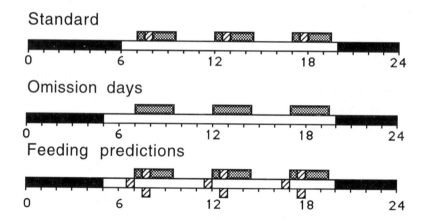

Figure 28. Procedure for the advanced houselight condition.

Now the dawn–initiated clock predicts feeding an hour earlier than the other two clocks. But the conflict between the dawn and cue clocks might be less clear than when dawn was delayed, since the gap between houselight onset and cue onset that is normally present is retained, though lengthened, whereas it was eliminated when dawn was delayed.

At breakfast (Figure 29, breakfast panels), when the conflict should be strongest, the peak in responding as measured in time since cue onset is shifted only slightly, if at all, toward the time predicted by the dawn clock.

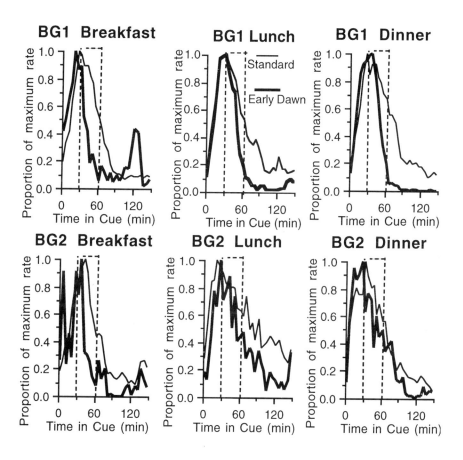

Figure 29. Responding in standard and advanced dawn omissions shown separately for each meal.

At lunch and dinner for both animals, peak times are nearly identical to the standard condition, so that a dawn–initiated clock appears to have no substantial influence on the timing of the peak for these meal positions.

It is noteworthy that the peak functions here on shifted days ·are somewhat narrower than for the standard condition. This was seen previously when dawn and cue light onsets were simultaneously delayed. In that condition the dissenting circadian clock indicated that feeding time had already passed when the cue light came on. In the present condition with dawn advances, the dissenting dawn clock also indicates that feeding time has passed. Since subjects can not show a peak at the time indicated by the dissenting clock because the keylight is off, the rapid decline from the peak––the narrowing of the food anticipation curves may be a manifestation of the influence of the dissenting dawn clock.

In summary, delaying dawn so that its onset is coincident with the onset of the keylight, produced a large delay in peak time for breakfast and diminishing or negligible delays in peaks for the later meals. In the present condition, advancing dawn by an hour produces very little conflict with the other two clocks as long as a discernible gap before the cue light onset is present. Then subjects "prefer" to rely upon the cue and time of day predictions for potential feeding. It is worth noting also that the large asymmetry between advances and delays here is not matched by a similar asymmetry when the cue light and the house light are both simultaneously advanced or delayed. This suggests that the cue light clock is more reliable and hence contributes more strongly to any compromise between dissenting clocks than does the dawn initiated interval clock. It was clear, however from earlier conditions (2.1 and 2.2) that a dawn initiated clock is certainly used to predict food when other cues are unavailable.

The phase shift experiments thus far have put the circadian clock in conflict with the other two, or the dawn initiated clock in conflict with the other two. In the final conditions described below, the cue initiated interval clock is put in conflict with the dawn and time of day clocks, which themselves agree.

3.5.   Cuelight delayed

In this condition, the onset of the cuelight was delayed by an hour (Figure 30). The dawn–initiated clock and circadian clock thus gave the same prediction for feeding time, while the cuelight–initiated clock predicted feeding an hour later. The cuelight clock indicates that feeding should begin a half hour after cue onset, while the two other clocks indicate that feeding time has already ended when the cuelight turns on.

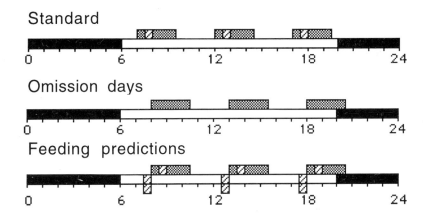

Figure 30. Procedure for the delayed cue condition.

In Figure 31 we present relative responding in time since cue for each subject in the top and bottom rows. This manipulation of the cue onset time alone is particularly interesting since it shows the operation of a compromise between the cue light prediction and that of the other two clocks, and also shows an anticipation of the next succeeding possible meal period within the delayed cue. Particularly for breakfast and to a lesser extent for lunch, there is an early peak which represents a compromise between the time that the standard condition peaks, close to 30 minutes into the cue, and the time predicted by the dawn and the time of day clocks, which is 1 hour earlier. The first peak in responding occurs about 15 minutes or so before that of the standard. The data have been normalized by this peak. One may also see, however, anticipation of the next succeeding meal period in the delayed cue. Particularly for the breakfast omission data, responding drops after the early peak and then begins to rise to a higher level towards the end of the cue. Note that when the cue light is delayed by 60 minutes, 150 minutes after it comes on in terms of time of day represents a bit more than an hour before the next meal period would be scheduled on standard days. Both subjects responded to this approach by an increase in rate late in the breakfast and lunch signals (the dinner signal of course would not be expected to show an anticipation of the forth coming meal since the houselight has already been turned off for the evening).

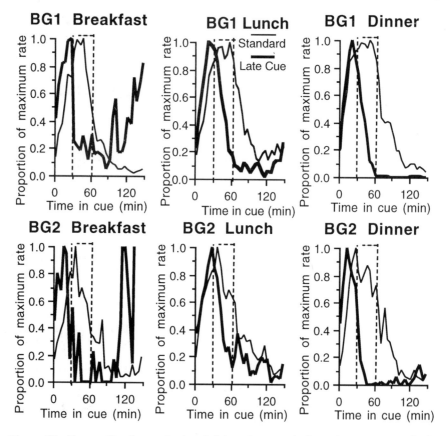

Figure 31. Responding in standard and delayed cue omissions shown separately for each meal.

The first peak, appropriate to a compromise between the time predictions shows about a 2/3: 1/3 weight for the cue initiated interval clock. The peak in the cue is advanced by about 15 minutes while the dawn and time of day circadian clocks would predict that food availability has just ended when the cue first appears. In terms of time since dawn (not shown), then, the food anticipation peak begins after the standard food availability period has passed and peaks about 15 minutes into the cue. Note also that the shifted peak is somewhat narrower than the standard in all cases, perhaps again reflecting the fact that the dissenting cue clock predicts food after the other two are well past their predicted feeding trials.

Another important feature of these data is that for this condition there is no need to hypothesize a small phase adjustment in the circadian clock, since times of dawn and dusk have not been changed. Yet, again a compromise in the predicted feeding time between the cue clock and the other two is seen.

3.6.   Cuelight advanced

In the final condition the times of cuelight onset were advanced by one hour on omission days (Figure 32).

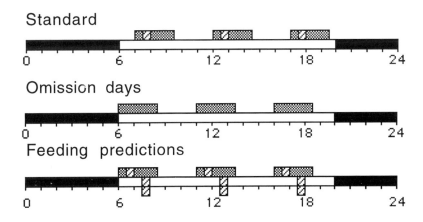

Figure 32. Procedure for the advanced cue condition

The dawn–initiated and circadian clocks still predict that feeding will start an hour and a half after cue onset, while the cuelight–initiated clock predicts feeding after a half hour.

At breakfast (Figure 33) the combined circadian and dawn–initiated clocks clearly predominate.

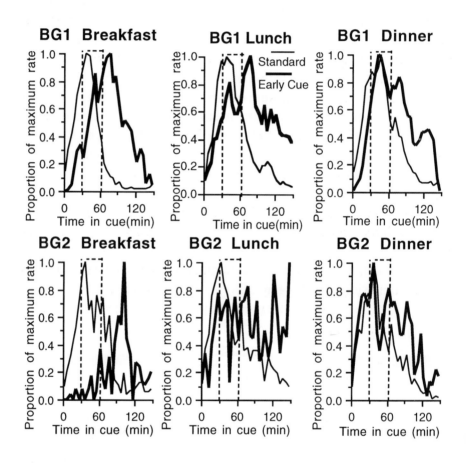

Figure 33. Responding in standard and advanced cue omissions shown separately for each meal.

For BG1, the peak is about 45 minutes later than the standard as measured by time since cue onset, and about 15 minutes earlier when measured in time of day or time since dawn. For BG2 the peak appears at the same time of day as the standard, and an hour later when measured by time since cue onset. The situation at breakfast is similar to that in phase 3.3, when dawn was delayed: the houselight and cuelight come on simultaneously. But in the dawn delay, the circadian clock agreed with the cuelight, whereas here it agrees with the houselight. In both cases peak responding comes at a compromise time, but that compromise time is nearer the houselight/TOD prediction when the houselight and circadian predictions agree, and nearer the cuelight prediction when the cuelight and circadian predictions agree. Both birds show this pattern for the breakfast meal. It is as though the compromise peak is contributed roughly equally in these conditions by all three clocks, so that when two of them agree, the compromise is weighted 2/3 in their favor and 1/3 in favor of the dissenting clock.

The pattern for lunch is similar for BG1 with a delayed peak in time since cue onset, while the lunch pattern for BG2 is unclear. For the dinner omission, however, both subjects show virtually identical curves with the standard peak for the time in cue. That is the cue clock for the dinner meal shows complete dominance.

These data are particularly instructive in their gradual shift of dominance from the dawn plus TOD clock at breakfast to the cue clock at dinner. It suggests to us a role for the lack of precision in a dawn initiated interval clock as contributing to "a lack of trust" of its prediction by dinner time.

General Discussion

The phases of Experiment 1 demonstrated interval timing with the scalar property in a discrete cue for anticipation of a meal at breakfast, lunch, and dinner. This finding extends the range over which we have seen interval timing to nearly four orders of magnitude--from seconds to hours. It also demonstrates the existence of interval timing when embedded within redundant circadian cues. While short interval work is also embedded within circadian cycles, it is unlikely that changes of the order of seconds would reflect a circadian anticipation of time of day at which reinforcement is to be expected. In the present case however, later manipulations do indeed show the existence of circadian food anticipation at the appropriate times within a 24 hour cycle.

The phases of Experiment 2 demonstrated control by one or another of three hypothesized clocks: a cue initiated interval timer, a dawn initiated interval timer and a circadian anticipation of the time of day of potential feeding episodes. This was done via eliminating one or another of these potential clocks by several different manipulations. When all clocks including the circadian clock are eliminated, no timing is seen. This was done by keeping a bright, constant light on at all times along with the keylight. Under these circumstances, the circadian rhythm is abolished and no timing is possible since there are no interval clock onsets from which to record feeding times. In contrast, when the circadian clock was restored (in dim–dim) some time of day anticipation of feeding was clear even in the absence of control by an external dawn signal or an external keylight cue. Thus, circadian anticipation of feeding opportunities when there are no other potential temporal cues is maintained.

In contrast, when the keylight signal is made redundant with the house light coming on at "dawn", a graded control by a dawn initiated interval clock was seen with strong control at breakfast and weak control at dinner which was better anticipated perhaps by the circadian TOD mechanism. It was not possible to eliminate circadian timing and timing from dawn, but it is noteworthy that when the keylight onset replaced the dawn houselight signal, timing briefly reverted to the half hour position for the breakfast meal, when a strong dawn zeitgeber was not present.

In summary, Experiment 2 demonstrated control by the cue light, control by circadian time of day and putative control by a dawn initiated interval timer as well.

The most important result emerged from Experiment 3. There, in several different phase advance and delay manipulations, one or another of the 3 putative clocks predicted food availability at a different time than the others. The result was a consistent tendency for a compromise position for food anticipation between the times predicted by the "dissenting" clock and the "other" clocks. In general, phase advance and delay conditions from this experiment showed a compromise that roughly weighted each clock equally, at least for breakfast when the dawn initiated clock is about as precise as the cue and TOD clocks.

This compromise or averaging of the time for potential feedings we regard as a fascinating and new finding. It means that some internal calculation must weight the differing predictions of feeding times and combine them to arrive at a time at which food anticipation is maximal – and such a time is often a time at which feeding has *never* occurred.

The strongest evidence of a "compromise" rule for interval clocks is perhaps that provided by data from the time left procedure in animal choice, and from titration procedures (Mazur, 1984, 1986, 1987), when a fixed delay is pitted against a set of variable delays. The point at which choice is indifferent between the two is the (harmonic) mean of the variable delays. This means that the rates experienced on the variable delay side are somehow averaged and this average then compared with a fixed delay. In the present experiments averaging was roughly equated between the predictions of the three putative food anticipation clocks, time since dawn, time since cue, and time of day.

The compromise prediction is *not* equivalent to a summation of responding at the prediction of each clock. In Figure 34, responding in the standard and early and late houselight and cue conditions (3.1 and 3.2) is represented by normal curves, thin lines for the houselight and cue clocks, which have the same prediction, and for the circadian clock, and a thick line for their sum. In each panel functions are normalized to the maximum of the sum. Panel (a) shows the standard condition: the narrower thin curve is responding to the houselight + cue clocks, the lower, broader one responding to the circadian clock. In panel (b), the houselight and cue come on an hour early, and responding to the circadian clock is centered at a point an hour later than houselight+ cue responding. The summation shows a second peak at the time of the circadian prediction, contrary to what is generally seen in the data, but virtually no change in the time of the major peak.

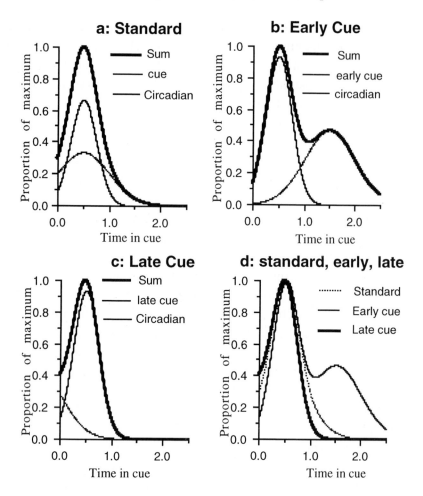

Figure 34. Building a response function by summation of responding to different clocks. In each panel the thin line with the higher, narrower peak represents responding to a combined dawn and cue clock, while the lower, broader thin line represents responding to a circadian clock. Their sum is the thick line. In each panel the values are normalized to the maximum of the sum. Panel (a) represents standard days. In panel (b), the cue and houselight come on hour earlier than standard, and in panel (c) an hour later than standard. In panel (d) the three sums are plotted on the same axes.

In panel (c), the houselight and cue come on an hour late, and only the right tail of circadian responding falls within cue–on time. Again, there is virtually no change in peak time. This is seen more clearly in panel (d), where the three summations are superimposed. Except for the second peak in the early cue condition, the three are virtually indistinguishable. The second peak can be eliminated by using a lower, broader curve for circadian responding (Figure 35), but peak time again remains virtually unchanged – again in contrast to the shifted peaks in the data for these conditions.

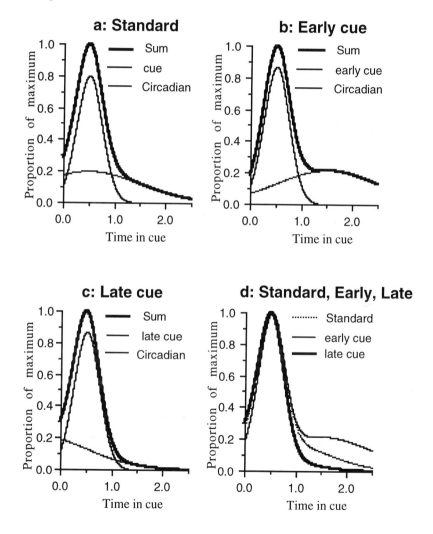

Figure 35. Building a response function by summation, using a lower, broader circadian function. The format and content of panels is the same as in Figure 34.

An alternative is that a compromise is computed, and then timed with a single interval clock. In Figure 36, normal curves with the same coefficient of variation are shown, with peaks at the standard time, 15 minutes before that time, and 15 minutes after it. These curves exhibit the scalar property, as they have the same coefficient of variation. If plotted against proportion of peak time, they will superpose. In panel (a) the curves are plotted against time in cue, in panel (b) against time of day.

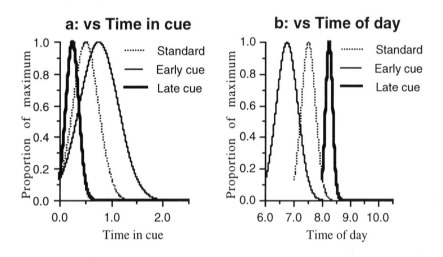

Figure 36. Response function with a single timer. Functions for the standard (dotted line) early cue and houselight (thin line) and late cue and houselight (thick line) have the same coefficient of variation. The peak time for the early cue and houselight function is 15 minutes later in the cue than the standard, the peak for the late cue and houselight is 15 minutes earlier than the standard. Each function is normalized by its maximum. (a) Responding plotted against time in cue. (b) Responding plotted against time of day.

These curves exhibit properties shown in the data: separation of peak times, and progressive broadening of the peaks with increasing peak time. In rare cases, the data exhibit a second peak like that seen in Figure 34b at the time of day or time since dawn predicted by a dissenting clock. Such a second peak could be produced by responding at that clock's prediction as well as at the compromise position. This would seem most likely to happen when the dissenting clock predicts later feeding, and the time of the compromise prediction has passed without reinforcement. Such additional responding would, as seen in Figures 34 and 35, produce virtually no change in the location of the main peak.

Some practical considerations should also effect the position of a compromise peak. In Experiment 3, subjects are never fed in the absence of the cue, and so should expect feeding only within cue–on time. In addition, subjects do not peck an unlit key, so that even if feeding was expected outside of cue–on time, no peak in pecking would occur to show that expectation. Thus the compromise prediction should always be sometime after cuelight onset.

On a few occasions, but they are rare, responding is seen to anticipate food at both of the two conflicting times arranged in the protocol. This is evidence, of course, that these conflicting predicted times are indeed remembered by subjects. It was seen in Experiments 1 and 2 that each of the three clocks can indeed be operative when they are, so to speak, the only game in town.

Timing the compromise

It was argued above that a compromise peak should always appear sometime after cue onset. It seems reasonable then that cuelight onset should trigger the timing of that prediction. Cuelight onset should almost always be a more salient cue for timing than the houselight onset, since it almost always occurs in closer proximity to the predicted feeding time than does houselight onset. And timing from keylight onset appears to be more accurate than circadian timing (Experiment 2), so that cue onset should be more salient than circadian phase. Thus timing an interval from cue onset would seem to be the preferred method. If this is the case, timing should exhibit the scalar property of superposition in relative time. In Figure 37, the standard, early, and late houselight and cue peaks, previously seen in Figure 25, are replotted relative to their peak times. Rough superposition is seen for both birds, suggesting scalar interval timing. As seen in Figure 25, in absolute time since cue onset, early peaks are narrower, later ones broader. When the dawn and/or circadian clocks give different predictions than the cue clock, peaks plotted at time since dawn or time of day show the reverse tendency – earlier peaks are broader, later peaks narrower. In sum, however the compromise is reached, it is timed as an interval from cuelight onset.

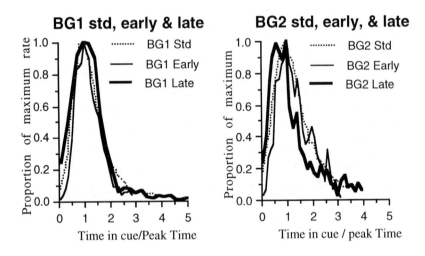

Figure 37. Data from Figure 25 (pools of standard omissions, early cue and houselight omissions and late cue and houselight omissions) replotted against proportion of peak time. The functions roughly superpose, indicating a scalar timing process.

The present data make an important advance on our knowledge base in showing that compromise, which has been documented in terms of rate averaging in operant choice behavior may now be seen to be a feature of conflicts between circadian and interval clocks as well. While it is still uncertain what the variance properties of these averages between a circadian oscillator system and interval clock systems might be, it is clear that the peak food anticipation times show averaging – not mixing.

Evolutionary pressure favoring averaging over mixing, in which all times are equally well remembered, is not obvious. Indeed in some of the experimental treatments of the rate averaging phenomenon in choice, subjects come to anticipate food at times at which it was *never* delivered. The ubiquity of Poisson scheduling of food in nature must favor averaging. Perhaps if variability dominates in nature an averaging rule might make evolutionary sense, because choice between resource sites should reflect some summary measure of resource quality. However periodic food sites are not uncommon. For example: bees harvesting flowers which bloom at different times of day, inter-tidal birds knowing when to go to the shore, and so forth, are a variety of examples of periodic, in time of day, feeding opportunities.

Similarly, it is rare that a cued food availability time of substantial duration is constant in nature, although it is possible that most of the time in the cue is spent "handling" the prey, which might be inherently variable. In terms of value, one might readily construe the "worth" of a resource site with variable delays as something like the mean of the rate of intake, but once in the site itself, with an ambiguous cue, why anticipate food at a time that has never been reinforced?

We end this report with the above question. In nature redundant and consistent clocks may be the rule rather than the exception. In our experiments we have put them in an arbitrary and unnatural context in which time of day and interval timing predictions of feeding conflict. Under these circumstances, subjects apparently use averaging mechanisms, perhaps most appropriate to foraging decisions in the field, with the result that experimentally manipulated conflicts can result in anticipation of feeding at times that have never been associated with a real feeding episode.

References

Aschoff, J. (1980). Ranges of entrainment: A comparative analysis of circadian rhythm studies. In *Proceedings of the XIIIth Conference of the*

International Society of Chronobiology, Pavia. F. Halberg, L. E. Scheving & E.W. Powell, Eds.: 105–112 Il Ponte. Milan.

Aschoff, J. (1984). Circadian timing. *Annals of the New York Academy of Sciences 423*, 470–487.

Biebach, H., Krebs, J.R. & Falk, H. (1994) Time–place learning, food availability and the exploitationof patches in garden warblers, *Sylvia borin. Animal Behaviour, 48*, 273–284.

Catania, A. C. (1970). Reinforcement schedules and psychophysical judgments: A study of some temporal properties of behavior. In W. N. Schoenfeld (Ed.), *The theory of reinforcement schedules*, (pp 1–42). New York: Appleton–Century–Crofts.

Gibbon, J. (1991) Origins of scalar timing. *Learning and Motivation, 22*, 3–38.

Gibbon, J. (1992). Ubiquity of scalar Timing with a Poisson clock. *Journal of Mathematical Psychology, 36*, 283–293.

Mazur, J. E. (1984) Test of an equivalence rule for fixed and variable reinforcer delays. *Journal of Experimental Psychology: Animal Behavior Processes, 10*, 426–436.

Mazur, J. E. (1986) Fixed and variable ratios and delays: Further test of an equivalence rule. *Journal of Experimental Psychology: Animal Behavior Processes, 12*, 116–124.

Mazur, J.E. (1987) An adjusting procedure for studying delayed reinforcement. In M. L. Commons, J. E. Mazur, J.A. Nevin, & H. Rachlin (Eds.) *Quantative analyses of behavior: Vol. 5. The effect of delay and intervening events on reinforcement value* (pp. 55–73). Hillsdale, NJ: Erlbaum.

Meck, W. H. (1986) Affinity for the dopamine D2 receptor predicts neuroleptic potency in decreasing the speed of an internal clock. *Pharmacology Biochemistry & Behavior, 25*, 1185–1189.

Meck, W. H. (1996a) [³H]–Choline uptake in the frontal cortex in mature and aged rats is proportional to the absolute error of a temporal memory translation constant. *The Journal of Neuroscience,* In press.

Meck, W. H. (1996b) Neuroanatomical localization of an internal clock: A functional link between mesocortical, mesolimbic, and nigrostriatial dopaminergic systems. *Behavioural Brain Reaserch,* In press.

Roberts, S. (1981) Isolation of an internal clock. *Journal of Experimental Psychology: Animal Behavior Processes, 7*, 242–268.

Time and Behaviour: Psychological and Neurobehavioural Analyses
C.M. Bradshaw and E. Szabadi (Editors)
© 1997 Elsevier Science B.V. All rights reserved.   *385*

CHAPTER 9

Factors Influencing Long–Term Time Estimation in Humans

Scott S. Campbell

Introduction

In the approximately 40 years since the establishment of human chronobiology as a scientific discipline, an enormous amount has been learned concerning the human circadian timing system. We know the anatomical locus of the clock and much about its pathways to and from other areas of the brain. A wide array of neurotransmitters used by the nervous system to dispatch temporal information have been characterized, and dozens of output measures of the endogenous circadian pacemaker have been identified and studied in detail. We know that external stimuli such as light exposure, exercise and exogenous melatonin administration can be used to reset the biological clock, and we know that certain medications can be made more effective by taking into account the circadian phase of administration.

Considering this sizable and diverse body of knowledge, it is somewhat surprising that we know so little about how humans use their internal clocks in the most fundamental way −− that is, to gauge the passage of time. The dearth of knowledge in this regard is particularly striking with respect to factors involved in the estimation of relatively long intervals (i.e., on the order of hours). For the most part, studies of long-term time estimation have been descriptive in nature, with a primary focus on the accuracy with which humans are able to gauge elapsed time. Thus, we know that most people tend to underestimate the passage of time, and that, by and large, they are capable of judging the passage of long intervals only in very general terms. Moreover, we know that time estimates following periods of sleep are generally less accurate than those following intervals predominated by wakefulness. Yet, we know virtually nothing about the underlying mechanisms that result in these features of human time estimation.

In the following paper, we will describe in greater detail the limited data addressing what we do know about human long-term time estimation, and we will speculate on possible mechanisms involved in the perception of the passage of long intervals. Before doing that, however, it seems appropriate to provide a brief background in the area of human circadian rhythms, since the very nature of studies into long-term time estimation frequently necessitates the use of experimental settings and methodologies unique to the study of human circadian rhythmicity. In addition, much of the data to be presented in this chapter strongly suggest that the human time sense, particularly as it applies to the estimation of longer intervals, is intimately tied to the functioning of the circadian timing system.

Human Circadian Rhythms

Human circadian rhythms, like those of all other organisms studied, are strongly influenced by the natural, 24–hour alternation of light and darkness, and to a lesser extent by the social and behavioral cues that structure our daily lives. Thus, in order to rule out the potential entraining influences of these external time cues, it is necessary to study putative endogenous rhythms under conditions in which such zeitgebers are eliminated or strictly controlled. The first objective studies designed to examine putative rhythmicity in human behavior and physiology took advantage of naturally occurring time isolation laboratories -- underground caves. Subjects volunteered to spend weeks, and even months, often in social isolation and always in a setting characterized by constant, if less than ideal, environmental conditions (˜55⁰ F and 100% humidity), and very primitive living facilities. For logistical reasons, artificial light was often limited, as was the opportunity for many work and leisure activities.

Because the conditions were so unpleasant, subjects were often the experimenters themselves, their unfortunate graduate students, or other professionals (e.g., geologists) with reasons of their own to occupy their time in this manner (see for example, Kleitman, 1963; Siffre, 1964). Despite these difficult circumstances, it was discovered early on that humans did, indeed, exhibit self–sustaining, near–twenty–four hour rhythms in a variety of measures, including sleep/wake behavior, cognitive performance, body temperature, and hormonal output.

By the late 1950's, more controlled studies of humans in temporal isolation were undertaken, most notably by investigators in Germany (Aschoff, 1965; Aschoff & Wever, 1962; Wever, 1979). Studies of hundreds of subjects over the next two decades confirmed, and added to, the earlier findings. It was determined that the so–called "free–running" period (*tau*) of the human endogenous clock was 25.1 hours, although some subjects exhibited behavioral rhythms (i.e. sleep/wake, or rest/activity patterns) with periods of 48 hours or longer. It was also established that in all but a small percentage of subjects, the various circadian rhythms remained synchronized with one another.

More recently, it has been proposed that this early assessment of endogenous period length is probably an overestimation, and may be a consequence of the conditions under which subjects were studied, rather than a true reflection of the endogenous pacemaker (Campbell, Dawson, & Zulley, 1993; Klerman, Dijk, Czeisler, & Kronauer, 1992). In traditional time isolation studies, subjects were instructed to continue to lead a "regular life" -- that is, to carry on their normal daily activities, to eat three meals in normal sequence, and to avoid napping (i.e, to sleep only when they are certain it is their major 'nighttime' sleep episode). When those instructions are altered only slightly, significant differences in the expression of the endogenous clock become apparent. Specifically, under conditions in which subjects are permitted to unstructure their days by eating and sleeping whenever inclined to do so, the average free–running period is not 25.1 hours, but rather, very close to 24 hours. This is the case for both sleep/wake (mean = 24.01h) and core body temperature rhythms (mean = 24.22h) (Campbell, et al., 1993).

Not only do the experimental instructions of the traditional free–run appear to alter *tau*, they also impose an artificial structure on the circadian timing system that probably serves to mask its natural lability. The degree of precision with which the underlying clock is expressed behaviorally, for example in sleep/wake or rest/activity patterns, is dependent largely on the degree of environmental structure within which it functions. In studies such as the one just described, in which subjects are permitted to sleep anytime throughout the circadian day, they do exactly as instructed. That is, sleep can, and does, occur at all phases of the circadian day, as illustrated in Figure 1. Under circumstances in which even fewer behavioral options are permitted, such as the bed rest design, or the "disentrainment" protocol, sleep and waking episodes become even more labile, rendering the circadian organization of behavior almost unrecognizable as such (see for example Campbell, 1984; Campbell & Zulley, 1985).

Yet, even under the most unstructured behavioral conditions, careful examination of certain physiological components of these behavioral variables reveals the continuing influence of the endogenous circadian pacemaker. For example, the proportions of both REM sleep and slow wave sleep comprising a given sleep episode are dependent on the time at which that sleep bout occurs. Likewise, as illustrated in Figure 1, the duration of a sleep episode is not only determined by whether a subject perceives the sleep to be a nap or a major sleep episode, but is also dependent on the circadian phase at which the sleep bout occurs (Czeisler, Weitzman, Moore–Ede, Zimmerman, & Knauer, 1980; Zulley, 1980; Zulley, Wever, & Aschoff, 1981).

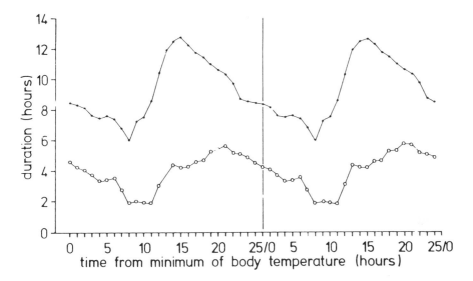

Figure 1. Average durations of sleep periods as a function of the circadian phase of onset, obtained from six subjects while living in a time-free environment (Zulley & Campbell, 1985). Subjects signaled their intention to initiate naps (open circles) and major sleep episodes (closed circles). In both cases, there was a clear circadian rhythm in the duration of the sleep bouts.

Finally, although sleep occurs throughout the circadian day, there are clearly identifiable "preferred" phase positions for the initiation of sleep (Campbell & Zulley, 1985; Zulley & Campbell, 1985), as well as the maintenance of wakefulness (Lavie, 1986; Strogatz, Kronauer, & Czeisler, 1987).

In terms of behavioral output then, the human circadian system may best be characterized as labile, or sloppy, but it is not completely without structure or form. Is this an accurate characterization of the clock driving the system, as well? Probably not. A functional analysis of the stability of the circadian pacemaker in mice revealed a remarkably precise clock, with day-to-day stability of the clock being approximately twice as good (standard deviation = 0.6% of period length) as the already precise activity rhythms which it governs (Pittendrigh & Daan, 1976). For their 150 subjects studied under time–free conditions, Aschoff and Wever found that the average within–subject standard deviation for the circadian rest–activity cycle was 5.2% of the period length. Using the same criterion for man as for mouse, this would make the day–to–day stability of the human circadian clock about 2.6% of period length, or a respectable ± 1.6 minutes per subjective hour.

It is clear, then, that the distinction between the lability of rhythms that we are able to measure and the neuronal mechanisms (i.e., the clock) that control those rhythms is an important one. If the estimation of the passage of long intervals of time is viewed as a "behavior", then we might expect subjects to be as sloppy in their estimates as they are in other behaviors, such as the alternation of sleep and wakefulness. Moreover, it might be expected that the degree of lability of estimates would be related to the degree of structure in the environment within which the estimates are made. On the other hand, if the judgment of time is a more direct reflection of the endogenous clock, then a capacity for more precise assessment of elapsed time might be hypothesized. The limited data that bear on this issue are the focus of the next section.

Characteristics Of Long Term Time Estimation

Underestimation of Elapsed Time

In many of their landmark studies using the time free environment, Aschoff and Wever asked their subjects to provide estimates of the passage of time throughout their stays. Most estimates were obtained by requesting subjects to signal by pressing a button each time that they believed an hour had elapsed, beginning when they awakened each day and continuing throughout their waking periods. Clearly, there are interpretational difficulties inherent in the data obtained using this procedure. If an estimate exceeds one hour it is impossible to know if the subject actually underestimated the interval, or simply forgot to respond at the proper time. Despite this, Aschoff's (Aschoff, 1985) analyses of these findings provided some interesting results, as shown in Figure 2. In all three examples, showing subjects' consecutive hourly estimates across their time

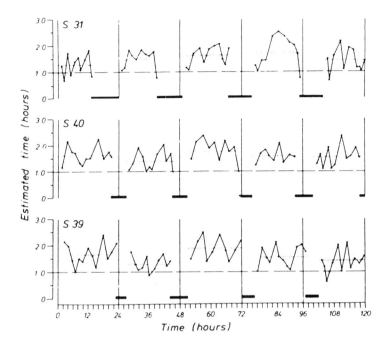

Figure 2. Consecutive estimates of one–hour intervals obtained from three subjects while living for extended periods in a time free environment (Aschoff, 1985). The dotted lines represent the duration of the mean "subjective hour". Black bars indicate times during which subjects were asleep.

in isolation, the large majority of interval productions exceed the 1–hr target interval. That is, the "subjective hour" routinely continues for longer than an actual hour (mean =1.14 h).

This tendency to underestimate elapsed time is one of the most consistent findings concerning the manner in which humans judge the passage of long intervals. All studies of long term time estimation have reported an average "subjective hour" of slightly longer than an hour, with a relatively small range across studies (1.02 to 1.47 hr). The single investigation in which the hour estimate is clearly discrepant (2.14 hr) was conducted under notably different, and substantially more difficult, circumstances than the other studies, as mentioned above (Siffre, 1964). As a result of living on a cave floor of melting ice, in constant, near freezing conditions and 100%

humidity, Siffre suffered from hypothermia ("a condition of semihibernation," p. 89) for most of the 2-month experimental period. As will be discussed later, it is little wonder that under such circumstances Siffre perceived time to pass so slowly; but perhaps for a different reason than that which seems obvious.

It might be argued that Siffre's slow clock was related to his isolation, his general physical discomfort and the substantial degree of boredom that characterized his days. After all, the expression, "time flies when you're having fun" is based on such common experience that it has become a truism. An interval filled with some activity, for example, solving anagrams or working problems in long division, is generally perceived to pass more quickly than an "empty" interval of the same duration, during which subjects simply sit. This result has been reported in virtually all studies investigating the perception of short intervals.

Yet, from his study of hourly estimations, in which he also examined the perception of shorter intervals, Aschoff (1985) concluded that short- and long-term time estimations are probably mediated by different mechanisms. As such, the composition of longer intervals may not have the same impact on estimation of duration as that of short time spans. In fact, a comparison of the results of two studies conducted in the same laboratory, but under very different conditions, suggests that in the case of long-term estimation, the degree to which an interval is "filled" or "unfilled" does not directly affect the manner in which it is perceived. In the first study (Lavie & Webb, 1975), subjects lived for 14 days in a time-free environment, during which they were permitted to continue their usual daily activities (i.e., reading, writing, studying, engaging in hobbies). In addition, a subgroup was required to maintain a daily exercise regimen involving brief intervals on a stationary bicycle. In the second study (Campbell, 1984; Campbell, 1986), subjects lived in the same time-free environment for 60 consecutive hours, but they were confined to bed for the entire period and were prohibited from engaging in virtually any activity. They were prohibited from reading, writing, listening to music, and so on. Thus, subjects in the first condition could be considered to be estimating filled time, whereas those in the bed rest condition were clearly judging the passage of empty time.

In both cases, subjects were asked to estimate the time of day at irregular intervals throughout the study. Under the assumption that filled time passes more quickly than empty time, it would be expected that subjects in the extremely monotonous bed rest condition would underestimate the passage of time to a much greater degree than those studied by Lavie and Webb. This was not the case, however. The average subjective hour in empty time continued for precisely the same duration -- 1.12 hr -- as that in filled time. Moreover, time estimates of subjects living under the even more drastic conditions of sensory deprivation resulted in an average subjective hour of a similar duration (Vernon & McGill, 1963). These findings strongly suggest that "having fun" does not necessarily make time fly.

Whereas the state of the environment may not affect the degree to which one underestimates the passage of long intervals, one's state of consciousness during the interval to be estimated may. Most (Campbell, 1986; Lavie & Webb, 1975; Tart, 1970; Zepelin, 1968), but not all (Boring & Boring, 1917; Noble & Lundie, 1974; Zung & Wison, 1971) studies have found that periods of sleep are underestimated to a greater extent than intervals of wakefulness, regardless of the degree of activity during the waking bout.

Subjects in the bed rest study mentioned above, for example, over- and under-estimated intervals comprised primarily of wakefulness to approximately the same degree (41% over-estimates, 35% under-estimates). Yet, for intervals in which the majority of time was spent asleep, there was a strong tendency for subjects to under-estimate elapsed time. Three-fourths of such intervals were under-estimated, while on only one occasion was the interval over-estimated (Campbell, 1986). Aschoff (1992) reported similar findings in a group of subjects who were asked to report the length of sleep and waking periods during their time in temporal isolation, and he concluded that "the passage of time slowed down during sleep, and speeded up during wakefulness" (page 10).

What is it about sleep that causes this apparent slowing in our perception of the passage of time? Does a change in our state of consciousness fundamentally alter the manner in which we read the clock? Or are there other physiological correlates of sleep that contribute to this apparent slowing in our biological clock? This issue will be addressed in a subsequent section, but first an additional, potentially more troubling, characteristic of human time estimation requires attention.

Lability of Time Estimation

In and of itself, under–estimation of the passage of time is not a fatal time–keeping flaw. If interval duration is consistently and reliably under–estimated, then a standard correction can be made and a reasonable approximation of real elapsed time can still be achieved. If, on the other hand, the general tendency for under–estimation is simply the average outcome of both over– and under–estimations (with a greater frequency of, or more extreme under–estimations) then the utility of the human time–keeping system might be called into question.

Aschoff (1985) identified "large intraindividual variability in consecutive estimates" (p. 42) as one of the most prominent features in the judgment of 1–hr intervals by subjects living for extended periods in isolation. The extreme variability in consecutive estimates of the three subjects' data presented in Figure 2 is quite evident. For the entire group, standard deviations of subjects' estimates ranged from 25% to 49% of the 1–hr interval. Although, as mentioned previously, the procedure that Aschoff used for obtaining interval estimates ran the risk of arriving at a somewhat liberal assessment of the degree of interindividual variability of 1–hour estimates, data from most other studies, using alternative methods of evaluation, are consistent with these findings. For example, Macleod and Roff (1936) found that two subjects (Macleod, himself being one of them) who estimated interval durations during periods in isolation of 86 and 48 hours, respectively, showed mean errors of 21.6% and 22.3% of interval duration.

The apparent sloppiness in time estimation is also seen in the low frequency with which the two subjects were actually accurate in their judgments of interval duration. In only 23.8% of all interval estimates were they able to come within ± 10% of the actual interval. Virtually identical results were reported in three studies conducted at the University of Florida. In the first two, estimates of interval durations ranging in length from 1 to 25 hr, were found to be "accurate" (i.e., within ± 10% of the actual interval) in only about one–fourth of total estimates made (Campbell, 1986; Lavie & Webb, 1975). In the third study, fewer than half of subjects asked to produce a judgment at each of four consecutive hours were accurate (± 10%) in their estimations after the first hour, and only one–fourth of subjects were accurate following the fourth hour (Webb & Ross, 1972).

In the studies by Lavie and Webb (1975) and Campbell (1986) the possible influence of sleep on the lability of interval estimation was also examined, with somewhat differing results. In the former study, subjects were relatively more accurate in gauging the passage of intervals spent asleep than of those spent awake. Again, defining as "accurate" those estimates within ± 10% of the actual interval, 52% of all estimates following major sleep episodes were accurate, while only 20% of those across waking periods were accurate. The latter study produced opposite results, with only 19% of intervals that were made up primarily of sleep being accurately estimated, but 31% of those with more waking being accurately estimated. Despite the relative differences, it should be emphasized that in both studies the likelihood of an accurate response was no better than chance.

Other researchers, using the different experimental approach of asking subjects to awaken at a specified time have arrived at essentially the same conclusion, though marked individual differences in this ability have been noted (see for example Boring & Boring, 1917; Zung & Wison, 1971). In that regard, an early French study cited by Kleitman (1963) should be mentioned: Vaschide (1911) studied a group of subjects who believed that they could awaken at a predetermined time. He found that a subgroup with "superior education" woke up, on average, within 25 minutes of the designated time; subjects with "rudimentary schooling", within 13 minutes; and those with no formal education, within 7 minutes.

To summarize, a generally consistent picture prevails from a rather limited database concerning fundamental properties of the human time–keeping system. Overall, people show a strong tendency to under–estimate the passage of time, more so when they are asleep than when they are awake. Contrary to a widely–held belief, the degree to which an interval is under–estimated does not appear to be dependent on whether that interval is "empty" or "filled".

Whether awake or asleep, whether bored or busy, a person's ability to judge the passage of long intervals is marked by a substantial degree of sloppiness. Boring and Boring (1936) calculated that their subjects were "able to estimate time with a degree of accuracy which was approximately one–half the accuracy that chance guesses would have given" (page 277). Similarly, von Skramlik (1934) concluded that the human physiological clock was approximately 400 times less accurate than the best mechanical clock (Macleod & Roff, 1936). Still others have questioned the existence of a human endogenous clock, or at the very least, the ability of an individual to put it to any useful purpose. Vernon and McGill (1963), for example, concluded that while the birds and the bees may have an internal clock that, among other things, allows them to accurately navigate long distances, humans probably shouldn't leave home without a map: "Chaos would result if solar navigation . . . was attempted using chronometers as poor as those demonstrated in the present study" (page 15).

It appears then, that something gets lost in the translation between a putatively precise internal clock and our capacity to use that clock to gauge the passage of time. In the next section, some of the factors that may help to shape the way in which we estimate long intervals will be discussed.

Factors Influencing Long Term Time Estimation

Behavioral Factors

In an earlier section, it was pointed out that the extent to which human sleep/wake patterns take on a coherent circadian profile seems to depend largely on the degree of behavioral controls in the environment. Indeed, a "sloppiness scale" of the human sleep system can be generated with position on the scale of various experimental designs and real life situations based primarily on the amount of behavioral structuring imposed. At the 'sloppy' end of the scale would be environments such as the bed rest protocol described earlier; at the 'precise' end would be any highly structured experimental condition, characterized for example, by the chronotherapy design in which bedtimes and wakeup times are strictly circumscribed, naps are not permitted during scheduled waking periods, and the subject is permitted to move about freely and generally maintain usual daily activities (Czeisler, Weitzman, Moore, Zimmerman, & Knauer, 1981).

There is suggestive evidence that the accuracy with which we estimate the passage of time is, in a sense, partially dependent on a structured environment, as well. For most of us, our daily lives would be placed far up on the 'precise' end of the sloppiness scale. The natural photoperiod, regular sleep, wake and meal times, job and family commitments, and the general tendency to schedule other routine activities all serve to partition the day into relatively uniform and consistent intervals. We have become accustomed to such structure, we have more experience with such a temporal framework, and therefore, we may be better able to judge the passage of time within such a setting. As the background against which estimation of the passage of time deviates from that prior experience, the likelihood of accurately estimating interval duration declines.

Consider the bed rest study described previously (Campbell, 1984; Campbell, 1986). For all of the subjects studied, sleep and waking episodes occurred throughout the 24–hr day, with a single episode of either sleep or wakefulness rarely exceeding 4 hr in length. In addition, meals consisted of a standard fare and they were served at irregular intervals. The accuracy with which subjects judged the passage of time under these conditions was directly related to the extent of behavioral deviation from their typical day. Specifically, the magnitude of error in subjects' estimates of time of day was significantly correlated with several measures reflecting alterations in the usual monophasic sleep/wake system – the number of sleep episodes taken (Spearman rank–order correlation =.63), the mean sleep/wake cycle duration (–.75) and the mean duration of wake periods (–.68). Simply, subjects had no familiar background against which to make a relative judgment of elapsed time.

Lavie and Webb (1975, see above) reached a similar conclusion, based on their finding that errors in time estimation were greater for the group of subjects who carried out a daily exercise regimen than for those who did not. The authors attributed this group difference to the more disrupted sleep/wake patterns of the exercise group, concluding that subjects' errors in time estimation were "a net result of rational problem–solving behavior" (p. 184). That is, they used prior experience about their own behavior to derive time–of–day values. Thus, when asked to give a time estimate on awakening from a sleep episode, a subject might reason, "I usually go to bed at midnight, and sleep for about 8 hours . . . . it must be 8AM". Rational enough, but of little value under "free–running" conditions in which the circadian clock ran at a frequency close to, but different from 24 hours, and in which sleep bouts occurred at any circadian phase, and lasted for a wide range of durations.

Such a reliance on prior experience is further underscored by Aschoff's (1992) finding that subjects estimating the duration of sleep and waking periods during their time in isolation showed a consistent trend toward overestimation of relatively short durations, and underestimation of relatively long durations, with the most accurate estimates centered around sleep and wake intervals that added up to 24 hours.

These data strongly suggest that as with the human sleep/wake system, the integrity of the human time–keeping system is largely dependent on the behavioral make–up of the environment within which it has to function. The more the environment deviates from that to which we are accustomed, the more our perception of the passage of time deviates from that which the earth's rotation dictates. Yet, in addition to the large impact of behavioral structuring on the human sleep/wake system, the continuing influence of the endogenous circadian pacemaker is still evident. Is there an equivalent underlying physiological drive associated with the human time–keeping system? And if so, in what way is it manifested?

Physiology Factors

Metabolic rate has long been hypothesized to influence the endogenous clock. For example, the medical team that examined Siffre (1963) upon his emergence from the Scarasson Cavern believed that his drastic underestimation of time was related to the slowing of metabolism that accompanied his extended bout of hypothermia. Aschoff has also hypothesized a link between metabolic rate and the degree to which elapsed time is underestimated (Aschoff, 1985; Aschoff, von Goetz, Wildgruber, & Wever, 1986; Aschoff, Wever, Wildgruber, & Wirz–Justice, 1984). This conclusion was based on the findings that both the timing of inter–meal intervals (presumably reflecting metabolic processes) and subjective estimation of 1–hr intervals lengthened with the lengthening of free–running rest/activity periods of subjects living in isolation.

Since daily metabolic rate is at its lowest during sleep, even when compared to bed rest without sleep (Shapiro & Moore, 1981), the studies described previously, in which slowing in the passage of subjective time was observed during periods of sleep, also support the argument that metabolism plays a role in long term time estimation. Finally, although Macleod and Roff (1936) rejected the possible influence of metabolism on the speed of the endogenous clock they, nevertheless, reported that a subject tended to overestimate the duration of those intervals during which exercise was the principal activity.

If metabolism does influence time estimation, and if it is assumed that the circadian course of body core temperature reflects metabolic processes, then a relationship might be expected between the daily oscillation in body temperature and the degree to which one over– or under–estimates the passage of time. As part of an ongoing study designed to examine the timing and composition of sleep taken *ad libitum*, we recently had an opportunity to test that hypothesis.

Estimations of time of day were obtained from 18 subjects, at irregular intervals, during 72 hours in a "disentrained" environment (see Campbell & Zulley, 1985; Campbell & Zulley, 1989, for complete description of disentrainment period), subjects ate and slept whenever inclined to do so, while living singly in an apartment isolated from cues to time of day. Behavioral options were highly restricted. Lighting was provided by floor and desk lamps (<100 lux) and was at the discretion of each subject. EEG and body temperature were recorded continuously throughout the study.

From sequential time–of–day estimates of each subject, we calculated a series of estimated interval durations across disentrainment. Those values were subtracted from the corresponding actual interval durations to arrive at a deviation score for each interval. Deviation scores were then plotted along with a time–series plot of body core temperature values obtained from each subject at the time of each estimate. The individual results from six of the subjects are summarized in the series of plots in Figure 3.

Visual inspection of the plots makes it clear that for most subjects, there was a striking correspondence between deviation scores of interval estimation and core body temperature. In general, on the rising portion of the temperature curve, the average "subjective hour" continued for a relatively short time, ie., subjective time ran fast relative to real time. Likewise, as temperature declined, the average "subjective hour" became longer, i.e., subjective time slowed down relative to real time. For the group, the mean subjective hour was significantly shorter at the temperature maximum (.98h) than at the minimum (1.41h).

Because of the relaxation of controls on sleep/wake behavior in the disentrainment protocol, sleep and waking bouts occurred at virtually all phases of the circadian day. Thus, it was possible to examine the effects of state of consciousness on time estimation, while holding circadian phase constant. Importantly, when intervals containing a majority of sleep were compared with same–phase intervals dominated by wakefulness, there was no significant difference in the duration of the subjective hour.

These data clearly support the notion of a time–keeping system modulated by the daily oscillation in metabolic rate. Yet, the existence of a "metabolic clock" seems to conflict with one of the basic assumptions concerning the biological timing system –– that of temperature compensation. Pittendrigh (1960; 1981), among others has made a convincing argument against any significant influence of changes in temperature (or any other perturbations likely to be normally encountered by the organism) on the inherent frequency of biological clocks. That is, the frequency of circadian oscillations is subject to "general homeostatic control." This argument was based on the logical premise that a temperature–sensitive clock mechanism would be of little adaptive value. "For unless the spontaneous frequency of a sense organ is compensated for temperature. . . it will be useless for anything except temperature sensing" (1960, p.179).

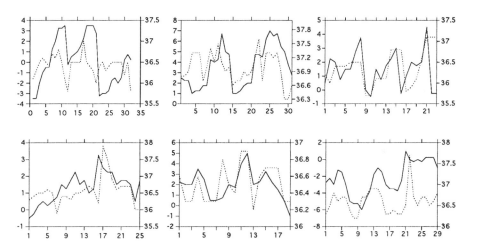

Figure 3. Relationship between the circadian course of body temperature (dotted line) and the degree to which subjects over- and under–estimated interval duration (solid line) during 72 hours in a disentrained environment. For most subjects, when temperature was relatively high, the subjective hour was relatively short; when temperature was low, the subjective hour was long.

How then, can our time estimation data be reconciled with the concept of a temperature compensated clock? Several possibilities can be envisioned. It is conceivable, for example that the human circadian clock itself is unaffected by changes in temperature (i.e., it is temperature compensated), but that the process by which we use the clock to estimate duration is not. Implied in this interpretation is the notion that the process of estimating elapsed time is either well "down stream", and influenced not only by the clock but by other factors, or alternatively, that the human "time sense" is subserved by physiological substrates completely unrelated to the circadian clock (see Pittendrigh & Daan, 1974).

Another possibility is that the relatively small daily excursion in body temperature that so significantly and systematically influences the duration of the subjective hour is a direct reflection of the circadian clock, remaining beneath the sensitivity of temperature compensation mechanisms. This would suggest that the circadian variation in the speed of the clock is, on one hand, an acceptable evolutionary flaw, or on the other, a vital component of the human biological clock.

These are clearly not the only possibilities, nor are they mutually exclusive. Indeed, a synthesis of the two may best account for the findings presented in this chapter. A time–keeping system comprised of a temperature compensated clock, the output of which is modulated by metabolic factors, and the utilization of which is influenced by prior experience would produce time estimates characterized by the features described above. The ultimate question, of course, is not how such estimates of time are accomplished, nor what mechanisms underlie them, but what their adaptive significance is, if any. This question is the focus of our final section.

Is a Sluggish, Sloppy Clock of Any Use?

The existence of the biological clock almost certainly has adaptive significance. The pervasiveness of circadian rhythms within individuals and across species strongly suggests that temporal organization in behavior and physiology is important to the overall well–being of the organism. Yet, it is conceivable that the capacity for humans to use their biological clocks to make general assessments of the passage of time has no adaptive value, whatsoever. By analogy, no one would argue that the opposable thumb does not hold distinct adaptive value for humans. Yet, the capacity to use that evolutionary gift to snap one's fingers in time with music, or as a gesture to gain attention, is probably worthless in terms of survival of the species.

Can a case be made for long term time estimation being anything more than a behavioral gimmick? The general lability observed in human long term time estimation under experimental conditions, if not terribly valuable from an evolutionary perspective, is almost certainly not damaging either. For all but the past century or so, the capacity to roughly determine suitable times for eating, sleeping, working and resting has been sufficient for the successful completion of one's day-to-day activities. Moreover, estimations of time-of-day, or interval durations, that are likely to be of importance for anything other than scientific inquiry are generally made not in isolation, but rather within a familiar temporal framework, characterized by a number of useful environmental cues. As such, normal utilization of the human "time sense" is likely to be marked by substantially less sloppiness than that revealed by controlled studies.

The fact that humans tend to overestimate the passage of time during their waking hours (on the rising portion of the temperature curve) and underestimate the passage of time during sleep (on the declining portion of the temperature curve) may have significant adaptive value. Many animals employ a time sense to more effectively make use of limited resources. In the laboratory, for example, numerous species exhibit anticipatory behavior shortly before food is available, after having been maintained on restricted daily feeding schedules for several days. Likewise, several species show the capacity to respond with behaviors that have been differentially conditioned to be rewarded at various specific times of day. In the wild, some predator species return daily to hunting sites at a time previously associated with the availability of prey. By having a time-keeping system that reliably overestimates the passage of time during the appropriate portion of the day, such anticipatory behavior is insured.

This overestimation of time would, of course, be equally valuable to humans in the role of prey as in the role of predator, insuring timely withdrawal from situations previously associated with danger. In the same way, underestimation of the passage of time on the declining portion of the temperature curve (i.e., at night) may have adaptive relevance. It has been proposed that one function of sleep is to enforce inactivity at times during which activity might be non-productive or even life-threatening. Yet, as cognitive, social animals, prolonged inactivity is anathema to the human species. By perceiving such time to pass more slowly than it actually does, that period of enforced sloth may be made more bearable by the perception that it is "over before you know it".

Summary and Conclusions

A wide array of human physiological processes and behaviors are mediated by an endogenous circadian pacemaker located in the suprachasmatic nuclei of the hypothalamus. There is overwhelming evidence that this neural structure functions as a self-sustaining oscillator that provides internal temporal organization among the various processes, and external temporal order within the natural 24-hour light/dark cycle. There is also good evidence to indicate that a large number of species utilize the endogenous clock to track the passage of time, often to a remarkably accurate degree.

The weight of the evidence presented here supports the notion that humans are also capable of estimating the passage of time, albeit with a certain degree of imprecision. There is no reason to believe that the human "time sense" is mediated by a mechanism, or mechanisms different than the circadian pacemaker, and in fact the human time-keeping system shares features common to other behavioral expressions of the circadian clock, such as the sleep/wake system. The functioning of both systems is governed to a large extent by the relative presence of behavioral structuring in the environment. At the same time, however, they both reflect an underlying influence by the circadian clock.

It is the circadian modulation of the subjective perception of the passage of time that allows us to anticipate events and act preemptively. Such a capacity to achieve a temporal headstart may have substantial adaptive advantage, both in our role as predator and as prey.

References

Aschoff, J. (1965). Circadian rhythms in man. *Science, 148,* 1427–1432.

Aschoff, J. (1985). On the perception of time during prolonged temporal isolation. *Human Neurobiology, 4,* 41–52.

Aschoff, J. (1992). Estimates on the duration of sleep and wakefulness made in isolation. *Chronobiology International, 9,* 1–10.

Aschoff, J., von Goetz G. C., Wildgruber, C., & Wever, R. A. (1986). Meal timing in humans during isolation without time cues. *J Biol Rhythms, 1,* 151–162.

Aschoff, J., & Wever, R. (1962). Spontanperidik des Menschen bei Ausschluss aller Zeitgeber. *Naturwissenschaften, 49,* 337–342.

Aschoff, J., Wever, R., Wildgruber, C., & Wirz–Justice, J. A. (1984). Circadian control of meal timing during temporal isolation. *Naturwissenschaften, 71,* 534–535.

Boring, L. D., & Boring, E. G. (1917). Temporal judgments after sleep. In *Studies in psychology: Contributions to the E.B. Titchener commemorative volume* (pp. 225–279). Worcester, MA.: Louis N. Wilson.

Campbell, S. S. (1984). Duration and placement of sleep in a "disentrained" environment. *Psychophysiology, 21,* 106–113.

Campbell, S. S. (1986). Estimation of empty time. *Human Neurobiology, 5,* 205–207.

Campbell, S. S., Dawson, D., & Zulley, J. (1993). When the human circadian system is caught napping: evidence for endogenous rhythms close to 24 hours. *Sleep, 16,* 638–640.

Campbell, S. S., & Zulley, J. (1985). Ultradian components of human sleep/wake patterns during disentrainment. In H. Schulz & P. Lavie (Eds.), *Ultradian Rhythms in Physiology and Behavior* (pp. 234–255). Berlin: Springer–Verlag.

Campbell, S. S., & Zulley, J. (1989). Evidence for circadian influence on human slow wave sleep during daytime sleep episodes. *Psychophysiology, 26,* 580–585.

Czeisler, C. A., Weitzman, E. d., Moore, E. M., Zimmerman, J. C., & Knauer, R. S. (1981). Chronotherapy: resetting the circadian clocks of patients with delayed sleep phase insomnia. *Sleep, 4,* 1–21.

Czeisler, C. A., Weitzman, E. D., Moore-Ede, M., Zimmerman, J., & Knauer, R. (1980). Human sleep: Its duration and organization depend on its circadian phase. *Science, 210*, 1264–1267.

Kleitman, N. (1963). *Sleep and Wakefulness.* Chicago: The University of Chicago Press.

Klerman, E. B., Dijk, D. J., Czeisler, C. A., & Kronauer, R. E. (1992). Simulations using self–selected light–dark cycles from "free–running" protocols in humans results in an apparent tau significantly longer than the intrinsic. In *Third meeting of the Society for Research on Biological Rhythms,* Amelia Island, Florida.

Lavie, P. (1986). Ultrashort sleep–waking schedule. III. 'Gates' and 'forbidden zones' for sleep. *Electroencephalogr Clin Neurophysiol, 63*, 414–425.

Lavie, P., & Webb, W. B. (1975). Time estimates in a long–term time–free environment. *Am J Psychol, 88,* 177–186.

Macleod, R. B., & Roff, M. F. (1936). An experiment in temporal disorientation. *Acta Psychologica, 1*, 381–423.

Noble, W. G., & Lundie, R. E. (1974). Temporal discrimination of short intervals of dreamless sleep. *Biol Psychiatry, 8*, 253–256.

Pittendrigh, C., & Daan, S. (1974). Circadian oscillators in rodents: A systematic increase of their frequency with age. *Science, 186*, 548–550.

Pittendrigh, C. S. (1960). Circadian rhythms and the circadian organization of living systems. *Cold Spring Harbor Symposia on Quantitative Biology, 25*, 159–184.

Pittendrigh, C. S. (1981). Circadian systems: Entrainment. In J. Aschoff (Eds.), *Handbook of Behavioral Neurobiology: Vol 4 Biological Rhythms* (pp. 95–124). New York: Plenum Press.

Pittendrigh, C. S., & Daan, S. (1976). A functional analysis of circadian pacemakers in nocturnal rodents: I. The stability and lability of spontaneous frequency. *J Comp Physiol, 106*, 223–252.

Shapiro, C. M., & Moore, A. T. (1981). Circadian heat transfer. In W. P. Koella (Eds.), *Sleep 1980* (pp. 340–343). Basel: Karger.

Siffre, M. (1964). *Beyond Time* (Briffault, H., Trans.). New York: McGraw–Hill.

Skramlik, E. von. (1934). Die Angleichung der Subjektiven Zeitauffassung an Astronomische Vorgaenge: Die Physiologische Uhr (Adaptation to astronomical events of subjective time estimation: The physiological clock). *Naturwissenschaften, 22*, 98–105.

Strogatz, S. H., Kronauer, R. E., & Czeisler, C. A. (1987). Circadian pacemaker interferes with sleep onset at specific times each day: role in insomnia. *Am J Physiol, 253* , 172–178.

Tart, C. T. (1970). Waking from sleep at a preselected time. *Journal of the American Society of Psychosomatic Dentistry and Medicine, 17*, 3–16.

Vaschide, N. (1911). *Sommeil et les Reves.* Paris: Ernest Flammarion.

Vernon, J. A., & McGill, T. E. (1963). Time estimations during sensory deprivation. *Journal of General Psychology, 69*, 11–18.

Webb, W. B., & Ross, W. (1972). Estimation of the passing of four consecutive hours. *Perceptual and Motor Skills, 35,* 768–770.

Wever, R. A. (1979). *The Circadian System of Man: Results of Experiments Under Temporal Isolation.* New York: Springer–Verlag.

Zepelin, H. (1968). Self–awakening and the sleep cycle. *Psychophysiology, 4,* 370.

Zulley, J. (1980). Sleep duration, sleep stages and waking time are related to circadian phase in young and older men during nonentrained conditions. *Trans Am Neurol Assoc, 105,* 371–374.

Zulley, J., & Campbell, S. S. (1985). Napping behavior during "spontaneous internal desynchronization": sleep remains in synchrony with body temperature. *Hum Neurobiol,* 4, 123–126.

Zulley, J., Wever, R., & Aschoff, J. (1981). The dependence of onset and duration of sleep on the circadian rhythm of rectal temperature. *Pflugers Archives, 391,* 314–318.

Zung, W. W. K., & Wison, W. P. (1971). Time estimation during sleep. *Biological Psychiatry, 3,* 159–164.

Time and Behaviour: Psychological and Neurobehavioural Analyses
C.M. Bradshaw and E. Szabadi (Editors)
© 1997 Elsevier Science B.V. All rights reserved.

CHAPTER 10

How Time Flies: Functional and Neural Mechanisms of Interval Timing

Sean C. Hinton & Warren H. Meck

Introduction

This purpose of this chapter is to show how psychological, biological, and mathematical analyses may be applied to particular patterns of data produced by rats, pigeons, and humans performing variations of a time perception task known as the peak–interval (PI) procedure. First, the origins of the PI procedure and its many variations will be discussed. Second, a number of models that attempt to account for the different observations of temporal performance will be described, although the primary focus will be on scalar timing theory, a derivative of Scalar Expectancy Theory (SET; Gibbon, 1977) and the information–processing (IP) model that grew out of it (Gibbon & Church, 1984). Using this theory, a quantitative description of the patterns of data obtained using the PI procedure can be linked to different psychological constructs such as the internal clock, memory, and attention, and to some of their biological underpinnings. Finally, research using PI procedure tasks will be discussed along with what it has disclosed about the substrates of the central nervous system that are involved in timing intervals in the seconds–to–minutes range.

Researchers who study the temporal control of behavior focus on several separate, though not absolutely defined, ranges of durations. The cerebellum is thought to be involved in the control of motor movements and classical conditioning through timing intervals in the millisecond range (e.g., Ivry & Keele, 1989; Raymond, Lisberger & Mauk, 1996). At the other end of the spectrum, the rhythmic oscillations of circadian (see Moore–Ede, Sulzman & Fuller, 1982) and circannual timing (see Gwinner, 1986) ensure appropriate behavior for diurnal or seasonal changes in the environment. In between these two extremes, interval timing refers to the estimation of temporal durations in the seconds–to–minutes range. It is a capacity required to determine rates of stimulus occurrence and to anticipate predictable events in an organism's near future.

History

Scalar timing theory and the information–processing model

An IP model of interval timing developed by Gibbon and Church (1984) allows a detailed analysis of the level at which various timing effects operate. The model, diagrammed in Figure 1, is hierarchically organized into three levels: a clock stage, a memory stage, and a decision stage. The clock stage consists of a pacemaker that transmits the pulses by which time is measured, a mode switch sensitive to attention that controls when and how the pacemaker pulses are gated, and an accumulator where the total number of pulses received is stored. The pulses in the accumulator represent the amount of time that has passed since the switch was closed and timing began. The accumulated pulses can be temporarily placed into working memory when timing must be briefly halted, and the accumulator value can then be copied into reference memory for later comparison. The comparator applies a decision rule to the value in the accumulator relative to the one in reference memory to determine when to respond. Changes in clock speed are produced by dopaminergic drugs (e.g., Maricq &

Church, 1983), attention may be manipulated by administering noradrenergic drugs to affect the operation of the switch (e.g., Penney, Holder & Meck, 1996), changes in the rate at which information is passed from the accumulator to reference memory are sensitive to cholinergic drugs (e.g., Meck & Church, 1987a), and strategic factors, such as how stringently the subject chooses to set response thresholds, can influence the decision process. The two hallmarks of scalar timing theory are the superimposition principle and the constant coefficient of variation (CV). Superimposition describes the property that when timing functions are plotted on relative axes, they will superimpose. The constant CV is a mathematical correlate of the superimposition principle which states that the standard deviation of a function increases in proportion to its mean value.

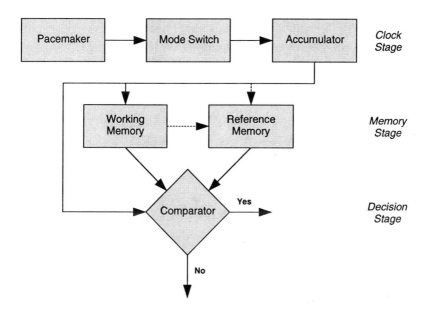

Figure 1. An information–processing model of timing after Gibbon, Church & Meck (1984). The clock stage consists of the pacemaker generating a continuous stream of pulses that are gated by the operation of the switch into the accumulator. The memory stage is composed of working memory for trial–specific information and reference memory for long–term storage. Comparison of the clock reading for the current trial in the accumulator with a sample from the set of representations of the previously learned durations determines when the subject responds.

Various investigators have developed working hypotheses about physiological substrates that may correspond to components of the IP model. The pacemaker cells of the substantia nigra pars compacta are thought to function as the internal clock (Meck, 1996b). These cells have major projections to the striatum, which is considered to serve as the accumulator (Meck, 1996b), and noradrenergic cells in the locus ceruleus modulating this projection could act as the switch by subserving selective attention (Penney et al., 1996). Animal experiments suggest that the hippocampus is the site of working memory (Meck, Church & Olton, 1984) and that divided attention and access to reference memory are properties modified by activity of the frontal cortex (Meck, Church, Wenk & Olton, 1987). Presumably, the executive functions of the frontal cortex also are involved in the decision stage represented by the comparator. Because the main symptoms of Parkinson's disease are due to degeneration of the substantia nigra pars compacta, patients with this disorder are a useful clinical population for testing some of the ideas of the IP model (Malapani et al., 1993).

Optimal foraging theory

According to utility theory as embodied in some optimal foraging models, an animal must be able to estimate the amount of time spent pursuing prey in a particular patch (e.g., Brunner, Kacelnik & Gibbon, 1992; Caraco, 1980) in order to determine a rate of reinforcement. This information, when compared to some representation of previously obtained rates of reinforcement stored in memory, can be used by the animal to assess when it becomes profitable to move to another patch. Human beings also benefit from the ability to time short intervals, such as when estimating how long it will take to cross the street or how soon oncoming traffic will arrive. These examples illustrate the inherent survival value of having an accurate internal clock for estimating the passage of time.

Traditional approaches to studying human time perception

There are three traditional classifications for timing tasks performed by humans: estimation, production, and reproduction (Bindra & Waksberg, 1956). The first two procedures tend to rely on verbal instructions or responses and require the subject to translate between performance and a verbal representation of a duration in seconds. For example, an estimation task might have an experimenter hold up her hand for a period of time and, upon lowering it, ask the subject how long her hand was up, requiring the subject to give an estimate in seconds. In a production task, the experimenter would instruct the subject to hold up his hand for a specified period of time in seconds. The confound exposed by these two tasks is that if the subject's verbal representation of times is distorted, so will be his verbal report, though this may have nothing to do with his intrinsic sense of time. A more reliable procedure for getting at time perception without having to rely on verbal translation is to ask the subject to reproduce an interval. For example, the experimenter might hold up her hand for a specific duration and then have the subject attempt to reproduce that interval by holding up his hand for the same period of time. Such a procedure ensures that any distortions of a subject's verbal representations don't influence the interpretation of their ability to perceive time. All three methods for examining time perception are limited in their usefulness because they don't allow a detailed analysis of how the brain's time perceptual system operates. In the following section, we will introduce some of the procedures used to study animal timing and argue for their usefulness when applied to human timing as well.

Origins and variants of the peak–interval timing procedure

A number of behavioral paradigms have been developed for assessing different aspects of subjects' ability to time short intervals. The methods in this chapter are all based on the PI procedure, which has at least three strengths: 1) memory is assessed with respect to its content (quantitatively) rather than its clarity (qualitatively); 2) timing effects are separable from other performance effects, such as might be due to motivational differences; and 3), analytical tools are available to associate aspects of timing performance with concepts like the internal clock, memory, and attention. What follows is some information about the PI procedure's antecedent, the fixed–interval schedule, as well as descriptions of how the PI procedure and its permutations are performed and the kinds of questions they were meant to assess.

Fixed–interval schedules

The PI procedure is derived from fixed–interval (FI) operant conditioning schedules in which reinforcement, usually the delivery of a small amount of food, is dependent upon a response made after some fixed period of time has passed. FI schedules can either be free–operant procedures, in which the inter–food interval serves as the only explicit discriminative stimulus and the interval requirement is reset after each reinforcement is obtained, or discrete–trial procedures, which are the basis for the PI procedure. In discrete–trial FI procedures, onset of the trial is signaled by the presentation of a stimulus that remains on continuously. The subject's first response after the FI duration has elapsed is followed by reinforcement, the signal is terminated, and an inter–trial interval (ITI) ensues in which the signal is absent. The signal may be of any modality, though it is important that it be salient to the subject. Typically, the presence of cue lights or the house light in the operant chamber are used as visual stimuli, while pure tones or white noise presented through a speaker may serve as auditory stimuli (e.g., Roberts & Church, 1978). Occasionally, studies have used tactile stimuli, for example, by passing a mild electric current through steel bars serving as the floor of the chamber (Meck, 1987).

One simple performance measure that can be derived from FI procedures is the post–reinforcement pause, defined as the amount of time the subject waits after signal onset before beginning to respond. Subjects will typically begin responding after a fixed proportion of the interval has passed, averaging about two–thirds of the FI value in well–trained subjects (Schneider, 1969). The post–reinforcement pause is a measure of the subject's ability to inhibit responding, and the later it occurs in the trial, the more precise the subject's temporal discrimination is said to be. The importance of the observation that the post–reinforcement pause is proportional to the FI duration is the implication that subjects have a threshold to begin responding and that this threshold is sensitive to relative rather than absolute time.

When responding on an FI schedule is either averaged across trials or measured using a cumulative recorder, for which each response increments the height of a line drawn horizontally across a drum with time, the resulting function sometimes appears as an "FI scallop", a curve gradually accelerating to a maximum at the time of reinforcement. This seemingly steady increase in responding develops a different form when the schedule is well learned. Especially in discrete–trial procedures, responding may be closely approximated by a horizontal line of low or no responding and an angled straight line corresponding to a constant high (Ferster & Skinner, 1957; Schneider, 1969). This means that on each trial, the subject's responding transitions abruptly from a low–responding state to a high–responding state, rather than accelerating smoothly to a high rate. The means and standard deviations of the time of the break point are proportional to the FI criterion (Gibbon, 1977), a property of scalar timing theory to be discussed later.

The peak–interval procedure

One of the problems with FI procedures is that because they only collect responses that are made before the criterion time, detecting changes in the animal's estimate of duration is difficult. This limitation is remedied by the PI procedure (Figure 2) which collects responses both before and after the criterion time. The task is derived from a discrete–trial FI procedure in which a subject is trained using a particular FI value, but during testing some proportion of trials are unreinforced and the signal continues long past the FI that it has learned (Catania, 1970; Roberts, 1981). Such "peak" trials allow the experimenter to assess within a single trial the subject's estimate of when reinforcement would have occurred. Responses summed and binned over many peak trials appear approximately normally distributed with some slight positive skew and are referred to as "peak functions" because of their characteristic Gaussian shape when the subject shows evidence of temporal discrimination (Figure 3). The mean response rate of a peak function starts low near the time of signal onset, increases to a maximum (*peak rate*) at about the time of the FI for which the animal has been reinforced, then gradually decreases to a low rate again in a fairly symmetrical fashion. The mode of the response distribution (*peak time*) is taken to represent the subject's remembered time of reinforcement, and the width (*spread*) of the peak function is a measure of the precision of temporal discrimination. The symmetry of peak functions on a linear time scale is taken as evidence that subjective time is also linear (Church, Miller, Meck & Gibbon, 1991).

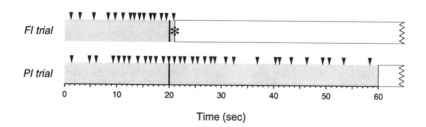

Figure 2. The peak–interval timing procedure consists of a mixture of fixed–interval (FI) trials in which the animal is reinforced (asterisk) for its first response (inverted triangles) after the criterion time (heavy vertical line) has elapsed since the onset of the signal (gray rectangle) and peak–interval (PI) trials in which the signal remains on for much longer than the criterion time and reinforcement is not made available. Both types of trials are followed by an inter–trial interval (open rectangle) during which the signal is absent.

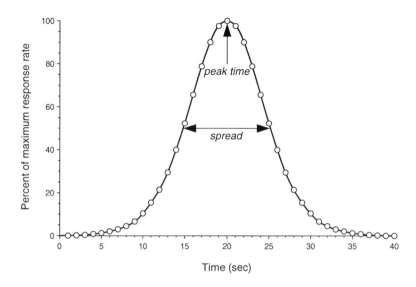

Figure 3. Idealized peak–interval timing function illustrating *peak time* (the maximum height of the function taken as a measure of timing accuracy, indicated by the vertical arrow) and *spread* (the width of the timing function at half the maximum height used as a measure of timing precision, indicated by the horizontal double-headed arrow).

The peak–interval gap procedure

A modification of the PI procedure introduces a retention interval, also called a "blackout" (Roberts, 1981) or "gap", during the presentation of the signal on peak trials to add a working memory component to the task. Consequently, the PI–GAP procedure allows for a test of working memory for signal duration. Working memory is a hypothetical cognitive construct conceived of as a temporary store for trial–specific information that needs to be retained only for a short period of time (Honig, 1978). During gap trials, a subject will typically record a clock reading how much time has elapsed when the signal goes off, store that quantity until the signal comes on again, and then continue timing the signal and respond at the appropriate time as though the signal had been continuous (Roberts, 1981). As in the standard PI procedure, the subject first learns the duration to be timed with simple FI trials, and then occasional peak trials are introduced. A testing session of the gap procedure then consists of a mixture of FI trials, peak trials, and peak trials with the gap inserted.

A rat with a working memory impairment will have difficulty holding in memory the amount of time that elapsed before the retention interval. Typically, a subject with such a deficit will completely ignore the part of the signal that occurs before the gap and will time the signal after the gap as though it were a new trial. The rat is said to "reset" its internal clock during the gap and start timing anew when the signal is restored (Meck et al., 1984). A normal rat is able to "stop" its internal clock during the gap, temporarily store the elapsed time in working memory, and start timing again when the signal resumes. Yet another alternative is for the subject to ignore the gap and "run" its internal clock throughout the retention interval (Meck, Church & Olton, 1984; Roberts & Church, 1978). Research on pigeon timing has found intermediate outcomes between these pure modes of operation (Cabeza de Vaca, Brown & Hemmes, 1994; Roberts, Cheng & Cohen, 1989), which the authors have interpreted as partial resetting of the internal clock or partial forgetting during the retention interval. Many studies, including those using other procedures such as spatial tasks, have demonstrated working memory deficits in rats with hippocampal lesions (e.g., Meck et al., 1984; Olton, Becker & Handelmann, 1979), suggesting that the hippocampus is the brain structure responsible for working memory.

The prior–entry method

The prior–entry (PE) method (Meck, 1984; Penney et al., 1996) is a form of the PI procedure in which selective attention is behaviorally manipulated. In this procedure, the subject is first trained with FI trials either of the same or different durations but presented in distinct modalities, such as a tone and a light. Then the subject experiences a series of PI procedure sessions with randomly presented FI and peak trials in either modality. In the final training phase, a brief warning cue of the same modality as the upcoming signal precedes every trial. The cue focuses the subject's attention and prepares it to respond to the signal of that modality. In such a standard PE trial, the latency to begin timing will typically decrease because the subject is prepared to expect a signal of that modality. If the warning cue is of the other modality, the subject's latency to start its internal clock is often increased because it requires some time to switch attention to the modality actually presented during the trial. This presentation of a timing signal in a modality that contradicts the subject's expectation generated by the cue is known as the prior–entry–reversal (PER) method.

Simultaneous temporal processing

The PI procedure may also be presented in a more complicated version known as simultaneous temporal processing (STP) in order to assess a subject's ability to divide attention between multiple signals that are presented concurrently and asynchronously (e.g., Olton, Wenk, Church & Meck, 1988). The subject is first exposed to FI trials in which each duration to be timed is associated with a distinct modality and then is trained with PI procedure sessions where the duration and therefore modality for each trial is randomly selected. Finally, in an STP session, the different stimuli are presented concurrently in mixtures of FI and peak trials. Concurrent FI trials and concurrent peak trials may be presented within a trial, or a trial might consist of an FI trial of one modality and a peak trial of another. Stimuli are typically presented with asynchronous onsets. Normal pigeons or rats can be trained in this fashion to time simultaneously up to three different durations presented in as many as three different modalities (Leak & Gibbon, 1995; Meck, 1987). In contrast, a subject that has difficulty dividing attention may either focus on only one of the signals and ignore any others present or attend to only one of the signals at a time (Olton et al., 1988).

Timing theories

Several theories have been developed to try to account for the various phenomena identified in timing experiments. The behavioral theory of timing (BeT; Killeen & Fetterman, 1988), for example, posits an internal pacemaker generating pulses according to a Poisson process. These pulses cause the animal to experience different states that correlate with adjunctive behaviors. When reinforcement is made available during particular adjunctive behaviors, they become discriminative stimuli that the animal learns to rely upon to produce the temporal discrimination. An interesting corollary of BeT is that reinforcement rate is what drives clock rate, so that higher reinforcement rates cause higher clock rates, although a recent study (Bizo & White, 1995) has refined this notion by arguing that the rate of the pacemaker varies inversely with the rate of reinforcement unrelated to the timing task.

    An entirely different approach is taken by Church and Broadbent's (1992) connectionist model which uses an internal clock composed of multiple oscillators having different periods. Each point in time is therefore associated with a unique state of the oscillators taken together. The oscillators begin cycling at their characteristic frequency at trial onset, and the time when reinforcement occurs is stored in an autoassociation matrix serving as reference memory. The time of reinforcement is represented as a multidimensional vector of status indicators recording the phase of each of the oscillators. The vector is combined linearly with the existing contents of the autoassociation matrix by modifying a series of weights that represents the others previously acquired. A weighted average vector can then be retrieved from reference memory for comparison with the vector representing the current time in order for the animal to decide when to respond.

    This neural network model was adapted from an IP model of interval timing developed by Gibbon and Church (1984) that will serve as the theoretical touchstone for this chapter. The IP model is hierarchically organized into clock, memory, and decision stages and allows a detailed analysis of the level at which various effects on timing may operate (see Figure 1). The clock stage consists of a pacemaker that transmits the pulses by which time is measured (as in BeT, the pacemaker is modeled with Poisson variability), a mode switch sensitive to attention that controls when and how the pacemaker pulses are gated, and an accumulator where the pulses are received. The accumulated pulses can be transmitted to working memory, which would be required, for example, by the PI–GAP procedure. When a response is reinforced, the value in the accumulator is copied into reference memory, adding to a distribution of values associated with that criterion that have been previously stored. The number of pulses in the accumulator is linearly scaled into reference memory, and the rate at which this occurs (the memory storage speed) may be biased in either direction. The comparator determines when to respond by continuously applying a similarity rule to compare the time for the current trial (the value in the accumulator) relative to one sampled from the distribution of previously reinforced times stored in reference memory.

A trial–by–trial analysis of performance may be accomplished by fitting data from a single trial of the PI procedure with a step function (Figure 4) using a least–squares fitting algorithm to determine a high state of responding flanked by two states of low responding (Cheng & Westwood, 1993; Church, Meck & Gibbon, 1994). One may then derive parameters from this fit, such as the break point into the high state ($S1$), the breakpoint out of the high state ($S2$), the duration of the high state (*spread*), and the midpoint of the high state (*peak time*). A threshold used by the comparator with a similarity rule determines when the subject decides to start and stop responding, with a more stringent threshold producing a smaller *spread* and therefore greater temporal precision (e.g., Cheng, 1992). The *peak time* may be either close to or far from the criterion time on any given trial and serves as an index of temporal accuracy.

The IP model grew out of a quantitative theory of timing behavior called Scalar Expectancy Theory (SET; Gibbon, 1977), also known as scalar timing theory (Gibbon, Church & Meck, 1984), which addresses the different sources of variance in an interval timing task. The theory allows specific predictions to be made about correlations between the trial parameters previously described, and the pattern of correlations can be used to make inferences about the forms and the relative contributions of memory or threshold variance to overall performance variability (Gibbon & Church, 1984).

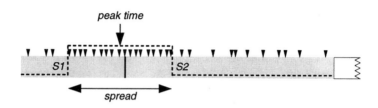

Figure 4. Individual trials may be fit by a step function in which the height of the function represents the response rate for that portion of the trial (dashed line). The following parameters may be derived from that least–squares fitted function: the break point into the high state is *S1*, the breakpoint back down to the low state is *S2*, the difference between these is the *spread*, and the midpoint of *S1* and *S2* is the *peak time*.

A series of trials with a lot of variability in the temporal placement of the *peak time* from trial to trial but with relatively little variability in the *spread* implies a preponderance of variability in the sample from reference memory with little contribution of threshold variability (Figure 5, left panel). The response distribution would be positively skewed because memory variability scales proportionally with increasing *peak time*. The resulting correlation pattern would be a moderate positive correlation between *S1* and *S2*, a moderate positive correlation between *S1* and *spread*, and a moderate positive correlation between *spread* and *peak time*. In contrast, if the *peak time* in a series of trials were centered reliably but the *spread* varied considerably from trial to trial, this would suggest a large contribution of threshold variability with little memory variability (Figure 5, right panel). The response distribution would be symmetrical because threshold variability is symmetrically distributed around one *peak time*. The corresponding pattern of correlations for this situation would be a highly negative *S1–S2* correlation, a highly negative *S1–spread* correlation, and a *spread–peak time* correlation near zero (Gibbon &

Church, 1990). In principle, the patterns of correlation can distinguish between variability produced by one threshold used to decide both when to start and stop responding versus a separate threshold for each (Cheng, 1992), and one versus multiple memory samples (Cheng, Westwood & Crystal, 1993). In practice, rats and pigeons typically show a mixture of both memory and threshold variability.

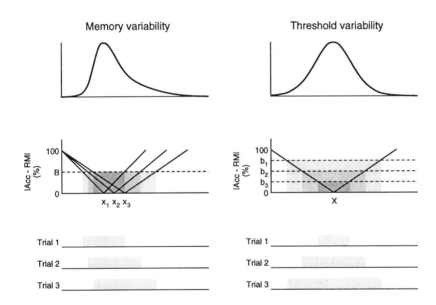

Figure 5. Schematic representation of memory and threshold variance. Subjective time (angled lines) accumulates linearly in real time (along the abscissa), and the state of high responding begins when the discrepancy between the value sampled from reference memory and the value in the accumulator crosses a lower threshold (dashed horizontal lines, plotted on a relative scale along the ordinate) and continues until it crosses the same upper threshold. When there is variability in the sample taken from reference memory (x_1, x_2, x_3) without variability in the threshold (B), the response distribution is positively skewed because of the scalar property. When there is no variability in the sample from reference memory (X) but variability in the threshold from trial to trial (b_1, b_2, b_3), the peak function is symmetrically distributed around the sampled value (adapted with permission from Gibbon & Church, 1990).

Peak functions generally obey the scalar property, a psychophysical principle of SET analogous to Weber's law which

S.C. Hinton & W.H. Meck

states that the variability around some psychophysical measure is proportional to its magnitude. When plotted on an absolute time scale (Figure 6, top panel), the mode of a peak function will usually occur near the FI value presented during training. One may assess the scalar property by plotting peak functions from different FI values on a relative time scale and as a proportion of the maximum level of responding, which should result in their superimposing (Figure 6, bottom panel). In other words, the scalar property is confirmed graphically when the width of a height–normalized peak function at half its maximum height is proportional to the mode of the same distribution obtained for a different time value. A mathematical equivalent of the scalar property is that the standard deviation of a distribution divided by its mean (the coefficient of variation) remains constant across signal durations.

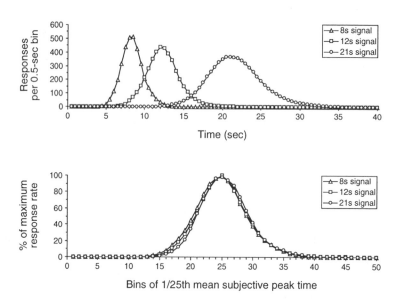

Figure 6. Pooled peak functions for seven human subjects trained with visual signals of 8, 12, or 21 sec on different blocks of trials. The upper panel shows absolute rate functions, while the lower panel shows the same functions plotted in bins proportional to the subjective peak time for each function. The lower ordinate has been scaled from minimum to maximum.

Temporal patterns reveal functional mechanisms of interval timing

One of the primary virtues of the PI procedure is that the patterns of data it produces can allow differentiation among the IP levels where a manipulation, such as a drug or lesion, affects the time perceptual system. Some of these patterns and their interpretation are discussed below.

The hedonic pattern

According to SET, hedonic value does not directly affect time perception (Gibbon, 1977). The total amount of expectancy or "hope" varies with deprivation and the amount and probability of reinforcement. Expectancy is modeled as a motivational parameter, H distributed over the value being timed, x, so that at the beginning of any particular trial, expectancy $h(0) = H/x$. Expectancy of reinforcement increases over the course of a trial in a hyperbolic manner $\{H/(x - t)\}$, because the expected time to reinforcement is linearly decreasing as $x - t$. Because responding depends on a ratio comparison between the local and overall expectancy and H enters into both values, it drops out of the equation, leaving $1/(1 - (t/x)) > b$, where b is the threshold used to determine when to start responding. Thus, the time to begin responding depends only on the threshold chosen, and the same is true of the time to end responding; expectancy determines only the strength of responding.

This theoretical account has set out to show that *peak time* and *peak rate* do not interact, according to the assumptions of SET. There is also experimental evidence that supports this view. A widely used measure of reinforcement value in operant procedures is response rate. A food–deprived subject will respond more frequently for food reward than a sated one, and this observation has been interpreted as the food–deprived subject being more motivated to perform the task. A perennial concern of experimental psychologists is whether observed changes due to a particular experimental manipulation indicate true cognitive or mere performance effects. This issue has been addressed in the PI procedure timing literature by a series of interval timing experiments in rats showing independence of *peak time* and *peak rate* (Roberts, 1981).

The first experiment of this study consisted of two parts to double–dissociate *peak time* from *peak rate*. The first part held the probability of reinforcement fixed at 80% with 20% peak trials for 20– and 40–sec signals counterbalanced across rats as either auditory or visual stimuli. The resulting *peak time* for each signal was different (about 20 or 40 sec), but the *peak rates* were identical. The second part used a 20–sec signal in each modality. For one group of rats, the visual signal was reinforced on 80% of the trials and the auditory signal was reinforced on 20% of the trials, and the assignment of reinforcement rates to signal modality was reversed for the other group. The *peak times* were equivalent at around 20 sec, but the *peak rate* was about four times higher for the 80% reinforcement group. This second part of the experiment showed that *peak rate* could be affected without disturbing *peak time*, thereby completing the double–dissociation.

Finding no difference in *peak rate* in the first part of the previous experiment is somewhat surprising given that each rat was exposed to both the 20– and 40–sec signals within a session. Ordinarily, one would expect a contrast effect to appear in such a within–subjects design, and responding would be lower for the longer signal duration because the effective rate of reinforcement is lower. However, a later experiment in the same paper using the same animals (Roberts, 1981) does appear more like the usual finding and shows some indication of a behavioral contrast effect.

Yet another experiment from this series examined the effect on *peak time* and *peak rate* of reducing the hedonic value of the reinforcement by pre–feeding the rats. On baseline days, the ten rats were run normally on a PI procedure in which they received occasional reinforcement on an FI–40 sec schedule. On pre–fed days, they received half their daily ration of food forty minutes before being tested on the same schedule. Although pre–feeding caused *peak time* to be somewhat later than under the baseline condition (a finding reminiscent of the effects of nutrients on clock speed, to be discussed below), the more relevant result for the current purpose was that pre–feeding reduced *peak rate* by about half. Like reducing the overall reinforcement rate, then, decreasing the hedonic value of the reinforcement causes a reduction in the amplitude of the peak function without having much effect on its horizontal placement.

A recent physiological study addressed one of the neural substrates for hedonic effects (Meck, 1996b). The experiment included one group of rats that had their nucleus accumbens (NAS) lesioned and another group of rats that received a control lesion, and both groups of subjects had been previously trained on the PI procedure with random presentation of 10– or 60–sec signals within a session. After the surgery, the control rats still showed contrast in the response rates for the two signals, whereas the maximal response rates for the 10– and 60–sec functions were identical for the NAS group (Figure 7). The inability of the NAS group to show contrasting response rates for the two signals suggests that the NAS is involved in assessing the hedonic value of a stimulus and therefore is at least one of the neural substrates that when damaged can produce the hedonic pattern.

Together, these experiments suggest that differences in response rate can be attributed to differences in reinforcement rate or hedonic value and can be largely separated from changes in peak time. In addition, these hedonic effects are associated with the normal functioning of the nucleus accumbens and, most importantly, can be dissociated from effects on other aspects of the IP model, such as the clock, memory, and attention.

The clock pattern

The clock pattern is represented by phasic proportional changes in the mean and variability of peak functions (Figure 8) and is usually obtained by administration of dopaminergic (DA) drugs. The subject is first trained to associate the criterion time with food delivery using an FI schedule. Testing then consists of a mixture of these training trials and unreinforced peak trials. When a DA drug is administered after training, an abrupt shift in the peak time is observed to a shorter (in the case of an agonist) or a longer (in the case of an antagonist) time value. If drug administration is continued over many days, the subject will gradually adjust to the new clock speed by changing the distribution of the values stored in reference memory. If the drug is then discontinued, an abrupt change in the opposite direction is observed followed by a gradual return to baseline as the subject readjusts to what was the original clock speed.

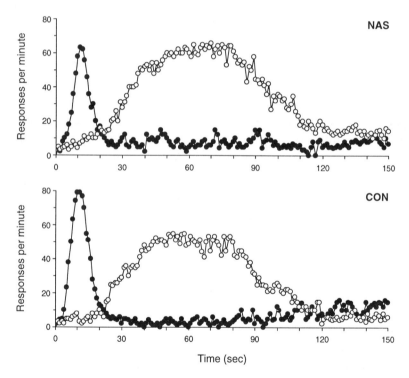

Figure 7. *The hedonic pattern*: Mean response rate (responses/min) as a function of signal duration (sec) for rats trained postoperatively on the peak procedure with both 10–sec (open circles) and 60–sec (filled circles) trial types presented in random order during the same sessions. Data for the nucleus accumbens (NAS) lesion group are shown in the top panel, and the data for the control (CON) lesion group are shown in the bottom panel (reprinted with permission from Meck, 1996b).

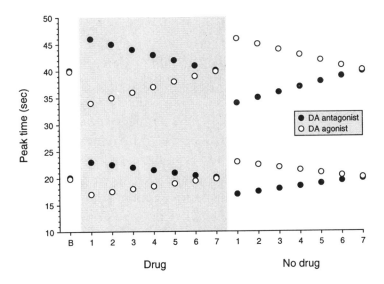

Figure 8. *The clock pattern.* Dopaminergic (DA) drugs produce an immediate decrease (agonists) or increase (antagonists) in peak time from baseline performance (B) that is proportional to the 20– or 40–sec signal durations. Over repeated drug administration days, the peak time gradually normalizes as the old values in reference memory are replaced by new ones appropriate to the altered clock speed. Discontinuation of the drug results in a rebound effect in the opposite direction, as the new clock speed without the drug produces an inaccurate time value. Repeated sessions without the drug allow renormalization of the time values stored in reference memory as they are replaced with veridical ones.

From the psychological perspective, the primary effect of DA agonists such as methamphetamine on the time perception system is to increase clock speed (Maricq, Roberts & Church, 1981), while DA antagonists such as haloperidol slow the clock down (Meck, 1996c). However, such changes in clock speed may only be observed behaviorally if the drugs are administered after training. If the drug is administered throughout training and testing, then the altered clock speed will be used both for learning and terminal performance, and no differences will be seen whether clock speed is increased or decreased. One minor exception to this principle is that a faster clock might allow for greater temporal precision (if variability were not also increased by the drug) while accuracy of timing should be unaffected.

Physiologically, administration of DA drugs has been shown to affect the output of the pacemaker cells of the substantia nigra pars compacta which project to the striatum (Mercuri, Calabresi & Bernardi, 1992) and are thought to represent the pacemaker in the IP model. Probably one of the primary mechanisms of action for DA agonists is to lower the threshold of stimulation in the striatum (for an example from the brain–stimulation reward literature, see Sarkar & Kornetsky, 1995). The net response is an increase in sensitivity to the DA input which corresponds to a faster clock speed. In contrast, DA antagonists (which decrease sensitivity to brain–stimulation reward) presumably decrease DA sensitivity in the striatum, which is manifested as a slower clock speed.

One of the earliest papers to demonstrate a clock effect showed a proportional decrease in *peak time* for 20– and 40–sec signals after methamphetamine administration in rats (Maricq et al., 1981), implying an increase in clock speed. The effect of the drug was immediate and decreased after repeated administrations. The authors argued that the decrease in the effectiveness of methamphetamine was not due merely to the development of tolerance to the drug but rather due to the rats' learning to perform the PI procedure while clock speed was altered by the drug. Methamphetamine also increased the overall response rate of the subjects without affecting their temporal discrimination, which is more evidence for the separability of timing from hedonic effects.

The rat forebrain has been shown to be involved in regulation of clock speed by a recent study examining the interaction between DA drugs and acetylcholinergic (ACh) lesions on timing behavior (Meck, 1996a). Five different groups of rats trained on the PI procedure task were given either control (CON), frontal cortex (FC), nucleus basalis magnocellularis (NBM: the main ACh input to the FC located in the basal forebrain), medial septal area (MSA: the major source of ACh input to the hippocampus, also in the basal forebrain), or fimbria-fornix (FF: the major output pathway of the hippocampus) lesions.

These groups were then exposed to challenge days where they were administered the dopamine antagonists haloperidol or sulfated cholecystokinin octapeptide (CCK–8S) in randomized order, with challenge days separated by six non–drug testing sessions to allow timing performance to return to baseline. In a later phase, the groups were challenged with dopamine agonists or behaviorally. Methamphetamine was given at doses known to produce a 10–20% horizontal shift in peak functions, or the rats experienced a "surprise" testing session that occurred during what would normally be a non-testing day on the subjects' alternating–day testing schedule. This behavioral manipulation was predicted to increase the subjects' arousal level and produced about a 10% leftward horizontal shift of the timing functions in control subjects, much as would a low dose of the dopamine agonist methamphetamine. Again, challenge days were presented in randomized order and were separated by six non–drug testing sessions. FF and MSA groups were no different in their response to the drug and behavioral challenges from the CON subjects. The methamphetamine and "surprise" treatments produced about 10–20% leftward horizontal shifts (a decrease in peak time and an increase in clock speed), while the haloperidol and CCK–8S treatments produced about 10–25% rightward horizontal shifts (an increase in *peak time* and a decrease in clock speed). However, neither the FC nor the NBM groups showed such a substantial response to any of the drug or behavioral challenges. Therefore it seems that frontal damage, such as may be caused either directly by a lesion or indirectly by reducing ACh input to the frontal cortex from the NBM, may abolish the sensitivity of the internal clock to the influence of either DA drugs or "surprise" that would ordinarily affect it profoundly.

Analogously to drug effects, clock speed may also be manipulated by supplying particular nutrients before a timing session. Snacks supplemented with casein (a protein found in milk and cheese) and given to rats twenty minutes prior to being tested on a PI procedure with an FI–20 sec schedule resulted in abrupt changes in the remembered time of reinforcement characteristic of clock effects (Meck & Church, 1987b). These protein–treated subjects exhibited an immediate leftward shift in their *peak time* that gradually returned to baseline after continued training and that rebounded to the right when protein administration was discontinued. This effect is qualitatively the same as that seen when subjects are treated with DA agonists, such as methamphetamine, which cause an increase in clock speed. The mechanism the authors use to account for this finding is that protein, which is high in the amino acid tyrosine, is metabolized and transported into the brain where it is then converted to DOPA by the enzyme tyrosine hydroxylase and thence to dopamine by DOPA decarboxylase (Cooper, Bloom & Roth, 1982). In contrast, sucrose snacks (a simple carbohydrate) caused an immediate rightward shift in *peak time* that normalized after repeated administration and rebounded to the left upon discontinuation. The explanation given by the authors for this observation is that sucrose causes an increase in secretion of insulin that preferentially concentrates tryptophan in the brain relative to other amino acids. Tryptophan is converted to serotonin (5–hydroxytryptamine), which may affect clock speed in a manner opposite to DA agonists and slows it down, as would a DA antagonist like haloperidol.

A set of lesion studies has focused specifically on the brain areas thought to represent the components of the clock stage (Meck, 1996b) and has demonstrated the important role played by the substantia nigra in determining clock effects. As mentioned earlier, neurons of the substantia nigra pars compacta (SNC) are thought to be the pacemaker cells used in interval timing. These neurons send DA projections to the striatum, also known as the caudate–putamen (CP), which is thought to serve as the accumulator by integrating these action potentials to provide a continuous sense of the passage of time. In this study, a group of rats was first trained to perform on a PI procedure with an FI–20 sec schedule. They were then divided into three groups and given either SNC, CP, or control lesions. The control subjects continued to accurately produce the 20–sec duration, but neither the SNC– nor the CP–lesioned groups showed a peak function. Rather, both groups showed a continuous moderate level of responding throughout the trial (their "peak functions" were flat) and therefore showed no evidence of temporal discrimination. All three groups were then given injections of L–DOPA, the metabolic precursor of dopamine. The control group showed a classic clock effect: an immediate leftward shift in the placement of the *peak time* because the increased levels of dopamine speed up the internal clock. The group with the CP lesion still exhibited a flat "peak function", although the average response rate in the trial was higher than it had been before the drug administration. Finally, L–DOPA actually restored temporal discrimination to the SNC–lesioned rats. Their peak functions were not as precise as they had been presurgically, their maximal response rate was not as high, and they did not show the leftward shift in *peak time* that the CON subjects did, but they definitely were able to show evidence of temporal discrimination. CP lesions presumably destroy the accumulator, and the SNC lesions damage the pacemaker, but some SNC neurons must survive the lesion so that the L–DOPA supplementation allows them to act effectively again.

A variety of converging studies, then, have provided evidence behaviorally, pharmacologically, and physiologically of clock speed effects. A clock speed manipulation produces an immediate proportional change in the *peak time*. This phenomenon depends on the integrity of the SNC, CP, and FC, and it can be evoked by increasing or decreasing the level of dopamine in the brain.

The reference memory pattern

In contrast to the rapid changes seen with manipulations affecting the pacemaker–accumulator system (the clock stage), the reference memory pattern is characterized by gradual proportional changes in the mean and variability of peak functions. The distribution of values associated with a time interval already stored in reference memory must be overwritten with new ones based on an altered memory storage speed. Changes in a multiplicative translation constant that modifies the value transferred from the clock stage to the memory stage can be obtained by administering acetylcholinergic (ACh) drugs (Figure 9). Functional agonists such as physostigmine (an acetylcholine esterase inhibitor) cause a gradual decrease in *peak time*, and antagonists like atropine produce a gradual increase in *peak time* (e.g., Meck & Church, 1987a). If the reference memory pattern is produced by drug administration, then discontinuing the drug causes a gradual readjustment back to the original *peak time*. A physiological approach for generating the reference memory pattern is to lesion either the FC or its ACh afferents, which produces effects that resemble those of ACh antagonists. This reference memory pattern is interpreted as a bias in the baud rate for transferring pulses from the clock stage to the memory stage, such that accumulator values are incorrectly scaled when they are transferred to reference memory. The underlying assumption of this storage mechanism is that the memory stage represents the duration of a signal according to the amount of time required to transfer the pulse accumulation from the clock stage to the memory stage. Consequently, an increase in memory storage speed would proportionally decrease the value of time represented in reference memory and therefore corresponds to a leftward shift in *peak time*, whereas a decrease in memory storage speed proportionally increases the value stored in reference memory and produces a rightward shift in *peak time*.

A rather indirect route to producing the reference memory pattern has been to treat subjects with pyrithiamine, a thiamine antagonist (Meck & Angell, 1992). Thiamine is required by pyruvate dehydrogenase, an enzyme that allows synthesis of a necessary ACh precursor. Two groups of ten rats each were trained on the PI procedure, one with a 20–sec and one with a 50–sec criterion. After their performance on the task stabilized, half the rats in each group were injected with pyrithiamine and the other half were injected with saline prior to every session for six sessions. The mean *peak times* for the pyrithiamine–treated subjects gradually increased over sessions and stabilized at a maximum value about 15% greater than those of the saline–treated subjects. One subsequent drug–free test day did not show any change in the mean *peak time* for the pyrithiamine–treated rats, thus confirming the reference memory pattern. The pyrithiamine prevented synthesis of new ACh in the brain and therefore reduced its effective level considerably. Memory storage speed was then decreased accordingly, and the new longer time values stored in reference memory gradually replaced those established during baseline training. The longer values were maintained through one test day without any drug injections, as is to be expected for the reference memory pattern.

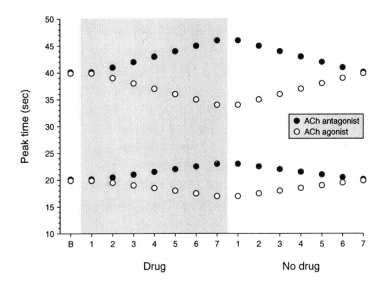

Figure 9. *The reference memory pattern.* Cholinergic (ACh) drugs produce a gradual decrease (agonists) or increase (antagonists) in peak time from baseline performance (B) that is proportional to the 20– or 40–sec signal durations. Over repeated drug administration days, the peak time gradually becomes more extreme as the old values in reference memory are replaced by new ones appropriate to the altered memory storage speed. Discontinuation of the drug results in a gradual return to the baseline peak time as the altered time values are replaced by veridical ones.

In the study of the effects of nutrients on interval timing mentioned previously in the context of the clock pattern (Meck & Church, 1987b), a reference memory effect was also observed. Administration of a snack containing lecithin to rats prior to testing them on the PI procedure produced a gradual leftward shift in their peak function which was maintained for two sessions after the lecithin snack was discontinued. This pattern of change corresponds to that seen with ACh agonists and is consistent with a mechanism in which choline contained in the lecithin is converted in the brain to acetylcholine and thereby increases memory storage speed, an effect opposite to that seen with pyrithiamine administration.

Lesion experiments offer another way to expose the reference memory pattern and also make it possible to identify the brain structures involved in temporal memory. Aspiration lesions of FC in trained rats produce rightward displacements of about 20% in the PI procedure using 10– and 20–sec FI signal values (Meck et al., 1987). In the same study, chemical lesions of the NBM also produced rightward shifts in *peak time*. The magnitude of the effect was smaller for NBM than FC lesions (closer to 10%) and it became smaller still with continued testing, whereas the *peak time* of the FC group was maintained at 20% greater than that of baseline performance. An intriguing outcome was observed in the same study for groups of rats with lesions of the MSA or the FF. These groups showed leftward displacements of *peak time*, about 10% for MSA and 20% for FF. Analogously to the NBM and the FC, the distorted performance of the MSA group diminished with continued training but that of the FF group did not. The authors interpret this finding to mean that damage to the MSA must produce a compensatory increase in ACh activity in other parts of the brain, presumably the NBM–FC system, that recovers to normal over time. This idea would have been stronger had they provided experimental evidence to back up the speculation, for example by assaying ACh in the FC directly after the FF lesion and at different time points after the MSA lesion.

In sum, the reference memory pattern is characterized by gradual proportional changes in *peak time* that remain stable over time. If the pattern is evoked by treatment with ACh drugs, then discontinuation of the drug will allow a gradual return of performance to baseline. A combination of drug and lesion studies has demonstrated the critical role the basal forebrain ACh system and the FC play in producing the reference memory pattern.

The working memory pattern

The working memory pattern is revealed by the PI–GAP procedure (e.g, Roberts & Church, 1978) a variant on the PI procedure that requires the subject to hold an interval of time in working memory during a retention interval and then to continue timing the signal when it resumes. A normal subject will spontaneously retain in working memory the duration of the signal that had elapsed before the retention interval and then continue timing when the signal comes on again as though the gap had not occurred. The *peak time* will therefore be greater by an amount equal to the duration of the gap, and the subject is said to have "stopped" its internal clock during the gap (Roberts, 1981). Two other possibilities also exist: the subject could continue to "run" its internal clock during the gap, producing a *peak time* near the FI value, or the subject might "reset" its internal clock and start timing when the signal resumes after the gap, as it would for the start of a new trial. Rats with hippocampal damage show this last pattern. They are unable to hold the first part of the signal in working memory and so they start timing anew when the signal comes on again. Their timing functions are shifted rightward by an amount equal to the gap duration plus the average amount of time the signal was on before the gap.

An example of this phenomenon is provided in one of the papers previously discussed with regard to the reference memory pattern (Meck et al., 1987). These authors showed that lesions of the MSA (which projects ACh afferents to the hippocampus) impair working memory in a PI–GAP procedure, whereas lesions of the NBM or FC had no effect. All rats were pre–trained on a PI procedure with FI–40 sec prior to surgery and retrained for one week after surgery before beginning the PI–GAP procedure portion of the experiment. During these testing sessions, trials were randomly determined according to the following probabilities: one–half were FI–40 sec, one–quarter were peak trials, and the remaining quarter were gap trials. Among the gap trials, the gap duration could either be 5 or 10 sec and the point in the trial where the gap was inserted was either 10 or 20 sec after signal onset, again with both parameters selected at random. The results were identical for the controls, NBM, and FC groups: the *peak time* was increased by the duration of the gap, either 5 or 10 sec, showing that the rats stopped their internal clock during the gap and continued timing when the signal resumed. The *peak times* of the MSA subjects, however, increased by the duration of the gap plus the signal before the gap, indicating that the gap caused them to reset their clocks and begin timing when the signal reappeared after the gap as though it were a new trial. From these data it is apparent that the ACh input from the MSA to the hippocampus is required for a rat to temporarily store in working memory the amount of time that has elapsed before an interposed retention interval.

A similar study comparing performance of rats with lesions of the FF and control rats on the gap procedure demonstrated the same characteristic impairments for the lesioned group (Meck et al., 1984). *Peak time* was accurate at 20 sec for the control group, but the FF group had a *peak time* of about 16 sec, 20% less than controls. This decrease in *peak time* has been hypothesized to result from compensatory activation of the ACh system that projects to the FC. For purposes of understanding the working memory pattern, however, one can compute a value called the peak shift by subtracting for each subject its *peak time* on gap trials from its *peak time* on regular peak trials without the gap. When a 5-sec gap was inserted 10 sec after signal onset, the control rats' average *peak time* was rightward shifted by about 5 sec, indicating they had stopped timing the signal during the retention interval. The FF group showed an average peak shift of about 15 sec, which was the duration of the gap plus the amount of signal that had transpired before the gap, indicating that these rats had reset their clock and started timing anew after the gap ended (Figure 10).

In a dissociation between the separate functions of nuclei in the basal forebrain ACh system, performance on the PI–GAP procedure has been shown to be sensitive to lesions of the hippocampus and its ACh afferents and insensitive to lesions of the FC and its ACh afferents (Olton et al., 1988). Five groups of rats were prepared for this study: sham lesion (CON), ibotenic acid lesions of the MSA or NBM, aspiration lesions of the FC, or radio frequency lesions of the FF. For the PI–GAP procedure, there are three possible outcomes: the subject can run, stop, or reset the internal clock during the retention interval. The CON, NBM, and FC groups were able to stop timing during the gap and their *peak times* were increased by the duration of the gap. The MSA and FF groups, in contrast, reset their clocks when the signal went off and started timing anew when the signal came on again because their *peak times* were shifted by the duration of the gap plus the amount of time elapsed by the signal before the gap. These data show that the hippocampus is necessary for working memory while the frontal cortex is not.

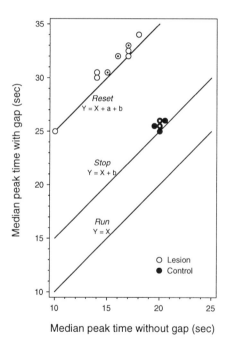

Figure 10. *The working memory pattern.* Median *peak times* with gaps as a function of median *peak times* without gaps for rats with fimbria–fornix lesions and control rats. The procedure used a 20-sec signal with a 5-sec gap inserted 10 sec after signal onset. Open circles with black dots represent multiple rats with lesions. Filled circles with white dots represent multiple control rats. (*a* = duration of the signal before the gap, *b* = the duration of the gap; adapted with permission from Meck *et al.*, 1984).

A dissociation has also been demonstrated between the roles of the hippocampus and amygdala in temporal memory in the rat (Olton, Meck & Church, 1987). All rats were trained preoperatively on a PI procedure with FI–50 sec and then ten were assigned to each lesion condition: sham, amygdala, and FF. The sham and amygdala rats were able to stop timing during a 5– or 10–sec gap, but the FF rats reset their clock when the gap appeared and started timing the signal when it continued as though it were a new trial. This study also provided evidence of the pharmacological basis of working memory impairments. Sham–treated subjects could be induced to perform like FF rats by administering 0.45 mg/kg atropine, an ACh antagonist, which caused them to reset their clocks after the gap, providing further evidence of the relevance of acetylcholine to working memory.

All the working memory pattern data discussed so far have come from research on rats. Experiments on pigeons suggest at first glance that they may respond differently than do rats to the insertion of a retention interval within a timing signal. On more careful examination, however, procedural differences from the rat studies allow alternative interpretations for the pigeons' peculiar performance.

One study trained pigeons on the PI procedure using both light and tone signals (Roberts et al., 1989). Eight pigeons were divided evenly into two groups: one group learned to associate a 15–sec FI with a tone and a 30–sec FI with a light, and the other group had the opposite assignment. When they were responding well to the FI schedules, testing began on peak trials (25% of each session) in which the signal remained on for 90 sec. Each trial was followed by a random ITI in the range of 10 to 40 sec. In experiment three, gap trials were introduced. Sessions consisted of 75% FI trials and 25% either peak or gap trials which alternated each day. Gaps always began 9 sec after the start of the trial and were 9, 3 or 1 sec in duration for blocks of 16 sessions. Most of the pigeons produced *peak times* during gap trials that clustered near the value expected if they had reset their internal clock and started timing the signal again from the end of the gap, with only one producing a *peak time* closer to the value expected if it had stopped timing during the gap as rats do.

There is a significant problem with the method of this study that makes the interpretation of the data less clear than it may appear. The ITI could be almost as short as the gap duration (10 sec vs. 9 sec), although from the description of the procedure it is not clear how often a 10-sec ITI occurred. The more similar that the ITI is to the gap, the more likely the pigeon will confuse the two and think that the end of the gap is really the end of an ITI signaling the beginning of a new trial. The fact that they were tested first on a 9-sec gap duration would contribute to such confusion and allow them to learn the reset rule and apply it in subsequent sessions when the gap decreased to 3 and then 1 sec.

A similar problem besets a more detailed analysis of pigeon working memory with the gap procedure (Cabeza de Vaca et al., 1994). This study attempts to dissociate the effects of gap duration from gap location by systematically varying both parameters alone and in combination. In Experiment 1, pigeons were trained for sixty sessions on a 30-sec FI in which peak trials lasted 90 sec plus a variable interval with a mean of 30 sec. The ITI consisted of a 15-sec blackout plus a variable interval with a mean of 45 sec. Each session started with five warm-up FI trials and was followed by 60 trials of which a random 30% were peak trials. Then pigeons experienced 14 blocks of three sessions in which a gap was inserted in half of the peak trials. The three sessions were chosen randomly without replacement from the following conditions: "early" had a 6-sec gap starting at 6 sec after signal onset, "late" had a 6-sec gap starting 15 sec into the trial, and "long" had a 15-sec gap at starting at second 6. If the pigeons stopped their internal clocks, they should produce *peak times* shifted by 6 sec for the early and late conditions and 15 sec for the long condition. If they reset their internal clocks, the increase in *peak time* should be 12 sec for the early condition and 21 sec each for the late and long conditions. The observed mean peak shifts were 8.1 sec for early, 13.8 sec for late, and 20.2 sec for the long condition. The long break produced a peak shift quite close to the value of 21 sec predicted by the reset hypothesis, whereas the other two *peak times* were intermediate between the two predicted values. Interestingly, the peak shifts produced in the first few sessions of the late condition began close to the reset value and gradually increased toward an intermediate value, suggesting some kind of learning process.

A similar account to the one provided for the findings of Roberts et al. (1989) can be applied here as well. The ITI could be as short as 15 sec, which was the duration of the long gap. If the pigeons were to confuse the gap with the ITI, they would reset their internal clocks, as it appears they did. The difficulty they had discriminating the 15–sec gap from the ITI may have generalized to the 6–sec gaps as well, so that *peak times* for both the early and late 6–sec gaps were shifted by an amount intermediate between the predictions of the stop and reset hypotheses. In effect, the birds were trained to partially reset their internal clocks because of the difficulty in discriminating the gap from the ITI. The gradual increase in the late group's peak shift away from the stop value provides support for this notion. Unfortunately, the design of the experiment does not allow a more stringent test.

If each of the six pigeons had received a counterbalanced order of presentation of 3 blocks of 14 sessions with each gap condition instead of 14 blocks of three sessions with gap condition randomized among the three, the proposed alternative explanation would predict peak shifts closer to the stop hypothesis except for pigeons that received the long condition first. They should learn to reset their internal clocks and show subsequent performance closer to resetting on the shorter gaps. A simple solution to this problem would be to increase the ITI so that it is more discriminable from the gaps. In this scenario, the pigeons should show evidence of stopping their internal clocks during a gap just as rats do.

The alternative explanation also applies to the data from Experiment 2 of Cabeza de Vaca et al. (1994). This procedure was a more elaborate variant of the previous one in which gap location, duration, or both varied across five levels while another parameter was held constant. In the fixed onset condition, the gap always began 6 sec after the trial began and could be either 3, 6, 9, 12, or 15 sec in duration. The fixed duration condition placed a 6–sec gap either 3, 6, 9, 12, or 15 sec after signal onset. Finally, in the fixed end condition the gap always ended at 21 sec but was either 3, 6, 9, 12, or 15 sec long.

In the fixed–onset and fixed–end conditions, the peak shift was a nonlinear function of gap duration that was closer to the stop hypothesis at shorter gap durations and increased to the reset hypothesis at a gap of 15 sec. This pattern looks much like a generalization gradient of gap duration. The more similar the gap is to the ITI, the closer the peak shift comes to the reset model. In the fixed duration condition, peak shift was a linear function of gap location so that earlier gaps were associated with smaller peak shifts. However, the magnitude of the peak shift was always intermediate between the stop and reset models. The authors present the evidence from Experiment 2 as support for partial resetting of the internal clock in pigeons, but the more parsimonious explanation is that they inadvertently trained their subjects to generalize from the value of the ITI to the gap duration. The pigeons provide evidence for an outcome intermediate between stopping and resetting the internal clock because that is what they learned to do.

We have seen that the gap procedure is a sensitive task for exposing working memory deficits and determining behaviorally how the internal clock operates. Research with rats has determined that it is the hippocampus that subserves the function of working memory and that its normal operation can be disrupted by lesioning its source of acetylcholine, the NBM, or by cutting the fibers of its output pathway, the FF. Normal rats will stop their internal clock during a gap interposed within a signal that they are timing and will continue timing when the signal resumes. Rats with damage to the input or output pathways of the hippocampus will be unable to retain in working memory the amount of time that has elapsed before a gap (or presumably even the signal's prior occurrence) and so will reset their internal clock and start timing from the resumption of the signal as though it were a new trial. Other research has demonstrated an outcome intermediate between stopping and resetting the internal clock in normal pigeons, but studies to date are flawed by making the gap duration similar to some ITI durations. This allows the subject to confuse a gap with an ITI and may produce a pattern that appears like partial resetting of the internal clock.

The general attention pattern

Probability of attention to the relevant stimulus dimension, symbolized by $p(A)$, to the relevant stimulus dimension has proven to be a useful measure for analyzing situations where animals are not always attentive to time. The $p(A)$ measure is typically applied in situations where a single stimulus is presented during any one trial. It has been shown that individual differences in temporal generalization can largely be accounted for by changes in the $p(A)$ parameter of Scalar timing theory (Church & Gibbon, 1982). Although the IP model described by Gibbon and Church (Gibbon & Church, 1984) characterizes many trials in a time discrimination task, on some trials the animal appears to respond independently of the signal duration. Quantitative fits of performance are often improved if it is assumed that with some probability the animal responds on the basis of signal duration and on the remainder of the trials, the animal responds without respect to signal duration. On these trials the animal selects the response on the basis of some constant bias (Heineman, Avin, Sullivan & Chase, 1969). In temporal generalization tasks, most of the between–subject variance can be accounted for by parameters representing the probability of attention and responsiveness given inattention. One can determine the proportion of responding controlled by time by partitioning the peak function into two components: a Gaussian curve representing the "signal" of the subject's temporal discrimination and a flat line which corresponds to "noise" or the trials when the subject's responding is only minimally controlled by time, if at all (e.g., Church, 1982; Roberts, 1981; see Figure 11).

The selective attention pattern

Selective attention describes a subject's ability to attend to a relevant

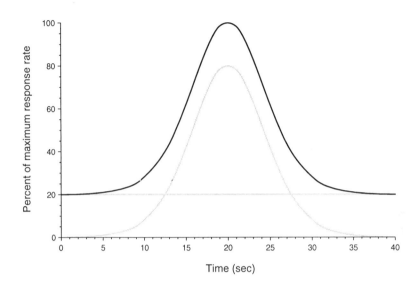

Figure 11. *The general attention pattern.* Idealized peak function for a 20-sec signal (black curve) can be separated into a signal component (gray curve) that represents the trials on which the subject shows evidence of temporal discrimination and a noise component (gray line) for trials in which responding is not controlled by time.

stimulus dimension while ignoring others. Selective attention may operate at different levels of the IP model and has been shown to affect performance dramatically in a prior–entry–reversal (PER) task by acting at the decision stage in rats (Meck, 1984). Five rats were trained on a 10–sec visual signal and a 30–sec auditory signal, and five rats learned the reverse association. Peak trials lasted twice as long as the FI value, and all trials were followed by a 130–sec ITI. Prior–entry (PE) training then followed in which a 1–sec cue in the modality of the next trial preceded it, thus predicting the signal to be timed. The cue was followed by a geometrically distributed delay interval with a mean of 15 sec and a minimum of 2 sec before the signal for the trial began. After rats were well–trained in the PE procedure, PER testing began. Three quarters of trials were standard PE trials as in the previous phase, but one quarter of the trials were preceded by a cue in the modality other than that of the trial: a visual cue preceded an auditory signal or an auditory cue preceded a visual signal. The *peak times* for PE trials were close to the value associated with the signal, either 10 or 30 sec. The *peak times* for the PER trials were also close to 10 or 30 sec, but they corresponded to the values associated with the modality of the cue, not the signal. Under PER conditions, the rats were prepared by the cue not merely to expect a signal of the same modality but also to respond for the duration associated with that value, so that even when the opposite modality appeared, they still produced the criterion time corresponding to the cue modality. This experiment shows that the decision stage may be manipulated by focusing the subject's attention on a cue in advance of the signal to be timed and preparing it to compare the passage of time during the signal with a memory sample of the criterion duration drawn from those associated with the modality of the cue.

Alternatively, rather than affecting the decision stage, selective attention may act at the level of clock. Attentional effects operating on the mode switch appear as changes in the latency required for the subject to start timing a signal. If the subject is attending to one modality when another appears, it may require some time to switch attention to the proper modality. In order to distinguish between the multiplicative (proportional) shifts one would expect from an effect at the level of the pacemaker or accumulator from the additive (constant) shifts that are characteristic of effects on the switch, one must test subjects using at least two different signal durations.

This was the approach taken in Penney et al. (1996) in which rats were first trained on the standard PI procedure with 10- or 30-sec criterion times associated with either an auditory or visual stimulus and counterbalanced between rats. They were subsequently tested using the PE and PER procedures. The 1-sec warning cue presented in the same modality in PE decreased the *peak time* by an additive constant for both signals, while the PER with an opposite modality cue caused *peak time* to increase by an additive constant. Clonidine, a noradrenergic functional antagonist (specifically, an α_2 autoreceptor agonist), was administered during some sessions of each behavioral condition and generally increased the latency to begin timing (Figure 12) similarly to PER. These results show that selective attention impairments can act by increasing the latency to close the mode switch which produces a constant shift in *peak time* for different signal durations and may be induced by decreasing the activity of the noradrenergic system in the brain.

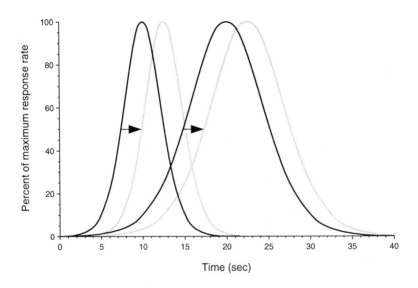

Figure 12. *The selective attention pattern*. Idealized peak functions for 10– and 20– sec signals at baseline (black curves) and after a manipulation impairing selective attention (gray curves). In contrast to the clock and reference memory patterns which appear as shifts proportional to signal duration, impaired selective attention would cause a rightward additive shift of the entire response distribution by an amount that is equivalent for different signal durations.

The PE and PER procedures have shown that selective attention may act either at the decision stage or the clock stage. The criterion time used for responding to a signal may be predicted by a cue, and sometimes the modality of the cue overrides the modality of the signal in determining the interval produced by the subject. Selective attention acting at the level of the clock produces subtler effects. The constant shifts in *peak time* that are independent of the duration being timed are very different from the proportional changes in *peak time* seen in clock speed or memory manipulations. These changes can be induced either by manipulating a subject's expectation for a signal of a particular modality by presenting it with a warning cue before each trial or by giving a drug that blocks the catecholamine neurotransmitter, noradrenaline. Because the locus ceruleus in the midbrain is the primary source of noradrenergic innervation throughout the brain, this region probably modulates the mechanism of selective attention.

The divided attention pattern

Divided attention is the capacity a subject uses when it must pay attention to multiple events simultaneously, for example, two different signals that must be timed concurrently. A technique used to assess divided attention in interval timing is the PI procedure version known as simultaneous temporal processing (STP) in which multiple signals of different modalities are timed independently. Probably the earliest study to assess this ability in rats used a mixture in equal proportions of single FI, compound FI, single peak, and compound peak trials (Meck & Church, 1984). Specifically, the authors trained the subjects to associate a 50–sec criterion with the presentation of the house light and sometimes compounded that signal with 1–sec bursts of white noise occurring every 10 sec. The white noise bursts began at the onset of the light signal but did not occur at the 50–sec criterion time. On FI trials the white noise was therefore predictive of reinforcement 10 sec later every fifth time it occurred. When tested on the light signal only, the rats produced a smooth and symmetrical peak function centered at 50 sec. Data from the compound peak schedule appeared as a 10–sec FI scallop superimposed on the 50–sec peak function. Even when the noise did not appear at the 50th sec, the average response rate of the peak function decreased abruptly for a few seconds, indicating that the subjects were timing the 10–sec signal as well as the 50–sec signal and anticipated its occurrence. This experiment provides strong evidence that rats have multiple pacemaker–accumulator systems that can independently time two concurrent signals.

A physiological study previously discussed in the context of working memory (Olton et al., 1988) also addressed the issue of divided attention. To review, the experiment included five groups of rats: control surgery (CON), ibotenic acid lesions of MSA or NBM, aspiration lesions of the FC, or radio frequency lesions of the FF. The NBM provides ACh input to the FC, the MSA is the major source of acetylcholine to the hippocampus, and the FF is the primary output pathway of the hippocampus. The reference memory pattern is apparent in the data (a leftward shift of *peak time* for MSA and FF, and a rightward shift for NBM and FC), but the question of divided attention is addressed by examining the difference between *peak times* for single peak and compound peak (STP) trials. During STP trials in which 10– and 20–sec durations were signaled by either an auditory or a visual stimulus, the NBM and FC groups were unable to divide attention between the short and long signals. Instead, they timed the long signal until the onset of the short signal, which apparently captured all their attention until its criterion time had elapsed, at which point they were able to switch attention back to the long signal and continue timing it. Effectively, they treated the long signal as if the short signal had been a gap, and they stopped timing it while the short signal was present. The net result was that their long signal *peak times* were longer by about 10 sec, the duration of the short signal. They were unable to time both signals simultaneously, although they were able to hold the time of the long signal in working memory until they could again devote attention to it when the criterion for the short signal had elapsed (Figure 13).

Observations that arginine vasopressin (AVP), a hypothalamic hormone and neuropeptide, enhanced memory in aversively motivated tasks such as passive avoidance, led Meck (1987) to study its effects on an appetitively motivated task, namely simultaneous temporal processing. The rats were trained to associate three different signal

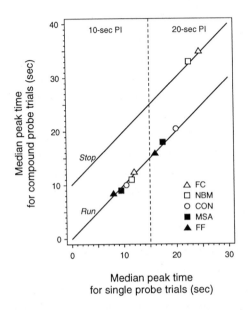

Figure 13. *The divided attention pattern.* These data summarize the behavioral results for the *peak time* in probe trials with a single stimulus (horizontal axis) and in compound probe trials with both stimuli (vertical axis) for the short stimulus (left side) and the long stimulus (right side). Points on the *Run* line indicate that rats in these groups timed both stimuli when they were presented together just the same as when they were presented separately. Points on the *Stop* line indicate that rats in these groups timed the second, short stimulus during the compound probe trials the same as during the probe trials with a single stimulus. However, they were unable to time the first, long stimulus simultaneously when the second stimulus was presented, a failure of simultaneous temporal processing (adapted with permission from Olton *et al.*, 1988).

modalities (light, sound, and footshock) with three different durations (15, 30, and 60 sec), and the assignment of modality to duration was randomized between subjects. All three signals began asynchronously and after they had been presented concurrently for between 5 and 15 sec, two of the signals were randomly chosen to terminate and the rat was left with only one signal remaining. This procedure required that the subject attend to all three signals because it had no way of knowing which of them it would ultimately be required to time. Fitting parameters from scalar timing theory to the peak functions showed that AVP produced a proportional leftward shift in all functions of about 20% as well as a proportional increase in the probability of attending to each of the signals. The increased probability of attention is manifested graphically by a decrease in the spread of each function greater than what would be expected from the scalar property and corresponds to an increase in timing precision. The gradual and proportional leftward shift observed with AVP administration indicates an effect on memory storage speed, which is consistent with other evidence that AVP increases or facilitates maintenance of ACh levels in the brain. The main effect of interest, however, is that AVP enhanced divided attention and improved timing precision in a simultaneous temporal processing task.

The divided attention task known as the simultaneous temporal processing version of the PI procedure depends upon a functional cortical ACh system. Lesions of the FC or its ACh afferent, the NBM, prevent subjects from simultaneously timing two signals at once. Instead they time one signal until the onset of the second one appears, store the value of the first one in working memory until they have finished timing the second signal, then continue to time the first signal treating the intervening second signal as a gap. Further evidence for the ACh hypothesis of divided attention comes from a study of the neuropeptide AVP and its enhancement of both ACh function and simultaneous timing in normal rats.

Circadian oscillations in interval timing processes

Fluctuations in the sensitivity to signal duration as a function of the time of day is apparently mediated by different mechanisms depending upon signal modality (Meck, 1991). Time–of–day oscillations in the psychophysical temporal bisection functions obtained by testing rats at various times during the light/dark (LD) cycle after they had initially been trained at a fixed time of day could best be explained by variation in the $p(A)$ for visual signals. In contrast, the same type of oscillations for auditory signals could best be explained by variability in temporal memory. In addition to this *between* modality difference, it was observed that sensitivity to signal duration as a function of the LD cycle varied *within* a signal modality according to the baseline phase of training. For example, rats initially trained in the morning (8 AM) typically demonstrated lower sensitivity to signal durations than rats initially trained in the evening (8 PM) for both signal modalities. This morning versus evening difference could best be explained by variation in the $p(A)$ for visual signals and by variability in temporal memory for auditory signals. At both training times, rats were observed to show greater sensitivity to time for auditory signals than for visual signals. Interestingly, the explanation for this light/sound difference varied as a function of circadian training phase. For rats trained in the morning, the light/sound difference was due primarily to changes in $p(A)$, whereas the light/sound difference for rats trained in the evening was due primarily to changes in the variability of temporal memory.

During the circadian test phase, it was observed that although the midpoint of the light cycle represented the "low point" for sensitivity to signal durations, this effect was dependent upon the interaction between signal modality and the baseline phase of training. Rats initially trained in the morning (i.e., "day–active" subjects) showed an increase in sensitivity to auditory signals and a greater than expected decrease in sensitivity to visual signals at the midpoint of the light cycle. In contrast, rats initially trained in the evening (i.e., "night–active subjects") showed an increase in sensitivity to visual signals and a greater than expected decrease in sensitivity to auditory signals at the midpoint of the light cycle. Thus, it appears that some sort of stimulus selection process is being invoked during this period of relative insensitivity to signal durations that is dependent upon the baseline circadian phase of training. Presumably this selection process is used to allocate resources whose limits fluctuate as a function of the LD cycle. The observed modality differences are particularly intriguing. Perhaps it is adaptive in terms of an optimal foraging strategy for "day–active" subjects to focus their mental abilities toward the detection and processing of auditory signals when their temporal discrimination sensitivity in that modality is at its lowest point, whereas it may be equally adaptive for "night–active" subjects to focus their mental abilities toward the visual channel when their temporal discrimination sensitivity in that modality is at its lowest point. This idea is based on the premise that the two signal modalities will alternate in being the most distinguishable from background stimuli given different LD contexts and that this will guide the adaptive nature of foraging strategies (Daan, 1981).

Summary and conclusions

In summary, a variety of patterns of temporal data generated using the PI procedure can be diagnostic of the level where a manipulation, either behavioral, pharmacological, or physiological, is acting in an IP model of interval timing. Hedonic effects may appear as changes in response rate that do not affect timing directly but can be manipulated by controlling the deprivation state of the subject and cause changes in *peak rate*. Other patterns of PI procedure data can be used to infer correlations between the components of the IP model and particular physiological and pharmacological manipulations. Lesions in SNC or the striatum, or DA drugs that act on these areas, produce the clock pattern that appears as abrupt, proportional changes in *peak time* that normalize over a period of several sessions and rebound in the opposite direction if the DA drug is discontinued. Lesions of the FC or its cholinergic input from the NBM generate the reference memory pattern in the standard PI procedure. In contrast to the clock pattern and its sensitivity to DA drugs, the reference memory pattern may be produced by ACh drugs and manifests as a gradual proportional change in *peak time* with equally gradual recovery to baseline performance if the ACh drug is discontinued. Damaging the hippocampus, its ACh input from the MSA, or its primary output pathway, the FF, produces the working memory pattern when rats are tested using the PI–GAP procedure. They will reset their internal clock after the retention interval and will treat the second part of the signal as though it were a new trial. Finally, attentional patterns of deficit come in several forms: a lack of general attention is posited when a well-trained subject shows no evidence of temporal discrimination during a peak trial; directed attention impairments are assessed using the PER procedure and may be characterized by either a reversal of the assignment of duration to modality if the decision stage is affected or by a constant additive change in *peak time* required to switch attention to an alternative modality; and divided attention is assessed by having the subject time two or more signals simultaneously and relies on an intact frontal cortex with its ACh input. Circadian oscillations affect the probability of attending to visual signals and temporal memory for auditory signals but do not affect the clock stage itself, a noteworthy observation given the pervasiveness of circadian rhythms.

References

Bindra, D., & Waksberg, H. (1956). Methods and terminology in studies of time estimation. *Psychological Bulletin, 53,* 155–159.

Bizo, L. A., & White, K. G. (1995). Reinforcement context and pacemaker rate in the behavioral theory of timing. *Animal Learning and Behavior, 23,* 376–382.

Brunner, D., Kacelnik, A., & Gibbon, J. (1992). Optimal foraging and timing processes in the starling, *Sturnus vulgaris*: Effect of inter–capture interval. *Animal Behavior, 44,* 597–613.

Cabeza de Vaca, S., Brown, B. L., & Hemmes, N. S. (1994). Internal clock and memory processes in animal timing. *Journal of Experimental Psychology: Animal Behavior Processes, 20,* 184–198.

Caraco, T. (1980). On foraging time allocation in a stochastic environment. *Ecology, 61,* 119–128.

Catania, A. C. (1970). Reinforcement schedules and psychophysical judgements: A study of some temporal properties of behavior. In W. N. Shoenfeld (Ed.), *The Theory of Reinforcement Schedules,* (pp. 1–42). New York: Appleton–Century–Crofts.

Cheng, K. (1992). The form of timing distributions under penalties for responding early. *Animal Learning and Behavior, 20,* 112–120.

Cheng, K., & Westwood, R. (1993). Analysis of single trials in pigeons' timing performance. *Journal of Experimental Psychology: Animal Behavior*

Processes, 19, 56–67.

Cheng, K., Westwood, R., & Crystal, J. D. (1993). Memory variance in the peak procedure of timing in pigeons. *Journal of Experimental Psychology: Animal Behavior Processes, 19*, 68–76.

Church, R. M., & Gibbon, J. (1982). Temporal generalization. *Journal of Experimental Psychology: Animal Behavior Processes, 8*, 165–186.

Church, R. M., Meck, W. H., & Gibbon, J. (1994). Application of scalar timing theory to individual trials. *Journal of Experimental Psychology: Animal Behavior Processes, 20*, 135–155.

Church, R. M., Miller, K. D., Meck, W. H., & Gibbon, J. (1991). Symmetrical and asymmetrical sources of variance in temporal generalization. *Animal Learning and Behavior, 19*, 207–214.

Cooper, J. R., Bloom, F. E., & Roth, R. H. (1982). *The Biochemical Basis of Neuropharmacology.* New York, NY: Oxford University Press.

Daan, S. (1981). Adaptive daily strategies in behavior. In J. Aschoff (Ed.), *Handbook of Behavioral Neurobiology,* (Vol. 4, pp. 275–298). New York: Plenum Press.

Ferster, C. B., & Skinner, B. F. (1957). *Schedules of Reinforcement.* New York, NY: Appleton–Century–Crofts.

Gibbon, J. (1977). Scalar expectancy theory and Weber's law in animal timing. *Psychological Review, 84*, 279–325.

Gibbon, J., & Church, R. M. (1984). Sources of variance in information–processing theories of timing. In H. L. Roitblat, T. G. Bever, & H. S. Terrace (Eds.), *Animal Cognition,* (pp. 465–488). Hillsdale: Erlbaum Associates.

Gibbon, J., & Church, R. M. (1990). Representation of time. *Cognition, 37*, 23–54.

Gibbon, J., Church, R. M., & Meck, W. H. (1984). Scalar timing in memory. *Annals of the New York Academy of Sciences, 423*, 52–77.

Gwinner, E. (1986). *Circannual Rhythms: Endogenous Annual Clocks in the Organization of Seasonal Processes.* New York: Springer–Verlag.

Heineman, E. G., Avin, E., Sullivan, M. A., & Chase, S. (1969). Analysis of stimulus generalization with a psychophysical method. *Journal of Experimental Psychology, 80*, 215–224.

Honig, W. K. (1978). Studies of working memory in the pigeon. In S. H. Hulse, H. Fowler, & W. K. Honig (Eds.), *Cognitive Processes in Animal Behavior,* (pp. 211–248). Hillsdale, NJ: Erlbaum Associates.

Ivry, R. B., & Keele, S. W. (1989). Timing functions of the cerebellum. *Journal of Cognitive Neuroscience, 1*, 136–152.

Killeen, P. R., & Fetterman, J. G. (1988). A behavioral theory of timing. *Psychological Review, 95*, 274–295.

Leak, T. M., & Gibbon, J. (1995). Simultaneous timing of multiple intervals. *Journal of Experimental Psychology: Animal Behavior Processes, 21*, 3–19.

Malapani, C., Deweer, B., Pillon, B., Dubois, B., Agid, Y., Rakitin, B. C., Penney, T. B., Hinton, S. C., Gibbon, J., & Meck, W. H. (1993). Impaired time perception in Parkinson's disease is reversed with apomorphine. *Society for Neuroscience Abstracts, 19*, 631.

Maricq, A. V., & Church, R. M. (1983). The differential effects of haloperidol and methamphetamine on time estimation in the rat. *Psychopharmacology, 79*, 10–15.

Maricq, A. V., Roberts, S., & Church, R. M. (1981). Methamphetamine and time estimation. *Journal of Experimental Psychology: Animal Behavior Processes, 7*, 18–30.

Meck, W. H. (1984). Attentional bias between modalities: Effect on the internal clock, memory, and decision stages used in animal time discrimination. *Annals of the New York Academy of Sciences, 423*, 278–291.

Meck, W. H. (1987). Vasopressin metabolite neuropeptide facilitates simultaneous temporal processing. *Behavioral Brain Research, 23*, 147–157.

Meck, W. H. (1991). Modality–specific circadian rhythmicities influence mechanisms of attention and memory for interval timing. *Learning and Motivation, 22*, 153–179.

Meck, W. H. (1996a). Frontal cortex or nucleus basalis magnocellularis lesions, but not hippocampal or medial septal area lesions, occasion the loss of control of the speed of an internal clock. (Submitted for publication)

Meck, W. H. (1996b). Neuroanatomical localization of an internal clock: A functional link between mesocortical, mesolimbic, and nigrostriatal dopaminergic systems. *Behavioral Brain Research*, in press.

Meck, W. H. (1996c). Neuropharmacology of timing and time perception. *Brain Research. Cognitive Brain Research, 3*, 227–242.

Meck, W. H., & Angell, K. E. (1992). Repeated administration of pyrithiamine leads to a proportional increase in the remembered durations of events. *Psychobiology, 20*, 39–46.

Meck, W. H., & Church, R. M. (1984). Simultaneous temporal processing. *Journal of Experimental Psychology: Animal Behavior Processes, 10*, 1–29.

Meck, W. H., & Church, R. M. (1987a). Cholinergic modulation of the content of temporal memory. *Behavioral Neuroscience, 101*, 457–464.

Meck, W. H., & Church, R. M. (1987b). Nutrients that modify the speed of internal clock and memory storage processes. *Behavioral Neuroscience, 101*, 465–475.

Meck, W. H., Church, R. M., & Olton, D. S. (1984). Hippocampus, time, and memory. *Behavioral Neuroscience, 98*, 3–22

Meck, W. H., Church, R. M., Wenk, G. L., & Olton, D. S. (1987). Nucleus basalis magnocellularis and medial septal area lesions differentially impair temporal memory. *The Journal of Neuroscience, 7*, 3505–3511.

Mercuri, N. B., Calabresi, P., & Bernardi, G. (1992). The electrophysiological actions of dopamine and dopaminergic drugs on neurons of the substantia nigra pars compacta and the ventral tegmental area. *Life Sciences, 51,* 711–718.

Moore–Ede, M. C., Sulzman, F. M., & Fuller, C. A. (1982). *The Clocks That Time Us: Physiology of the Circadian Timing System.* Cambridge: Harvard University Press.

Olton, D. S., Becker, J. T., & Handelmann, G. E. (1979). Hippocampus, space, and memory. *Behavioral and Brain Sciences, 2,* 313–365.

Olton, D. S., Meck, W. H., & Church, R. M. (1987). Separation of hippocampal and amygdaloid involvement in temporal memory dysfunctions. *Brain Research, 404,* 180–188.

Olton, D. S., Wenk, G. L., Church, R. M., & Meck, W. H. (1988). Attention and the frontal cortex as examined by simultaneous temporal processing. *Neuropsychologia, 26,* 307–318.

Penney, T. B., Holder, M. D., & Meck, W. H. (1996). Clonidine–induced antagonism of norepinephrine modulates the attentional processes involved in peak–interval timing. *Experimental and Clinical Psychopharmacology, 4,* 82–92.

Raymond, J. L., Lisberger, S. G., & Mauk, M. D. (1996). The cerebellum: A neuronal learning machine? *Science, 272,* 1126–1131.

Roberts, S. (1981). Isolation of an internal clock. *Journal of Experimental Psychology: Animal Behavior Processes, 7,* 242–268.

Roberts, S., & Church, R. M. (1978). Control of an internal clock. *Journal of Experimental Psychology: Animal Behavior Processes, 4,* 318–337.

Roberts, W. A., Cheng, K., & Cohen, J. S. (1989). Timing light and tone signals in pigeons. *Journal of Experimental Psychology: Animal Behavior Processes, 15,* 23–35.

Sarkar, M., & Kornetsky, C. (1995). Methamphetamine's action on brain–stimulation reward threshold and stereotypy. *Experimental and Clinical Psychopharmacology, 3,* 112–117.

Schneider, B. A. (1969). A two–state analysis of fixed–interval responding in the pigeon. *Journal of the Experimental Analysis of Behavior, 12,* 677–687.

Time and Behaviour: Psychological and Neurobehavioural Analyses
C.M. Bradshaw and E. Szabadi (Editors)
© 1997 Elsevier Science B.V. All rights reserved.

CHAPTER 11

On the Human Neuropsychology of Timing of Simple, Repetitive Movements

Donald J. O'Boyle

Introduction

For most of us, moving around in the everyday world is, like many of the extraordinary things of which we are capable, a pretty simple business. Although it is often important *when* we get out of bed in the morning, and the question of *why* we should do so might sometimes exercise our minds, we seldom concern ourselves with the question of *how* it is that we are able to do so, or give a moment's thought to the nature of the internal workings which allow us to perform any of the enormous variety of complex, yet commonplace, actions which unfold during a typical day: we just do them, and we take it for granted that we can do them. As has often been remarked, it is only when the underlying systems and mechanisms break down, perhaps as a consequence of ageing or of neurological insult, or if we try to build a system to simulate even a small fraction of what we can do, that we realise the magnitude of the problems which are solved with such astounding ease by our nervous systems.

While a few of our everyday actions may be very simple, involving little more than basic reflexes, most of them require the effective planning and control of co–ordinated sequences of movements of limbs and limb–segments; sequences which can be described, at one and the same time, in terms of spatial trajectories of limbs and of joints, of patterns of muscle contractions and of activity of central and peripheral neurones. Thus, one major, general problem in understanding the mechanisms underlying effective action concerns the *level* at which planning and control are mediated. Furthermore, sequences of movements unfold not only in space, but in time, and in a particular serial order: a movement trajectory is defined with respect to both its spatial and temporal characteristics. As a consequence, and irrespective of the level of analysis, all movement sequences will display some form of ordinal and temporal structure. Thus, two additional, fundamental problems in motor control, the problem of *sequencing* (famously addressed by Lashley in 1951) and the problem of *timing*, concern the characterisation of the processes and mechanisms which determine the ordinal and temporal structure of action.

Sequencing and Timing

    In some instances, and at some level of description (e.g. limb movement), the successful completion of a task will not be dependent upon the adoption, or control, of either a particular serial order, or of a specific pattern of absolute or relative timing, of successive movements. In such instances, the ordinal and temporal structure of behaviour will *emerge,* as an *implicit* property, from a variety of sources, including the dynamic and biomechanical characteristics of the system. But note that even in such cases, the movements will themselves each consist of 'sub–components' (e.g. muscular contractions) which must be sequenced and timed appropriately. This again highlights the issue of level of control, in that some decision must be made about what we consider the controlled components or 'units' of action to be. More commonly, however, the success of action will be critically dependent upon the accurate control, and the *explicit*, internal representation, of one or both of these processes, as in writing, conversing, playing table–tennis or cricket, gymnastics, playing a musical instrument or driving a car. Indeed, precision of sequencing and of the timing of sequence components lie at the heart of skilled performance and, depending upon the task, attainment of the necessary level of control may require many hundreds or thousands of hours of practice. At least at the level of description, then, sequencing and timing are intimately related and considerable attention, particularly in the psychological literature, has been devoted to the questions of whether or not the underlying mechanisms are dependent and of the extent to which possible dependence might vary as a function of the nature and demands of the task (e.g. see Collier & Wright, 1995; MacKay, 1987; Summers and Burns, 1990).

Explicit timing of simple repetitive movements

One fruitful strategy for addressing the problems posed by sequencing and timing has been to try to examine each process, so far as is possible, in isolation from the other. In the case of motor timing, the two components of a simple, repetitive movement task, originally introduced by Stevens (1886), have proved to be particularly useful in this respect. During the first phase of this task, referred to as *synchronization* or *temporal tracking* (Michon, 1967) subjects are required to synchronise each of a series of repetitive movements (usually finger–taps) to the beat of a metronome. When synchronization performance has stabilised, the metronome is switched off, and there ensues the second phase, referred to as *continuation* or *serial, repetitive interval production*, during which subjects are required to continue tapping at the same rate. Thus, the essential procedural difference between the two phases is that while, during synchronisation, performance may be timed or paced with respect to an external (exogenous or extrinsic) pacemaker, subjects are required, during continuation, to provide their own internal (endogenous or intrinsic) timing based on the immediately–prior experience with the metronome.

These tasks have proved to be popular in the study of motor timing for several reasons. First, they are simple: they involve the same repetitively–produced response, and the primary and explicit task–demand (controlled variable) is movement timing with minimal competing processing demand for sequencing, the generation or modulation of force, or for spatial accuracy. Such properties are especially attractive when studying patients with neurological disorder and, as Vorberg & Wing (1996) have remarked, the fact that performance on these tasks yields a sequence of elements which can be considered functionally equivalent facilitates both data collection and statistical modelling. Third, despite their simplicity, the tasks ask interesting questions of the nervous system and may be seen as simple analogues (or model systems) of more natural, complex tasks in which explicit movement timing must be provided intrinsically, as when playing a solo instrument, or when, instead or in addition, performance is regulated by extrinsic timing cues, as when playing the same instrument in an ensemble.

The scope of this chapter

Until quite recently, performance on these tasks has been studied by psychologists with the intention of developing mathematical (systems–analytic) and functional (information–processing) models of normal movement timing, with relatively little concern about the degree to which the hypothetical processes or black boxes contained within the functional models might be mapped onto, or realised physically within, the nervous system. During the past 10 years, however, there has been an increasing interest in attempting to characterise, within the empirical and theoretical context provided by the normal psychological literature, the performance of neurologically–damaged individuals. It is with this body of literature that I shall be concerned primarily in this chapter. In particular, I shall try to place the study of abnormality within the context of what is known or theorised about normal performance in addressing two main, related questions. First, what neural structures, systems and mechanisms are implicated in the control and production of timed, repetitive movements? Second, what might such studies tell us about the functions of different parts of the brain?

As any review is, as a matter of necessity, selective, I should make it clear at the outset that I shall be concerned almost exclusively with research which has been conducted within the general conceptual framework of cognitive, or information–processing, psychology. This approach to the study of timing of repetitive movements, which was given a great deal of quantitative and theoretical impetus by the seminal work of Michon (1967) has involved, in large part, the use of auto– and cross–covariance functions to capture the sequential structure of statistical dependencies between successive response intervals and between response intervals and other salient events (e.g. extrinsic timing cues). The form of the observed statistical dependencies is seen to provide clues about the nature of the underlying processes, and constitutes a statistical description of the temporal characteristics of performance which are to be explained by any model within which hypothesised processes are instantiated. Such models are linear and have included, almost invariably, a hypothetical timekeeping, or 'clock', process. Given that little neuropsychological work has been conducted outside of this general perspective, no more than passing reference will be made in this chapter to the non–linear, oscillator–based accounts of motor co–ordination (for reviews, see Heuer, 1996a; Pressing, 1995) provided within the dynamical frameworks of synergetics (e.g. Kelso et al., 1981) and 'ecological physics' (Kugler & Turvey, 1987), or to linear models in which time is represented without recourse to an hypothesised 'clock' mechanism (e.g. Todd & Brown, 1996).

Continuation: Explicit movement timing in the absence of extrinsic timing cues

Normal continuation performance

During continuation, then, subjects are required to re–produce serially a particular inter-tap target interval, the duration of which is established during an immediately-prior 'entrainment' period of tapping in synchrony with the beats of a regular metronome. Following Stevens's (1886) original work, performance on the task was studied intermittently (e.g. Ehrlich, 1958; Gottsdanker, 1954; Michon, 1967), but it was not until 1973, when Wing & Kristofferson (1973a,b) published a model of continuation performance which provided the means to estimate the precision of an hypothesised 'clock' process underlying motor timing, that it became of significant theoretical interest and, as a consequence, more widely used.

Wing & Kristofferson's (1973b) two-process model of continuation performance

Wing and Kristofferson (1973b) proposed that the statistical structure of a series of intervals produced during continuation is determined by two hypothetical processes, a 'clock' and a 'motor-implementation system'. They supposed that the clock, having become entrained to the frequency of the pacing cues during the initial period of synchronised movement, ticks subsequently, with successive clock-intervals (C_j, with mean μ_C), which are subject to random temporal variation (σ^2_C; clock variance, CV). Each tick of the clock initiates the production of a response by activating the motor-implementation system which imposes a delay (D_j, with mean μ_D), which is subject to independent, random temporal variation (σ^2_D; motor-delay variance, MDV), before the eventual occurrence of the response (R_j). The interval (I_j) between two successive responses (R_{j-1} and R_j) is therefore given by,

$$I_j = C_j + D_j - D_{j-1}$$

and the mean inter-response interval (IRI) is μ_I ($= \mu_C$), with variance

σ^2_I. Assuming that C_j and D_j are random, independent variables, it can be shown that,

$$\sigma^2_I = \sigma^2_C + 2\sigma^2_D$$

that is, the total variance (TV) of response intervals (reflecting the *precision* of timing about some mean level of *accuracy*) is constituted by the sum of the variance of the clock intervals (CV) plus twice the variance of the delays (MDV) in the motor–implementation system: TV = CV + 2MDV.

Now, in the hypothetical case of a clock which ticks at a perfectly constant rate (CV = 0), a response (R_j) which follows a motor delay interval (D_j) which is randomly long (or short) will close a response interval (I_j) which is longer (or shorter) than the mean (μ_1). Given that the delay interval (D_{j+1}) is equal to μ_D, the next response interval (I_{j+1}) which is opened by R_j will be necessarily shorter (or longer) than the mean. This negative covariation between neighbouring response intervals, which is thus predicted to arise as a consequence of random variation in the delay imposed by the motor–implementation system, is not predicted, on the other hand, in the hypothetical case in which the motor–implementation system imposes a perfectly constant motor delay (MDV = 0), following clock intervals which are subject to random variation.

The model predicts, therefore, that neighbouring response intervals (intervals at lag 1) will tend to covary negatively, and this negative covariation is attributed to random variation ('noise') in the duration of the delay within the motor–implementation system. Calculation, across the series of response intervals, of the autocovariance, γ, at lag 1 provides, therefore, an estimate of the variance attributable to the motor delays (MDV). Thus,

$$\text{MDV } (\sigma^2_D) = - \gamma \quad (1)$$

The variance attributable to the clock (CV) can then be obtained indirectly (but not independently) by elimination in the equation, TV = CV + 2MDV, and it follows from the assumptions underlying the model that the lag 1 auto*correlation*, $\rho(1)$, is bound between theoretical limits of 0 and -0.5 and that, for lags (k) >1, the predicted value of autocovariance (γ_k) is zero (Wing & Kristofferson 1973b).

Validity of the two–process model of continuation in normal individuals

a) the model prediction that $-0.5 < \rho(1) < 0$.
The general validity of this prediction has been confirmed repeatedly in studies of normal subjects (e.g. Wing & Kristofferson, 1973b): although violations of $\rho(1) < 0$ and, much less frequently, of $\rho(1) > -0.5$ do occur during some runs, they have been attributed usually to sampling error, consequent upon the analysis of what is usually a small number of response intervals during each run.

b) the model prediction that $\gamma_k = 0$, for k >1.
The validity of this prediction has been assessed formally only rarely. Wing (1977a, 1979) showed that adequate fitting of the model to data from normal subjects (among whom the mean autocovariance at lag 2 was negative and significantly larger than predicted) required relaxation of the subsidiary assumption of statistical independence of successive motor delays. It may be of significance, however, that in the experiments from which these data were derived, subjects were provided with a brief feedback tone 5 ms following each tap. Other patterns of marginal violation of this prediction of the model among neurologically–intact individuals have also been reported (e.g. Helmuth & Ivry, 1996; O'Boyle et al., 1996a).

c) the assumption of independence between the two hypothetical processes.
Several lines of evidence address the question of the general validity

of this basic assumption which, I should point out, mandates that timing control is open–loop. First, Wing (1980) found that, within a range of tapping target intervals of 220 ms – 490 ms, *i)* CV increased as a reasonably–linear function of the target interval[1], whereas MDV remained relatively constant (also see Ivry & Hazeltine, 1995), and *ii)* individual differences in TV were attributable to individual differences in CV rather than in MDV. Second, Wing (1977a, 1980) found that, with changes in effector, CV was more stable than MDV. However, one should note that Wing obtained this result using a form of the model involving three parameters (the third reflecting a correlation between successive motor delays) in order to fit the data adequately (see *b)* above): when analysed in terms of the basic two–process model, CV varied more than MDV as a function of the effector employed. Third, Sergent et al. (1993) reported that the performance of a verbal, anagram–solving task, concurrently with continuation at a target interval of 400 ms, resulted in an increase in CV, but not MDV, irrespective of the hand used. Note that while these data indicate that the two hypothetical processes are dissociable, they also suggest that some component(s) of the verbal anagram–solving task competed with motor timing for common processing and neural resources. Fourth, neuropsychological evidence suggests that the two hypothetical processes can be mapped onto discrete neural systems, and thereby dissociated. For example, Ivry & Keele (1989) reported that four patients with peripheral neuropathy (resulting in some degree of difficulty in making finger movements) tapped, by comparison with controls, with an elevated TV which was attributable, in the absence of a difference in CV, to a significantly elevated MDV (also see Keele & Ivry, 1987, pp. 204–206). The analysis of tapping performance following damage within the CNS will be discussed in detail below. Finally, while these studies do provide support for the notion that the two hypothesised processes are independent, there is some direct evidence to the contrary: that is, during continuation performance, normal subjects have been shown to be sensitive to occasional perturbations in the delay between the time of occurrence of a tap and the time of occurrence of auditory feedback provided shortly following each tap (Wing 1977b; McIntosh & O'Boyle, 1995). These data suggest that subjects do monitor available feedback and are able to use it to adjust the clock interval, and thereby make compensatory adjustments to the durations of response intervals, on occasions when the difference between the desired response interval and feedback interval exceeds some threshold value. Thus, under these conditions, motor timing during continuation appears to be subject to some degree of closed–loop control.

Another measure of the general validity of the model is that it has been applied, with a good deal of success, to a wide variety of issues in the study of motor timing[2]. Here, I concentrate on its application to the study of motor timing in subjects with disorders of the CNS.

Abnormal continuation performance: the effects of neurological damage

During the past twelve years, there has been considerable interest in studying, within the theoretical context afforded by the basic two–process model, the timing of repetitive movements in neurological patients of different types. This work has been motivated by the same symbiotic aims which drive neuropsychology in general. First, and as indicated above, the way in which movement timing breaks down as a consequence of damage to one or other part of the nervous system might inform us about the validity of the structure of, and of some of the assumptions underlying, the model. Second, the explanation of normal movement timing which is provided by the model might help us to understand and explain the characteristics of abnormal movement timing, such as that observed following neurological insult to certain parts of the brain. Third, by attempting to map the hypothesised processes contained within the model onto the nervous system, we should hope to learn something about the nature and organisation of the neural systems and processes underlying normal movement timing. The most influential work of this nature has been conducted by Keele and Ivry and their colleagues who have argued forcefully that the cerebellum plays a primary role in explicit timing computations. To their work, then, I turn first.

Studies of patients with cerebellar lesions

a) The work of Keele & Ivry: the 'cerebellar timing hypothesis'
In an early case study of a patient with a unilateral cerebellar lesion, Keele et al. (1985a) observed that the variance (TV) of response intervals produced during continuation when she tapped with the hand ipsilateral to the lesion was larger than that associated with tapping with the hand contralateral to the lesion[3], and that this elevation of TV appeared to be attributable specifically to an elevation of CV. Furthermore, the patient appeared to have, by comparison with older controls, a raised threshold for the discrimination of duration of an unfilled interval. These suggestive findings were pursued subsequently in two substantial papers (Ivry & Keele, 1989; Ivry et al. 1988) which, because of their seminal position in the development of these authors' ideas about the role of the cerebellum in timing, I shall describe in some detail.

First, Ivry & Keele (1989) studied the performance, on the same two tasks, of different groups of patients including those with, respectively, Parkinson's disease (PD), or lesions of the cerebral cortex (extending into the posterior frontal lobe) or cerebellar lesions. They reported that for both cortical and cerebellar groups, by comparison with both elderly controls and PD patients, *i)* TV was significantly larger, and *ii)* while both CV and MDV were larger, the obtained pattern of statistically–significant contrasts was dependent upon the method chosen for dealing with the disproportionate incidence, among the cerebellar group, of violations of the model prediction of negative lag–1 covariance. On the one hand, by comparison with elderly controls, and irrespective of the method of dealing with model violations, CV of the cerebellar group was significantly larger, whereas CV of the cortical group was not (although $p <0.10$). On the other hand, whereas MDV of the cerebellar group was significantly larger than that of both controls and PD patients if violating subjects were omitted from analysis, the contrast in neither case was significant if, instead, all subjects were retained in the analysis and MDV, in case of violating runs, was set to zero (else, CV >TV). In addition, within-subject comparisons of performance between hands, in groups of subjects whose motor signs were clearly bilaterally asymmetrical, revealed that while mean TV, CV and MDV during continuation were all larger for the more–affected hand than for the less–affected hand in case of both cortical and cerebellar groups, within–subject comparisons for a group of PD patients revealed that an observed elevated TV on the more affected side was attributable specifically to elevated CV. This latter finding, however, was not consistent with the results of a within–subject comparison of PD subjects' performance when 'on' L–dopa medication with that when 'off' such medication: no difference between these medication conditions was observed for obtained values of TV, CV or MDV. Finally, the authors reported that while the cortical group alone were impaired on a test of loudness perception, only for the cerebellar group was the mean threshold for the detection of difference in duration significantly larger than that of elderly controls.

In a second study, Ivry et al. (1988) compared continuation performance between hands in seven patients with unilateral, focal lesions of the cerebellum. They reported the remarkable findings that while all subjects tapped with a larger TV with the hand ipsilateral to the lesion, the source of this elevated TV in each case differed according to the primary locus of the lesion within a hemi–cerebellum: in those four patients in whom the lesion implicated primarily the lateral (hemispheral) region, the elevated TV was attributable to elevated CV, and there was little difference between hands in MDV. In the case of those patients with lesions restricted primarily to the medial region, however, the elevation of ipsilateral TV was attributable to elevated MDV, and there was little difference between hands in CV. From these results, the authors concluded, in line with Wing & Kristofferson's (1973b) model, that the elevated TV in those with lateral lesions reflected a central impairment of motor timing, whereas the elevated TV in those with medial lesions reflected an impairment of motor implementation.

In earlier studies of normal individuals, the authors had observed, first, that the precision of continuation performance correlated, across subjects, significantly between finger and foot, and with acuity of perception of duration (Keele et al., 1985b) and, second, that subjects' ability to control the timing of taps during continuation appeared to be independent of the ability to control their force (Keele et al., 1987). On the basis of these results, and of those of the studies of neurological patients described above, the authors argued that the cerebellum (in particular, its lateral, hemispheral regions) constitutes the primary neurobiological substrate of an intrinsic, horizontally–modular, timekeeping (clock) system or mechanism which is largely independent of mechanisms involved in the computation of other movement parameters (e.g. force, or sequence), and which supplies the explicit temporal computations which are required for the temporal control of movement *and* which underlie temporal perception (Ivry & Keele, 1989; Ivry et al., 1988; Keele & Ivry, 1987; Keele & Ivry, 1990).

In more recent experiments, Ivry and his colleagues have accumulated a good deal of experimental evidence which bears directly on one or other component of this argument (Ivry, 1993). First, Ivry & Diener (1991) reported that cerebellar patients were impaired, by comparison with controls, in the perception of the velocity of apparent movement of stimuli, but not in the perception of their vertical position. Second, Ivry & Gopal (1992) have presented evidence that cerebellar patients show temporal impairments in the production, but not in the perception, of voicing of consonant–vowel syllables. Third, Lundy–Ekman et al. (1991) and Williams et al. (1992) compared the performance of groups of 'clumsy'[4] children with that of normal children on the continuation task and on tests of the perception of duration, of loudness and of force control. They reported that whereas clumsy children with 'soft' neurological signs indicative of cerebellar dysfunction tapped with a larger TV on the continuation task and showed an impairment of the perception of duration (but not of loudness), clumsy children with 'soft' signs indicative of basal ganglia dysfunction were not impaired significantly on any of these three tasks, but showed significant impairment of force control. Fourth, Papka et al. (1995) reported that, in normal subjects, performance of the continuation task interfered significantly with the concurrent acquisition of eye–blink conditioning (which is known to depend upon the integrity of the cerebellum both in humans [e.g. Daum et al., 1993; Lye et al., 1988] and other animals [e.g. Hardiman et al., 1996; Thompson & Krupka, 1994]), whereas no effect on acquisition of conditioning was observed during concurrent performance of either a task of recognition memory (in which the cerebellum is not known to be involved and for which the integrity of the hippocampus has been shown to be essential [e.g. Squire et al., 1988]), or of a choice reaction time (control) task. Fifth, Franz et al. (1996) studied the tapping performance, during unimanual and bimanual continuation, of a group of four patients who were selected according to the two criteria that neuroimaging revealed evidence of focal, unilateral damage restricted to the cerebellum, and that tapping TV when using the hand ipsilateral to the lesion was significantly higher than that associated with tapping with the contralateral hand. The authors observed that, *i)* in the bimanual condition, tapping TV using the hand ipsilateral to the lesion declined to the same level as that of the contralateral hand which, in turn, remained unchanged from the value observed in the unimanual condition, and *ii)* the reduction, during bimanual tapping, of TV associated with the hand ipsilateral to the lesion was attributable to a reduction in CV and not MDV. Sixth, Clarke et al. (1996) have presented some preliminary evidence that bilateral lesions aimed at the cerebellar dentate and interpositus nuclei (which receive projections from lateral regions of the cerebellar cortex) in rats produced transient changes in the psychometric characteristics of performance of temporal bisection (Church & DeLuty, 1977), within a temporal range of about 200 ms – 850 ms, but not within a range of 20 s – 45 s, and not in the characteristics of performance of an equivalent task involving auditory stimulus 'intensity bisection'.

b) The work of others
While there have been, to my knowledge, no published studies of continuation performance of cerebellar patients other than those originating in the laboratories of Ivry and Keele (but see below),

several other researchers have examined the role of the human cerebellum in temporal perception. In a recent report of a PET study of normal subjects engaged in either a task of temporal discrimination or in a similar control task, Jueptner et al. (1995) argued that whereas observed activation of inferior regions of a cerebellar hemisphere reflected ipsilateral finger movements involved in the tasks, observed activation of superior regions of the cerebellar vermis and of the cerebellar hemispheres (bilaterally) reflected specifically the operation of a cerebellar timing process. The appropriateness of this conclusion has been challenged, however, by Maquet et al. (1996) who conducted a separate PET study of normal subjects engaged in a task of either temporal generalisation (D), or stimulus intensity generalisation (I), or in a similar control task: they observed a similar pattern of cerebellar activation during the D and I tasks, and argued that such activation was, therefore, attributable "to processes shared by discrimination tasks in general or to ... casual timing aspects" (p. 124) which, although implicit in the tasks, do not require temporal judgement on the part of the subject. Nichelli et al. (1996), on the other hand, have reported recently that a group of 12 patients with cerebellar degeneration showed some degree of impaired performance of temporal bisection: whereas patients' temporal discrimination was normal, by comparison with controls, within a temporal range of 100 ms – 325 ms, it was impaired within ranges of 100 ms – 600 ms and 8 s – 32 s. Finally, Daum and her colleagues have reported that temporal reproduction, within a temporal range of 1s – 8 s, is impaired in subjects with cerebellar damage (Schugens et al., 1994b) and that subjects with bilateral cerebellar atrophy, but not those with unilateral lesions, show (contrary to the results of the similar experiment conducted earlier by Ivry & Gopal (1992)) a temporal impairment of categorical speech perception (Ackermann et al., 1996).

Several issues in this substantial body of work invite discussion which, however, I shall postpone until I have described an analogous series of studies of continuation performance, and of temporal perception, by patients with neuropathology involving the basal ganglia.

Studies of patients with disorders of the basal ganglia

a) Parkinson's disease
Parkinson's disease (PD) is the most common disease of the basal ganglia and is considered to be the best available human model of basal ganglionic dysfunction. Conventionally, the disease has been thought of as one which results primarily in impairments of the control of movements, as manifest in the classical triad of neurological signs of akinesia/bradykinesia, rigidity and tremor. Recent, rapid advances in our understanding of the way in which the component nuclei of the basal ganglia inter–connect with each other, and with other parts of the brain (see Alexander & Crutcher, 1990), suggest that the functional architecture of the basal ganglia is such that they are organised into several structurally and functionally distinct (hence, parallel) circuits, each of which involves a loop extending from cortex to basal ganglia to thalamus to frontal cortex, on a different region of which each circuit is focused separately. Each circuit is thought to involve, in turn, two parallel pathways through the basal ganglia: first, the 'direct' pathway, which projects from cortex to striatum (caudate nucleus and putamen) to the internal globus pallidus and pars reticulata of the substantia nigra (GPi/SNr) to thalamus to cortex and, second, the 'indirect' pathway in which the striatum projects to GPi/SNr indirectly by way of the external globus pallidus (GPe) and subthalamic nucleus. The primary neuropathology in PD, to which the major symptoms are attributed, is a progressive degeneration of dopaminergic neurones in the pars compacta of the substantia nigra (SNc), which results in a concurrent loss of dopaminergic innervation of the striatum. The hypokinetic motor signs of PD are thought to reflect disturbances of normal functioning within the 'motor' circuit: loss of dopaminergic innervation of the striatum is believed to enhance the output activity of the 'indirect' pathway (which is excitatory to GPi/SNr) and decrease the output activity of the 'direct' pathway (which is inhibitory to GPi/SNr), the net result being excessive excitatory drive of GPi/SNr which, in turn, gives rise to excessive inhibition of the thalamus and pre–frontal cortical motor fields (see Albin et al., 1989; DeLong, 1990). An admirable tutorial review of current understanding of normal and

abnormal basal ganglionic functioning has been provided recently by Rothwell (1995; also see Chesselet & Delfs, 1996).

The first neurological patient studied using the continuation task suffered from PD. This patient, M.F., had predominantly right–sided signs and was tested on seven occasions over a one–year period (Wing et al., 1984; Wing and Miller, 1984), during which time she took inconsistently a low, daily dose of L–dopa. On average, over the period of testing, M.F.'s TV and CV were significantly larger when tapping with the right hand than when tapping with the left: MDV did not differ between hands. Since these initial reports, a number of studies of PD performance on the task have been published (Duchek et al., 1994; Freeman et al., 1994; Keele & Ivry, 1987; Ivry & Keele, 1989; O'Boyle et al., 1996a; Pastor et al., 1992b) and the apparent inconsistencies, which I have described above, in the patterns of observed impairment reported by Ivry & Keele (1989) in their different between– and within–subject comparisons involving PD patients have largely been resolved. Thus, while both Ivry & Keele (1989) and Duchek et al. (1994) observed no differences in TV, CV or MDV between controls and PD patients who were tested while maintained 'on' their normal L–dopa medication, Pastor et al. (1992b) and O'Boyle et al. (1996a) reported significant differences in all three variables in comparisons between controls and PD patients tested while 'off' dopaminergic medication. In addition, O'Boyle et al. (1996a) showed, in within–subject comparisons, that resumption of normal dopaminergic medication in patients tested initially following 12–15 h abstinence from medication, produced a significant reduction of TV, CV and MDV and that, in patients with bilaterally asymmetrical motor signs and tested while 'on' normal medication, all three variables were elevated significantly during tapping with the more–affected hand, by comparison with values observed when tapping with the less–affected hand. In several reports, moreover, evidence has been presented that the elevated TV observed among PD patients may be attributable, on occasion, to a preferential elevation of CV (Freeman et al., 1994; Ivry & Keele, 1989; Keele & Ivry, 1987, pp. 207–208; O'Boyle, 1996a; Wing et al., 1984; Wing and Miller, 1984). The single experimental finding which is inconsistent with this overall pattern of results is Ivry

& Keele's (1989) failure to find significant differences in TV, CV or MDV in seven PD patients between 'on' and 'off' dopaminergic medication conditions. Possible reasons for this discrepancy have been discussed by O'Boyle et al. (1996a).

Aside from Ivry & Keele (1989), none of these authors tested PD patients on a task of temporal perception, in addition to testing them on the continuation task. Nevertheless, there is some evidence, from other studies, that PD patients are impaired on perceptual tasks requiring the explicit representation of temporal information. For example, Artieda et al. (1992; also see Lacruz et al., 1992) reported that intra–modal (tactile, auditory and visual), sensory, temporal discrimination thresholds were elevated significantly, by comparison with controls, among a group of PD patients tested while 'off' normal dopaminergic medication, and that this impairment was both attenuated substantially by a single dose of L–dopa/carbidopa and attributable to an abnormality of central, rather than peripheral, underlying mechanisms. In a second study, Pastor et al. (1992a) found that, relative to controls, PD patients tested while 'off' dopaminergic medication were impaired on tasks of both verbal time estimation and temporal reproduction and that, in certain experimental sub-conditions, PD patients' performance was improved significantly following administration of a single dose of L–dopa/carbidopa. Other human studies bearing on the issue of impairment of temporal perception in PD patients have been reviewed recently by Pastor & Artieda (1996).

Finally, a separate body of relevant evidence derives from the animal work of Meck who, working within the theoretical context of scalar timing theory (Gibbon, 1977; Gibbon & Church, 1984), has shown that the psychometric characteristics of time estimation (within temporal ranges of 2 s – 8 s and 4 s – 16 s) in rats can be manipulated systematically by the administration of dopamine receptor agonists and antagonists (Meck, 1983, 1986).

b) *Huntington's disease*
Whereas PD is a basal ganglionic disease seen to be characterised by

poverty of movement (hypokinesia), Huntington's disease (HD) isone of a number of neurological conditions which involve *excessive* involuntary movement (hyperkinesia), including chorea. In the early stages of the disease, the primary neuropathology is a loss of the GABA/enkephalin neurones which give rise to the 'indirect' pathway projecting from the striatum to the external globus pallidus. In terms of transmission within the proposed 'motor' circuit described above, the hyperkinetic symptoms of HD are thought to reflect a reduction of inhibition of GPe by striatum, which will result in reduced excitatory drive, by the 'indirect' pathway, of GPi/SNr and, therefore, excessive disinhibition of thalamus and cortex (DeLong, 1990).

The only published study of continuation performance by HD patients is that of Freeman et al. (1996), who reported that tapping TV, CV and MDV were all significantly higher among a group of five HD patients, than among controls. In the light of our current understanding of the differences in the pathologies underlying PD and HD, it is interesting that the two diseases should give rise to the same pattern of impairment on the continuation task. However, the theoretical significance of this finding is not yet clear as it is known that the pathological processes in HD progress to affect widespread areas within the basal ganglia and cerebral cortex (Albin et al., 1989) and that bradykinesia accompanies chorea frequently in HD (Thompson et al., 1988).

Finally, it is of interest in this context that while the pattern of acquisition of eye–blink conditioning appears to be unimpaired in patients with PD (Daum et al., 1996) and in patients with HD (Woodruff–Pak & Papka, 1996), the optimisation of the timing of conditioned responses has been reported by the latter authors to be abnormal in HD.

Conclusions: the status of the 'cerebellar timing hypothesis', and the possible role of extra–cerebellar neural systems in explicit temporal representation

On the basis of their own early work, and in addressing the question of what the precise role of the cerebellum might be, Keele &

Ivry developed the ideas of previous researchers (e.g. Braitenberg, 1967; Pellionisz & Llinas, 1982) in proposing the 'cerebellar timing hypothesis', which was described earlier and according to which the cerebellum "can be conceptualised as a relatively task–independent timing mechanism" (Clarke et al. 1996). Accordingly, the authors argued that impairments in the performance of a variety of tasks observed previously to be associated with cerebellar damage (e.g. locomotor ataxia, hypometria during rapid movements, and impaired acquisition and retention of classically–conditioned motor responses) might all be explained in terms of a common underlying deficit to a modular timing system (see Ivry & Corcos, 1993; Keele & Ivry, 1990). During the past six years, the basic hypothesis has undergone some degree of elaboration by Ivry and his colleagues in that, *i)* the cerebellar timing system is thought to be non–oscillatory in nature (Clarke et al., 1996; also see Ivry & Hazeltine, 1995) and, as a consequence, its operation is assumed to be restricted to relatively brief durations within a range of several milliseconds to several seconds (Ivry, 1993; Clarke et al., 1996), and *ii)* different regions of the cerebellum are thought to provide separate timing computations for different effectors (e.g. the two hands) and different tasks, and outputs of these separate timing processes are hypothesised to be constrained by a common output gate (Franz et al., 1996; Helmuth and Ivry, 1996).

It is undoubtedly the case that a considerable body of experimental evidence indicates that the integrity of the cerebellum is necessary for the explicit representation of temporal information during both normal continuation performance and during normal temporal perception. Nevertheless, the claims which lie at the heart of the 'cerebellar timing hypothesis', that is that the cerebellum provides the *primary* neurobiological substrate or mechanism underlying motor timing *and* temporal perception, rest primarily on two critical pieces of evidence, one of which has yet to be replicated and a second which appears to apply to neural systems involving the basal ganglia, as well as those involving the cerebellum.

The first piece of evidence is Ivry et al.'s (1988) demonstration, among seven patients, of a double dissociation of the two processes hypothesised by Wing & Kristofferson (1973b): lateral cerebellar lesions were associated with a specific elevation of CV, whereas medial cerebellar lesions gave rise to a specific elevation of MDV. This finding is of critical import because without it, there is no reason to assign a special role to the cerebellum in explicit motor timing during continuation: the evidence that an elevation of both CV and MDV, or a specific elevation of CV, is associated with damage to basal ganglionic systems (Freeman et al., 1994, 1996; Ivry & Keele, 1989; Keele & Ivry, 1987; O'Boyle, 1996a; Pastor et al., 1992b; Wing et al., 1984; Wing and Miller, 1984) is equally as good as that which derives from studies of patients with undifferentiated unilateral or bilateral cerebellar damage (Ivry & Keele, 1989; Franz et al., 1996). It is of considerable importance for the 'cerebellar timing hypothesis', therefore, that Ivry et al.'s (1988) findings be replicated. This will not be a straightforward matter, however, as patients with focal, unilateral lesions of the cerebellum are uncommon, while those in whom the lesion is restricted to either the lateral or medial aspect of one hemi-cerebellum are very rare. We have examined the tapping continuation performance of one patient, B.P., with a unilateral cerebellar lesion, and some preliminary, unpublished data[5] from our experiments are presented in Figure 1. B.P., who was the subject of the first published report of impaired acquisition, in humans, of classical eye–blink conditioning following a cerebellar lesion (Lye et al., 1988), suffered a spontaneous right cerebellar hemisphere infarction (nine years prior to testing on the continuation task) resulting in a lesion which, according to CAT/MRI scans and operation notes, was restricted to the posterior two thirds and lateral aspect of the right cerebellar hemisphere, with considerable sparing of the intermediate zone, and preservation of the anterior third of the hemisphere and of the vermis. B.P. was tested repeatedly on the continuation tapping task using each hand separately at a target interval of 550 ms. While, by comparison with that of eight, age–matched controls, his performance with each hand was characterised by substantially elevated TV and CV and a slightly elevated MDV, values of none the three variables varied substantially according to whether B.P. used the hand ipsilateral or contralateral to his lesion (Figure 1). These data will be presented in full at a later date (O'Boyle et al., 1996b).

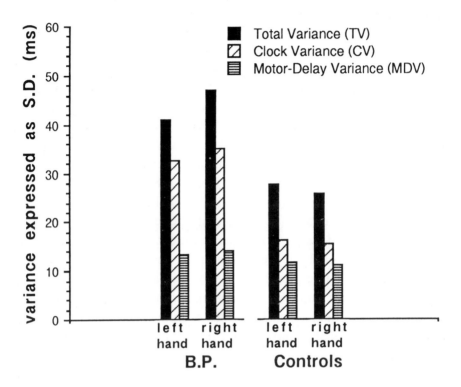

Figure 1. Mean TV, CV and MDV during continuation performance with each hand, for the cerebellar patient B.P., and group mean values for eight age–matched controls.

The second fundamental building block of the 'cerebellar timing hypothesis' was provided by the demonstration that cerebellar damage can, in the same subjects, give rise to impairments not only of motor timing, but also of performance on tasks of temporal perception requiring explicit temporal representation. This latter finding, which has been replicated, constitutes the critical evidence supporting the argument that the essential function which is localised uniquely to the cerebellar system is the provision of timing computations whenever they are needed across a variety of tasks, irrespective of their nature.

This argument has been challenged, however, by the demonstration that impairments of both motor timing during continuation and temporal perception accompany damage to basal ganglionic systems, although it should be noted that this has yet to be demonstrated in the same subjects, and the data concerning impairments of temporal perception in parkinsonian patients are not as extensive as is the case for patients with damage to the cerebellum. In cognisance of such evidence, Ivry has suggested that the cerebellar system, having evolved originally in the service of the co-ordination of action and motor control, might operate specifically within a temporal range bound by an upper limit of two or three seconds (Ivry, 1993; Clarke et al., 1996) and that beyond this limit, other neural systems (including those involving the basal ganglia) may play a more important role in timing operations. Such an argument appears to be weakened, however, by observations that impairments of motor timing during continuation at target intervals ranging between 400 ms and 2000 ms accompany damage to basal ganglionic systems (Freeman et al., 1996; O'Boyle et al., 1996a; Pastor et al., 1992b), and the report that tactile, auditory and visual temporal discrimination (in the low ms range) is impaired centrally in parkinsonian patients (Artieda et al., 1992). In addition, impairments of temporal reproduction within a temporal range of 1–8 s (Schugens et al., 1994b) and of temporal perception within a range of 8–32 s (Nichelli et al., 1996) have been reported following cerebellar damage, although the degree to which such impairments might reflect a specific disorder of timing, rather than of other cognitive processes including attention, is not clear.

These issues await clarification in future research: it is important to know whether or not the data of Ivry et al. (1988) can be replicated, and more information is required about the nature of the impairments of temporal perception which accompany basal ganglionic dysfunction. More generally, there is a need to generate bodies of data, about the performance of the same patients on a variety of motor and perceptual tasks requiring the explicit representation of temporal information, which are equivalent for patients with damage to different neural systems, so that similarities and differences between observed patterns of impairment which are associated with different neuropathologies can be compared more systematically. An important contribution, in the pursuit of such a strategy, might be expected to result from the detailed dissection of temporal impairments on different tasks in individual patients, as well as from group studies. More information is also required about the performance of patients with primary neuropathology located outside of the cerebellum or basal ganglia. Relatively little attention has been paid in this literature, for example, to the potential role of regions of the cerebral cortex, despite reports that, during continuation performance, unilateral cortical lesions extending into pre–motor regions can give rise to significantly elevated values of CV or MDV, "but rarely both", when tapping with the hand contralateral to the lesion (Ivry & Keele, 1989), and that, by comparison with the performance of controls, CV, but not MDV, is elevated among patients with diffuse cortical damage associated with mild senile dementia of the Alzheimer type (Duchek et al., 1994). One potentially–promising, yet rarely used, approach to these problems might involve the use of appropriate dual–task methodology with both neurological patients and neurologically–intact individuals, to examine the degree to which the processing resources and, by implication, the underlying neural mechanisms employed in different timing tasks overlap.

Before turning, in the next section, to consideration of the timing of repetitive movements in the presence of extrinsic timing cues, brief mention should be made of several other issues in this literature which require closer attention. First, the predictions of Wing & Kristofferson's (1973b) model that, during continuation, lag–1 covariance should be negative and covariance at lags > 1 should be zero appear to be violated more commonly in the performance of neurological patients than in that of normal individuals. Such violations (which may reflect primary violations of model assumptions and/or violations of the statistical assumption of stationarity in the series of response intervals), and the manner in which they are addressed, can have significant theoretical consequences, some of which have been discussed by O'Boyle et al. (1996a; also see Helmuth & Ivry, 1996; Heuer, 1996a; Vorberg & Wing, 1996). Another, and very difficult, problem concerns the possible influence of neural mechanisms of adaptive change to neuropathology, including those underlying recovery of function and compensation (e.g. see Latash & Anson, 1996): there is some evidence, for example, that impairments of temporal perception observed following cerebellar damage may be short–lasting (Ivry et al., 1988; also see Clarke et al., 1996) and, in patients with clinical signs indicative of early and/or mild cerebellar damage, of the possible operation of compensatory mechanisms in the cerebral cortex which may play a role in meliorating impairments of temporal reproduction and reaction time (Schugens, 1994a,b). Finally, the degree to which impairments of temporal perception are dependent upon the extent and location of cerebellar damage is also unclear (see Ackermann et al., 1996; Ivry et al., 1988; Schugens et al., 1994b).

Synchronization (temporal tracking): Explicit movement–timing in the presence of extrinsic timing cues

In the continuation experiments described above, the synchronization phase served merely to establish a reference interval which had to be re–produced by subjects during the continuation phase. Synchronization performance, however, has been studied in its own right for at least 95 years (Dunlap, 1910; Miyake, 1902; Scripture 1899), especially since the publication of Michon's (1967) remarkable monograph. Normal subjects are able easily to synchronize on–the–beat tapping to an isochronous, auditory metronome and, but for a small average anticipatory error of the order of tens of ms (see below), accurately. This means, among other things, that a tap timed to synchronize with the next metronome beat must be initiated in advance of the beat. Assuming that subjects use the timing information provided by the extrinsic timekeeper to regulate the accuracy of performance, one might suppose, therefore, that effective synchronization requires that subjects be able to, *i)* evaluate, represent and re–produce the reference interval provided by the metronome, *ii)* predict the times of occurrence of both metronome beat and associated tap, *iii)* perceive any temporal asynchrony between metronome beat and tap and, *iv)* regulate subsequent performance on the basis of that perception. Thus, viewed in this way, the characteristics of synchronization performance will reflect the operation of underlying mechanisms for the accurate perception and re–production of metronome intervals, for the perception of temporal simultaneity, temporal succession and temporal order of metronome beat and tap, and for feedback correction of timing on the basis of perceived synchronization error.

Normal temporal–tracking performance: experimental data[6]

Four dependent variables (DV) are of interest: the *metronome interval* (MI), and a subject's *inter–tap interval* (ITI), *delay interval* (DI: the interval between a given metronome beat and the subsequent tap intended to synchronize with the next metronome beat) and *synchronization error interval* (SE: by convention, the error is negative if the tap precedes the metronome beat and positive if beat precedes tap). As in studies of continuation performance, experimental attention in the temporal–tracking literature has been centred upon both the static (in particular, the mean and variability of SE, and variability[7] of ITI) and dynamic (sequential) characteristics of performance which are reflected in the statistical dependencies between the durations of the various intervals defined above (characterised by auto– and cross–covariance and correlation functions).

Normal subjects start to tap in synchrony with an auditory metronome with reasonable accuracy within the first five or so beats, although it should be noted that it is not clear when, following the start of tapping, performance stabilises. The feature of stable performance which has received most attention is that subjects, although exhorted to tap on the beat exactly in time with the metronome, do so, on average, slightly in advance (about 20–60 ms) of the beat. While this mean anticipatory (negative) synchronisation (i.e. phase) error has proved to be a highly reliable finding, its magnitude (and, on occasions, its sign) have been shown to vary, as have the observed values of other static DVs, as a function of variation in the following independent variables (IV).

a) The duration of the metronome interval.

Although the literature is not entirely consistent, the range of auditory MIs within which a subject can synchronize reliably seems to lie between about 300 ms and 2000 ms (Fraisse, 1966; Mates et al., 1994; Peters, 1989). Within this range, the mean SE is reliably negative while its (negative) magnitude may increase with MI, and the distributions of ITIs and SEs are roughly symmetrical. In addition, while both Mates et al. (1994) and Peters (1989) observed that the mean absolute standard deviations (SD) of the SE and of the ITI increased gradually across the range, others have reported that the variability about the ITI is lowest within a region centred on an MI of between about 400 and 800 ms, either side of which the variability increases (e.g. Fraisse, 1982; Michon, 1967; Woodrow, 1932). At MIs of less than about 300 ms, which is larger than the ITI corresponding to the maximum rate of tapping, the mean SE is zero or positive and subjects report lack of a subjective impression of reliable synchronisation and that taps are not experienced individually (Peters, 1989). Above about 1800 ms to 2400 ms, SE distributions tend increasingly to become bimodal, reflecting the occurrence of both anticipatory and reactive taps, and subjects again report a difficulty in synchronizing (Mates et al., 1994). Peters (1989) has argued that a sudden increase in the mean SD of the ITI at 300 ms reflects a transition from automatic to controlled movement, while Mates et al. (1994) opined that the breakdown of synchronization at long MIs reflects the operation of a temporal integration process with a maximum capacity of about 3 s (also see Woodrow, 1932). This breakdown might also be interpreted to reflect the upper limits of accuracy and precision of an internal interval–timing mechanism (see Mates, 1994b).

b) The modality in which the metronome is presented.

Early observations that the negative SE tended to be smaller when tapping to a visual metronome than when using an auditory metronome (Dunlap, 1910; Miyake, 1902; also see Bartlett & Bartlett, 1959) have been confirmed and extended by Kolers and Brewster (1985). Using auditory (A), visual (V) and tactile (T) metronomes with MIs of 400 ms, 500 ms and 600 ms, these authors found that the magnitudes of mean negative SEs associated with each condition were ordered, A (most negative) >T >V (least negative), and only in the case of the visual metronome at an MI of 400 ms was the mean SE positive. Some preliminary data from my own laboratory also support this pattern of findings: in a study in which four subjects were required to synchronise tapping with a metronome with an MI of 550 ms, the mean negative SE was smaller when subjects tracked a visual metronome than when they tracked an auditory metronome (Figure 2).

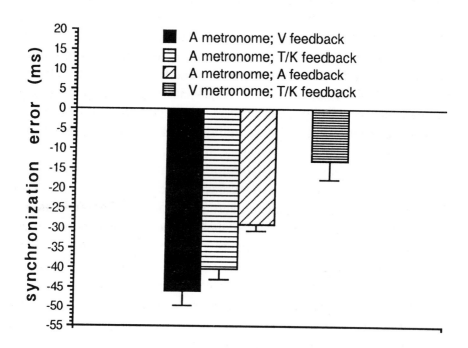

Figure 2. Group mean (+s.e.m.) synchronization error during tracking for each of four different combinations of modality of a regular metronome (MI =550 ms) and modality of feedback. A = auditory; V = visual; T/K = tactile/kinaesthetic only. Note that normal T/K feedback was present during all runs.

c) The nature of the available sensory feedback information.

It has been reported, for both finger– and toe–tapping in time with an auditory metronome, that the magnitude of the negative SE is reduced if subjects are provided with tap–contingent auditory feedback, in addition to the tactile/kinaesthetic feedback deriving naturally from the effector (Aschersleben & Prinz, 1995; Mates et al., 1992). Aschersleben & Prinz (1995) showed that this effect was not attributable to auditory masking and, in an additional experiment (Aschersleben & Prinz, 1996) in which they investigated the effect of imposing a delay between the tap and the contingent auditory feedback, they found that the magnitude of the negative SE increased as a linear function of the delay. Furthermore, in a preliminary report of an experiment involving an auditory metronome, and in which subjects were provided with either visual (V), auditory (A) or no tap–contingent feedback, in addition to the veridical tactile/kinaesthetic feedback (T/K) arising from the finger, O'Boyle & Clarke (1996) showed that the magnitude of the negative SEs associated with each condition was ordered, V (most negative) > T/K >A (least negative) (Figure 2).

d) The motor effector employed.

In 1949, Paillard reported that if subjects are required simultaneously, and in a self–paced manner, to raise the heel and to extend the ipsilateral index finger, the onset of heel movement precedes the onset of finger movement, whereas the order of movements is reversed when subjects respond in reaction to an auditory stimulus. This finding has been replicated by Bard et al. (1991, 1992) and Billon et al. (1996a). Bard and her colleagues (1992) also demonstrated that the order of movements, during the self–paced condition was reversed in a subject suffering from de–afferentation arising from specific, permanent loss of the large sensory, myelinated fibres in all four limbs (resulting in the total loss of the senses of touch, vibration, pressure and kinaesthesia, with preservation of pain and temperature sensation, and absence of tendon reflexes). These findings were then extended by Billon et al. (1996a) who observed that if normal subjects are required to track an auditory metronome simultaneously with repetitive finger extension and raising of the heel, movement of the heel preceded that of the finger which, in turn, preceded the metronome; that is, a larger negative SE was observed for heel raising than for 'simultaneous' finger extension.

What is the source of the negative synchronization error?

A number of different, but not necessarily mutually–exclusive, explanatory accounts of the anticipatory SE have been proposed. Here, I describe four of them (also see Aschersleben & Prinz, 1995).

1. The 'perceptual–latency' (or 'Paillard–Fraisse') hypothesis.
It has been suggested that the negative SE reflects primarily a difference between the perceptual latency in the modality in which the metronome information is presented and that in the modality in which the movement–contingent feedback is available. Furthermore, it is proposed that the difference in perceptual latencies will reflect, in turn and at least in part, the difference in conduction times in the particular sensory pathways involved. This hypothesis was first suggested by Paillard (1949) and was elaborated subsequently by Fraisse (1980). According to this hypothesis, subjects evaluate the temporal simultaneity of a metronome beat and associated tap on the basis of a central comparison of the time of arrival or 'availability' of sensory information derived from the metronome with that of the sensory feedback derived from the occurrence of the tap (c.f. Dennett & Kinsbourne, 1992). In the typical case of tapping in time with an auditory metronome, then, a mean negative SE will be observed because the perceptual latency associated with the tactile/kinaesthetic modality is longer than that associated with the auditory modality which, in turn, will reflect, to some extent at least, the fact that the tactile/kinaesthetic information has farther to travel than does the auditory information. Thus, if the two sorts are information are to arrive or be available centrally at the same time, the tactile/kinaesthetic information has to start first, and this can only be effected if the tap

occurs appropriately in advance of the metronome beat.

All of the effects described above concerning the effects, on the magnitude of the SE, of the modality of the metronome and of the nature of the available feedback are consistent with the supposition that, under the stimulus conditions pertaining during the described variations in the nature of the synchronization task, the perceptual latencies in the auditory (A), visual (V) and tactile/kinaesthetic (T/K) modalities are ordered in magnitude, V > T/K > A. The fact that the magnitude of the mean SE is reduced, but not to a value of zero, when a subject is provided with tap–contingent feedback in the same modality as the metronome (e.g. see Figure 2), is consistent with the idea that the auditory feedback information is incorporated into a "joint–event code" (Aschersleben & Prinz, 1995, 1996) with the concurrently–available, natural T/K information deriving from the tap. Several other pieces of evidence support the notion of the directional ordering of perceptual latencies described above. First, the mean simple reaction time to auditory stimuli has been observed to be about 20–50 ms faster than that to visual stimuli (Jaśkowski et al., 1990; Pöppel et al., 1990; Rutschmann & Link, 1964). Second, it has been reported that a visual stimulus must be presented in advance of an auditory stimulus in order for the two to be judged as occurring simultaneously (e.g. see Jaśkowski et al., 1990), However, it must be added that there is a good deal of inconsistency of data in the complex literature on the judgement of intra– and heteromodal simultaneity, succession and order (e.g. also see Hirsh & Sherrick, 1961; Rutschmann & Link, 1964), that it has been demonstrated that such judgements can vary considerably as a function of the degree to which attention is directed to one or other stimulus (e.g. Kristofferson, 1967; Stelmach & Herdman, 1991; Stone, 1926), and that it is unlikely that simple perceptual–latency models can provide the best explanation for the perceptual data (e.g. Sternberg & Knoll, 1973; Ulrich, 1987).

The reported precession of toe/heel raising over finger extension during self–paced movement and during simultaneous synchronization to a metronome may be explained, in terms of Paillard–Fraisse hypothesis, by the fact that the distance to the brain which feedback tactile/kinaesthetic information has to traverse is longer from the foot than from the finger. This explanation receives some support from the findings of Halliday & Mingay (1964) who found that electrical stimulation of the toe had to precede electrical stimulation of the finger, in order for the two stimuli to judged as simultaneous. However, the temporal separation of the two stimuli when judged to be simultaneous could not be explained entirely in terms of the difference in latencies of the cortical potentials evoked by electrical stimulation at the two sites.

Finally, some very recent evidence suggests that the role of sensory feedback during temporal tracking may be restricted to the setting of a mean SE, rather than the regulation of on–line correction of SE repeatedly during a run, as is hypothesised in several models of synchronization performance (see below). This evidence comes from a study by Billon et al. (1996b), of tracking by a de–afferented patient (A.N.) who, as a consequence of the loss of large–diameter, myelinated sensory fibres, has lost all sensation (other than pain and temperature) below the neck. In the absence of any sensory (including auditory and visual) feedback, A.N. proved to be able to track an isochronous metronome in much the same way, and with about the same mean SE, as normal, control subjects. Given that this remarkable finding has only just appeared in the literature, its reliability, and its full theoretical significance for understanding normal tracking performance, have yet to be evaluated.

2. The 'P–centre' hypothesis.

Other evidence suggests that the negative SE arises from, or is at least modulated by, a temporal asymmetry in the perceptual (P)–centres (Morton et al., 1976) of the metronome stimulus, which is usually short in duration, and of the tap, the duration of which is usually relatively much longer. That is to say, if, during synchronization, a subject's strategy is to align temporally the P–centres of metronome and tap, rather than their physical onsets, then the longer the relative duration of the tap (dwell time), the more the physical onset of the tap should be shifted to the left of the metronome beat and, hence, the more negative should be the SE. Vos et al. (1995) have summarised the evidence supporting this hypothesis, and they reported data which support the derived prediction that, with tap duration held constant, the magnitude of the negative SE should vary as an inverse function of duration of the metronome stimulus. However, if it is assumed that the P–centres of metronome stimuli and taps are located in the same positions relative to the times of their physical onsets (e.g. in the temporal centre), the P–centre hypothesis is not able to explain why a negative SE is observed when taps and metronome beats have the same duration (as in Vos et al., 1995, Figure 5, at a metronome–stimulus duration of 100 ms). It is important, therefore, that the question of the respective locations of the P–centres, relative to times of physical onset, of metronome stimuli and taps be examined.

3. The 'evaluation' (or 'SE cost–function') hypothesis.

Vos and Helsper (1992) have suggested that "..... the subject, aware of his inability to be perfectly accurate, prefers being too early to being too late"; that is, a negative SE is produced as a consequence of an (intentional) asymmetric evaluation of SEs during error correction. Koch (1992) has shown that, according to this hypothesis, the magnitude of the negative SE should increase with the variance of the timekeeper mechanism underlying tap–generation. If this is the case, then the mean and variance of the negative SEs should correlate positively. However, Aschersleben & Prinz (1996; also see Dunlap, 1910) found no support in their data for this hypothesis.

4. The 'undershooting' (or 'optimisation of timekeeper precision') hypothesis.

Vorberg and Wing (1996) have argued that, as the variance of timed response intervals (and, it is supposed, the variance of intervals produced by an underlying, internal timekeeper) tends to increase with the mean, a consistent negative SE might reflect an optimisation of timekeeper variance (precision) at the expense of a bias for anticipation. This hypothesis awaits further examination and analysis.

Normal temporal-tracking performance: models

In terms of a subject's processing capabilities, we should expect the characteristics and limits of temporal-tracking performance to be constrained by those of the sensory, perceptual and motor systems involved, and by the accuracy and precision of the internal mechanisms underlying both the temporal regulation and execution of movements. A number of linear, information-processing models of normal synchronization performance have been proposed (Hary & Moore, 1985, 1987a,b; Mates, 1994a,b; Michon, 1967; Michon & van der Valk, 1968; Schulze, 1992; Voillaume, 1971; Vorberg & Wing, 1996; Vos & Helsper, 1992). Each of these models contains, *i)* an hypothesised internal, interval timer (or 'clock') which can be set to the metronome period, and which may be reset to zero by some local event (e.g. the metronome beat, or the response) after each cycle, and *ii)* some hypothesised mechanism(s) for the on-line correction of response timing on the basis of the evaluation and feedback of the synchronization error (i.e. involving closed-loop control). On the other hand, the models differ according to the precise nature of the hypothesised underlying processes and mechanisms (e.g. whether both period- and phase-correction are involved) the extent to which distinction is made between variables which are either external or internal to the subject, and, as might be expected, the degree to which they are consistent with the psychological data. A succinct summary of the important functional characteristics of different models has been provided by Mates (1994a).

The validity of models of temporal tracking has been assessed primarily by analysis of their statistical properties and in terms of the success with which they are able to provide an explanation of, *i)* the observed mean anticipatory SE, *ii)* the observed patterns of sequential dependencies (characterised by the auto- and cross-covariances and correlations) between the various intervals defined at the beginning of this section, *iii)* the characteristics of subjects' performance in the face of sudden or gradual changes in metronome interval and, in some instances, *iv)* mathematical analysis of their stability (Hary & Moore, 1987b; Mates 1994b). For example, Hary & Moore (1987a), in a study of temporal tracking by trained musicians, observed a negative lag–1 cross–correlation between metronome intervals[8] and delay intervals, and a positive lag–1 autocorrelation between delay intervals. In similar vein, Michon (1967) has characterised the performance of subjects when tracking sinusoidal and ramped metronomes (also see Ehrlich, 1958; Gottsdanker, 1954), and when responding to the injection of a sudden, positive or negative step in metronome interval. The results produced by these authors have been used as 'benchmark' data against which to test model performance.

The currently most promising model of temporal tracking appears to be that of Mates (1994a,b) as, in addition to coping well with the tests described above, it is more highly specified in terms of realistic psychological processes than other models. Nevertheless, detailed description of this, or other, models here would be premature because, as yet, there have been no published attempts to map the hypothesised internal variables on to the nervous system nor, with the exception of the few studies described below, to analyse the effects of neurological damage on synchronization performance within the empirical and theoretical context provided by the psychological literature on normal performance. This is clearly an important, and potentially productive, area for future research.

Abnormal synchronization performance: the effects of neurological damage

It is surprising to find, given the long tradition of study of synchronization in psychology, the simplicity of the task, and the interesting nature of the processing capacities assumed to underlie performance, that the task has received relatively little attention in the neurological literature. Most of what little work that has been done has concerned the performance of parkinsonian patients.

Studies of patients with Parkinson's disease

a) The 'hastening' phenomenon

The only systematic body of work is that of Nakamura and Nagasaki and their colleagues, who appear to have been unaware of the considerable psychological literature on normal synchronization performance. In experiments in which PD patients and controls were required to synchronize tapping with an auditory metronome at frequencies ranging between 1 Hz (MI: 1000 ms) and 7.5 Hz (MI: 133 ms), Nakamura et al. (1976, 1978) found that, at a 'transition' metronome frequency of 2.5 Hz (MI: 400 ms), the tapping of many PD patients became desynchronized (as indicated by the rate of tapping) and abruptly increased to a frequency of 4–6 Hz (MI: 167–200 ms), which was usually independent of, and invariably lower than, subjects' maximal rate of tapping. This phenomenon, which they termed 'hastening', also occurred in some subjects (although less often) at other metronome frequencies, especially 4 Hz (MI: 250 ms), and was observed, on at least one side, in 66% of PD patients. In addition, while the SD of the PD patients' response intervals generally declined with increasing metronome frequency, an abrupt increase was evident at the metronome transition–frequency whereafter, during hastened tapping, the SD returned to a much lower value. On the grounds that, both among controls and among PD patients who did not show hastening, plots of the SD of response intervals against metronome frequency also revealed peaks (in a generally declining function) at 2.5 Hz and 5 Hz, and on the basis of the characteristics of the response–interval auto-correlelograms of controls and patients, the authors concluded that the hastened tapping represented an intrinsic oscillation of 5–6 Hz in the CNS which was masked in normal subjects, but was released in PD when the variance of subjects' response intervals exceeded some critical value (e.g. at 2.5 Hz). Furthermore, as such hastening was observed during continuation (in the absence of the metronome) and, therefore, was not dependent on feedback–based error correction, they attributed the phenomenon to the impairment of a mechanism of intrinsic "rhythm formation in the CNS", which they modelled in two ways.

First, they proposed a mathematical, 'forced–oscillation' model (Nagasaki et al., 1978), according to which the appearance of hastening is attributable to the impairment of an inhibitory mechanism which damps an inherent response instability of the hypothesised rhythm–formation system at certain resonant frequencies. Second, they modelled the phenomenon in terms of a purely peripheral oscillatory system, based on the skeletal muscle reflex pathway, in which hastening was attributed to impairment of a mechanism which serves normally to inhibit the gain of a feedback–loop, controlled by the gamma motoneurones, from muscle to alpha motoneurones (Nagasaki & Nakamura, 1982).

These authors also reported that hastening was observed during synchronization using effectors other than the finger (Nagasaki & Nakamura, 1982), that its appearance was related clinically to the phenomenon of 'freezing' or 'festination' commonly observed in parkinsonism (Nakamura et al., 1976), and that the frequency with which hastened tapping was observed among healthy subjects increased with age, especially after the age of 50 (Nagasaki et al., 1989). Finally, they argued that hastening is attributable to organic or functional impairment specifically of the striatum, as hastening was also observed in patients with Huntington's disease, striato–nigral degeneration, and olivo–ponto–cerebellar atrophy (OPCA, in which striatal lesions are implicated), but was rarely observed in patients with forms of spinocerebellar degeneration other than OPCA, or following lesions confined to the thalamus or cerebellar cortex (Kosaka et al., 1982).

Several features of this work invite comment. First, given that only isochronous metronome sequences were employed and that subjects were not required to accentuate particular taps, the authors' reference to underlying processes of "rhythm formation" might be interpreted more appropriately as reference to an intrinsic timing mechanism. Second, although the authors show that it is possible, at least in principle, to model hastening in terms of an impairment of a purely peripheral mechanism, one obvious implication of the work is that the normal functioning of this hypothesised timing mechanism is dependent upon the functional integrity of brain systems involving the basal ganglia. However, as many of the potentially theoretically-significant characteristics of the hastening phenomenon in patients with disorders of the basal ganglia do not appear to have been replicated by others (e.g. see Freeman et al. 1993, 1996; Pastor, 1992b), and the observed peaks of response interval variance, among controls, at metronome frequencies of 2.5 Hz (MI: 400 ms) and 5 Hz (MI: 200 ms) are not consistent with the normal psychological literature described above, it might be argued that the significance of this work is restricted to simply that of the characterisation of the upper response–rate limits of synchronization and continuation performance in PD and other patient groups, limits which might be expected to be determined by factors having nothing to do directly with timing. Thus, the theoretical significance of this body of work remains unclear.

b) Synchronization error

It has been observed repeatedly that, during continuation performance at a target frequency of around 550 ms, PD patients tap, on average, too fast, and faster than controls (e.g. see O'Boyle 1996a). Reasoning that this might reflect, among such patients, an impairment of the establishment of a reference interval (or the setting of an internal 'clock', or the entrainment of a neural oscillator) during the synchronization phase of the task, I and my colleagues, John Freeman and Fred Cody, have recently started to examine the synchronization performance of PD patients in detail.

In our initial experiments (Freeman, 1994; O'Boyle et al., 1995), we examined the performance of six PD patients (while taking their normal L–dopa medication) and six control subjects (matched with respect to age and general cognitive state) who were required to synchronize finger tapping during each of six series of 120 isochronous metronome intervals of 700 ms. When asked, at the end of each run, both control and PD subjects usually expressed the opinion that, on average during the run, they had tapped "on" the metronome beat, indicating that subjects in neither group were aware of a mean anticipatory error. As shown in Figure 3, however, we found that the PD patients tapped significantly further in advance of the metronome than controls; that is, the mean negative SE was larger for the patients than for controls, a finding which we have very recently confirmed among a separate group of PD patients. How might this be explained?

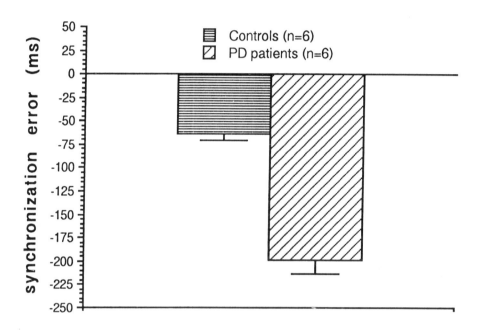

Figure 3. Group mean (+s.e.m.) synchronization error during tracking of a regular metronome (MI = 700 ms) by patients with Parkinson's disease (PD) and controls.

We are currently conducting experiments to determine the degree to which this finding might be understood in terms of one or other of the various hypotheses which have been proposed to explain the smaller mean negative SE observed in neurologically–intact individuals, and which were described earlier in this chapter. For example, based on the 'perceptual latency' hypothesis, one explanation for the larger mean SE of PD patients might be that the difference in perceptual latencies between the auditory and tactile/kinaesthetic (T/K) modalities is larger for PD subjects than for controls. As it seems unlikely that the auditory perceptual latency among PD subjects is shorter than that among controls, this explanation requires that T/K information, or its central 'availability' be delayed. Such a notion is not new: in two, rarely–quoted papers published more than 30 years ago, Dinnerstein proposed that delayed tactile/proprioceptive feedback might be the essential impairment underlying parkinsonian motor signs (Dinnerstein et al., 1962, 1964; also see Marsden, 1982), and the more general idea that the basal ganglia play a role in the sensory regulation of movement has been around even longer (e.g. see Martin, 1967; Schneider & Lidsky, 1985).

It is also possible that the larger negative mean SE of PD patients reflects primarily a mean displacement of their tap P–centres (hence, tap onsets) to the left, relative to the P–centres of metronome beats, which would occur if their tap durations (dwell times) were longer than those of controls. On the other hand, the abnormally large SE of our PD patients might best be explained in terms of their having, in the presence of normal tactile–kinaesthetic delay, an abnormally high threshold for the perception of temporal order between metronome beat and tap. Impairments of the perception of temporal order have been associated with damage to a variety of brain regions (reviewed by Nichelli, 1993) and, in the current context, it is of particular interest that impairments of (intramodal) temporal discrimination (Artieda et al., 1992) and (heteromodal) temporal order judgements (Dinnerstein, 1964) have been reported to attend parkinsonism (also see Lacruz et al., 1991; Pastor & Artieda, 1996). Why it should be that, in the case of our PD patients, this hypothesised increase in threshold should be asymmetrical about simultaneity might be explained by recourse to an hypothesis couched in terms of 'attentional bias'. Although it has been known for at least 70 years (Stone, 1926) that the perceived temporal order of heteromodal stimuli may vary according to which stimulus subjects direct their attention and although, as mentioned earlier, the phenomenon has been the subject of continued interest in the literature on temporal perception, it does not seem to have received serious consideration in the temporal–tracking literature, other than by Mates (1994a). According to this hypothesis, the negative SE observed in normal subjects might be attributable to subjects' attention being directed preferentially to, or 'captured' by, the metronome stimulus. This, in turn, might result in the subjective impression of delay, relative to the metronome, of the corresponding tap. As a consequence, subjects might be expected to advance the relative time of the tap in order to compensate. According to this hypothesis, then, the fact that PD patients tap further in advance of the metronome than controls might be explicable in terms of an abnormally large effect of the metronome as an attentional 'attractor', or of a reduction in the salience of sensory feedback. Such a notion is consistent with clinical observations that the initiation of movements in PD patients with a freezing gait can be facilitated by visual or auditory stimuli (Martin, 1967), with experimental evidence which suggests that PD patients may be impaired in the provision and/or use of internal, as opposed to external, cues for the control of attention (Brown & Marsden, 1988) and for the generation of movement (e.g. Georgiou et al., 1993), and with the observation that PD patients' response timing is more accurate during synchronization than during continuation (Freeman et al., 1993).

Finally, I should like to mention one other interesting piece of evidence from our preliminary experiments. Discussion of these data so far has been restricted to that of the static DV of mean SE. In extending our analysis to the question of the sequential characteristics of SEs during a run, we have observed that the magnitudes of successive SEs of PD patients often tend to vary in a cyclical fashion (while remaining negative throughout), sometimes with high amplitude and regular period. An example of this phenomenon is portrayed in Figure 4. It remains to be seen whether the cycling of SE reflects a magnification of a similar phenomenon in controls (c.f. Dunlap, 1910) and whether it reflects potentially an impairment of mechanisms underlying the regulation of ITI (period) and/or the evaluation and correction of SE (phase).

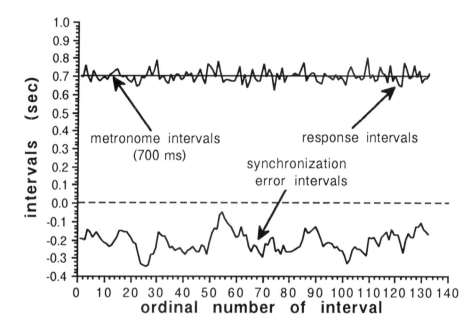

Figure 4. Successive metronome intervals (700 ms), response intervals and synchronization–error intervals during a single tracking run, for a patient (J.S.) with Parkinson's disease. Note the cyclical variation of the synchronization error during the run. This run does not represent the most pronounced example of such variation observed among PD patients.

Studies of other patient groups

I am aware of only two other studies of abnormal temporal-tracking performance. First, it may be inferred from an abstract published by Abell (1962) that his schizophrenic subjects tapped further in advance of an auditory metronome (MI of 1000 ms) than did controls. However, no full account of this work seems to have been published subsequently. Second, Najeson et al. (1989) studied a group of six patients, with either cortical (n=4) or subcortical (n=2) lesions sustained following unilateral cerebrovascular accidents, on synchronization to auditory, visual and tactile metronomes at isochronous MIs of 1 s, 2 s, 3 s and 4 s. The authors reported that the patients, unlike controls, failed to produce "predictive" responses with either one or both hands. However, for a number of reasons (e.g. the reported SE was invariably positive, even among controls at MI of 1s, and the authors defined "predictive" responses as those occurring within 150 ms following the metronome beat), the significance of these data is difficult to assess.

Conclusions

    The study of the ways in which temporal–tracking performance may be impaired by damage to different neural systems would seem to be a highly promising area for future research, although it may be, as intimated by Vorberg & Wing (1996), that such studies will reveal more about the neural mechanisms underlying sensory–motor and/or sensory–sensory synchronization than about those underlying response timing. However, this remains to be seen. Currently, we are engaged in extending our work on the effects of damage to basal ganglionic systems beyond a concern with simple static performance variables to the study of response kinematics, to the analysis of the sequential dependencies between intervals produced by metronome and subject, and to the modelling of patients' performance. In addition, we are extending these studies to include the examination of neurological patients with damage to other neural systems.

General summary and conclusions

Although there may be circumstances under which it is difficult to determine whether the temporal structure of action is emergent or controlled, it seems reasonable to suppose that the successful completion of many sequences of movements, including those, for example, which are highly constrained by external temporal considerations, depends upon the operation of internal mechanisms for the generation of the explicit representation of time in a way which, within certain limits, is isomorphous with real time. If this is the case, what neural form might such representations take, and what neural systems are involved in their provision? One general approach to such problems has been described in this chapter. As stated at the outset, the research which I have discussed has been conducted from a general perspective of information–processing psychology, within which a dominant theoretical role has been played by the notion that explicit timing computations are provided by an internal, general–purpose, task–independent 'clock' or timekeeper. This notion derived from the idea that certain important kinematic and kinetic characteristics of voluntary movements are controlled by central 'motor programs' (e.g. Keele, 1968). Furthermore, the neuropsychological work conducted within this framework suggests unequivocally that the normal control of explicit timing of repetitive movements, and the explicit temporal representations underlying important aspects of normal temporal perception, are dependent upon the functional integrity of the cerebellum and of the basal ganglia. As yet, however, we have little idea of the precise roles played by these and other structures in the explicit processing of temporal information, of the nature of the underlying mechanisms or, given the highly–complex connectivity of the nervous system in general and of the motor systems in particular (e.g. see Alexander et al., 1992), of the degree to which timing functions might be localised to specific neural structures or systems. It is far from clear, for example, how abstract 'motor programs' (Alexander et al., 1992; Summers, 1992) or the metaphorical, black–box 'clocks' contained within information–processing theories of motor timing or temporal perception might be realised physically in the nervous system, despite the evidence in mammals, including humans, for the existence of spinal and brainstem central pattern generators underlying the temporal organisation of certain rhythmic behaviours (e.g. Calancie et al., 1994; Nakamura & Katakura, 1995), of neocortical–thalamic 40 Hz oscillators (e.g. Jeffreys et al., 1996) and of one or more entrainable, circadian pacemakers located within the suprachiasmatic nucleus of the hypothalamus (e.g. Ishida, 1995; Miller, 1993). Indeed, there is no general consensus among members of the information–processing fraternity about the nature and properties of 'clocks' hypothesised within one or other model, or even about how many separate 'clocks' there should or might be.

These are difficult issues. Nevertheless, the problem of the complex connectivity of the nervous system and the associated problems of neuropsychological inference (e.g. see Farah, 1994) are not, of course, restricted to this literature and there is good reason to believe that significant advances in our understanding of how and where the nervous system represents time will accrue, as in related areas of research in neuropsychology and cognitive neuroscience, from a convergence of information derived from the use of a diversity of experimental and theoretical approaches, including functional imaging in combination with the temporal resolution afforded by EEG recording (e.g. Jahanshahi et al., 1995), computational modelling (e.g. Miall, 1996) and dynamical analyses (e.g. Kelso et al., 1992). Time, one must assume, will tell.

Notes

1.   The question of whether a linear function better describes the relationship between target interval and, on the one hand, the variance of 'clock' intervals or, on the other hand, the standard deviation of 'clock' intervals is not pursued further here. The question is of theoretical interest because the exact form of the relationship is potentially informative about the functional characteristics of the hypothetical clock or timekeeping process (see, for example, Gibbon, 1977; Ivry & Corcos, 1993; Killeen & Weiss, 1987; Ivry & Hazeltine, 1995). A related issue concerns what exactly CV is the variance of: as it is calculated indirectly, and represents all variance in the system which is not attributable to MDV, Ivry & Corcos (1993; also see Ivry & Hazeltine, 1995) have argued that it should be thought of as constituting 'central' variance, of which 'clock' variance will be an important component. There are, however, problems with this formulation in which MDV is attributed entirely to peripheral (or non–central) processes. For example, and as explained later in the main text, this view would require that the medial regions of the cerebellum be considered as peripheral structures. Either way, the lack of specification in the model beyond two processes does constrain severely the possible scope of neuropsychological analysis.

2. The model has been applied, for example, to the analysis of rhythmic performance (Vorberg & Hamburg, 1978; Vorburg & Wing, 1996) and of synchronous and alternate two–handed tapping (Helmuth & Ivry, 1996; Wing, 1982; Wing et al, 1989; Vorberg & Wing, 1996), to the reproduction of tonal sequences (Vos & Ellerman, 1989; Mates, 1991), to the issue of relative–timing invariance (Heuer, 1996b; Vorberg & Wing, 1996), to the modelling of temporal tracking (Vorberg & Wing, 1996), and to the analysis of the effects of normal ageing (Greene & Williams, 1993).

3. The distal motor effects of a unilateral cerebellar lesion are manifest primarily on the side ipsilateral to the lesion. This is because the output from the intermediate and lateral zones of each cerebellar hemisphere undergoes a double crossing of the midline before accessing spinal motor neurones. The motor effects of unilateral cortical and basal ganglionic lesions are, on the other hand, manifest primarily contralaterally. This organisation of the motor projections in these three groups allows, in patients with unilateral lesions or bilaterally–asymmetrical motor signs, the use of powerful within–subject designs involving comparison of performance between effectors on the two sides.

4. 'Clumsy' children were defined as "children who have developmental apraxia and agnosia but who are otherwise normal" (Williams et al., 1992, p. 165).

5. The model prediction of negative lag–1covariance was violated in 13.9% of all trials for control subjects and 15.3% of all trials for B.P. In the decomposition of TV into CV and MDV, and in the calculation of the mean values of these variables portrayed in Figure 1, the uncorrected, positive values of lag–1 covariance from such runs were employed. In this preliminary analysis, in addition, possible violations of the model prediction of $\gamma_k = 0$, for k >1 were ignored.

6. Discussion of studies of temporal tracking in this chapter is restricted primarily to those in which a single isochronous series of metronome intervals was employed, and in which subjects were required to produce a series of unaccentuated taps of similar nature as, except in the study of the de–afferented patient A.N. (Billon et al., 1996b), and in my own preliminary work with John Freeman and Fred Cody, only such series have been used in the investigation of neurological patients. By way of an introduction to studies of performance of more complex forms of synchronization task as used, for example, in the study of rhythm, the interested reader is directed to Auxiette et al. (1992), Billon & Semjen (1995), Deutsch (1983), Franěk et al. (1991, 1994) and Mates et al. (1992).

7. Note that, aside from its use in calculating serial covariances and correlations, the mean ITI during temporal tracking is usually of little interest because if, as should be the case during synchronization performance, subjects produce one tap to each metronome beat, any difference between mean MI and mean ITI will simply reflect

a difference in the magnitudes of the SEs to the first and last taps of the series. This arises as a trivial consequence of the way in which mean MI and mean ITI are calculated (total duration/number of intervals): so long as the number of taps in a series is equal to the number of metronome beats, and if the SEs to the first and last taps are equal, the mean ITI will be exactly equal to the mean MI, irrespective of how the taps are distributed throughout the series. Thus, if the mean ITI is not almost identical to the mean MI, then subjects are not tapping on a one-to-one basis with the metronome beats; in other words, they are not temporally–tracking. Mean ITI is of interest, therefore, only as an index of the breakdown, or limits of synchronization, as in studies of the way in which normal or abnormal performance varies as a function of MI at very short or long durations.

8. The way in which a subject might use the temporal information provided by the metronome beat in the regulation of synchronization behaviour has been characterised by the cross–covariance and cross–correlation functions between MI and ITI, DI and SE. As these functions cannot be calculated in case of a perfectly isochronous metronome, investigators have used a 'random metronome', in which successive intervals vary fractionally in duration, around some pre–determined mean, to an extent which is not detectable by the subject. Such intervals might, for example, be drawn randomly from a normal distribution with a given mean and SD. The use of such a metronome was introduced by Michon (1967).

References

Abell, A.T. (1962). Sensory–motor synchrony and schizophrenia: a study of temporal performance. *Dissertation Abstracts, 23*, 1776-1777.

Ackermann, H., Gräber, S., Hertrich, I. & Daum, I. (1996). Categorical speech perception in cerebellar disorders. *Brain & Language*. In press.

Albin, R.L., Young, A.B. & Penney, J.B. (1989). The functional anatomy of basal ganglia disorders. *Trends in Neuroscience, 12*, 366-375.

Alexander, G.E. & Crutcher, M.D. (1990). Functional architecture of basal ganglia circuits: neural substrates of parallel processing. *Trends in Neuroscience, 13*, 266-271.

Alexander, G.E., DeLong, M.R. & Crutcher, M.D. (1992). Do cortical and basal ganglionic motor areas use 'motor programs' to control movement? *Behavioral and Brain Sciences, 15*, 656-665.

Artieda, J., Pastor, M.A., Lacruz, F. & Obeso, J.A. (1992). Temporal discrimination is abnormal in Parkinson's disease. *Brain, 115*, 199-210.

Aschersleben, G. & Prinz, W. (1995). Synchronizing actions with events: the role of sensory information. *Perception & Psychophysics, 57*, 305-317.

Aschersleben, G. & Prinz, W. (1996). Delayed auditory feedback in synchronization. Manuscript submitted for publication.

Auxiette, C., Drake, C. & Gerard, C. (Eds.), (1992). *Proceedings of the Fourth Rhythm Workshop: Rhythm Perception and Production*. Bourges, France.

Bard, C., Paillard, J., Teasdale, N., Fleury, M. & Lajoie Y. (1991). Self-induced versus reactive triggering of synchronous hand and heel movement in young and old subjects. In J. Requin & G.E. Stelmach (Eds.), *Tutorials in motor neuroscience* (pp. 189–196). Dordrecht: Kluwer.

Bard, C., Paillard, J., Lajoie, Y., Fleury, M., Teasdale, N., Forget, R. & Lamarre, Y. (1992). Role of afferent information in the timing of motor commands: a comparative study with a deafferented patient. *Neuropsychologia, 30*, 201–206.

Bartlett, N.R. & Bartlett, S.C. (1959). Synchronization of a motor response with an anticipated sensory event. *Psychological Review, 66*, 203–218.

Billon, M., Bard, C., Fleury, M., Blouin, J. & Teasdale, N. (1996a). Simultaneity of two effectors in synchronization with a periodic external signal. *Human Movement Science, 115*, 25–38.

Billon, M. & Semjen, A. (1995). The timing effects of accent production in synchronization and continuation tasks performed by musicians and nonmusicians. *Psychological Research, 58*, 206–217.

Billon, M., Semjen, A., Cole, J. & Gauthier, G. (1996b). The role of sensory information in the production of periodic finger–tapping sequences. *Experimental Brain Research, 110*, 117–130.

Braitenberg, V. (1967). Is the cerebellar cortex a biological clock in the millisecond range? *Progress in Brain Research, 25*, 334–346.

Brown, R.G. & Marsden, C.D. (1988). Internal versus external cues and the control of attention in Parkinson's disease. *Brain, 111*, 323–345.

Calancie, B., Needham–Shropshire, B., Jacobs, P., Willer, K., Zych, G. & Green, B.A. (1994). Involuntary stepping after chronic spinal cord injury. Evidence for a central rhythm generator for locomotion in man. *Brain, 117*, 1143–1159.

Chesselet, M.-F. & Delfs, J.M. (1996). Basal ganglia and movement disorders: an update. *Trends in Neurosciences, 19*, 417–422.

Church, R.M. & DeLuty, M.Z. (1977). Bisection of temporal intervals. *Journal of Experimental Psychology: Animal Behavior Processes, 8*, 165–186.

Clarke, S., Ivry, R., Grinband, J., Roberts, S. & Shimizu, N. (1996). Exploring the domain of the cerebellar timing system. In M.A. Pastor & J. Artieda (Eds.), *Time, internal clocks and movement*. Amsterdam: Elsevier Science.

Collyer, G.L. & Wright, C.E. (1995). Temporal rescaling of simple and complex ratios in rhythmic tapping. *Journal of Experimental Psychology: Human Perception and Performance, 21*, 602–627.

Daum, I., Schugens, M.M., Ackermann, H., Lutzenberger, W., Dichgans, J. & Birbaumer, N. (1993). Classical conditioning after cerebellar lesions in humans. *Behavioral Neuroscience, 107*, 748–756.

Daum, I., Schugens, M.M., Breitenstein, C., Topka, H. & Spieker, S. (1996). Classical eyeblink conditioning in Parkinson's disease. *Movement*

Disorders, 11, 639–646.

DeLong, M.R. (1990). Primate models of movement disorders of basal ganglia origin. *Trends in Neuroscience, 13*, 281–285.

Dennett, D.C. & Kinsbourne, M. (1992). Time and the observer: the where and the when of consciousness in the brain. *Behavioral & Brain Sciences, 15*, 183–247.

Deutsch, D. (1983). The generation of two isochronous sequences in parallel. *Perception & Psychophysics, 34*, 331–337.

Dinnerstein, A.J., Frigyesi, T. & Lowenthal, M. (1962). Delayed feedback as a possible mechanism in parkinsonism. *Perceptual and Motor Skills, 15*, 667–680.

Dinnerstein, A.J., Lowenthal, M., Blake, G. & Mallin, R.E. (1964). Tactile delay in parkinsonism. *Journal of Nervous and Mental Disease, 139*, 521–524.

Duchek, J.M., Balota, D.A. & Ferraro, F.R. (1994). Component analysis of a rhythmic finger tapping task in individuals with senile dementia of the Alzheimer type and in individuals with Parkinson's disease. *Neuropsychology, 8*, 218–226.

Dunlap, K. (1910). Reactions to rhythmic stimuli, with attempt to synchronize. *Psychological Review, 17*, 399–416.

Ehrlich, S. (1958). Le mécanisme de la synchronisation sensori–motrice: étude expérimentale. *L'Année Psychologique, 58*, 7–23.

Farah, M.J. (1994). Neuropsychological inference with an interactive brain: a critique of the 'locality' assumption. *Behavioral & Brain Sciences, 17*, 43–104.

Fraisse, P. (1966). L'anticipation de stimulus rythmiques vitesse d'établissement et précision de la synchronisation. *L'Année Psychologique*, 66, 15–36.

Fraisse, P. (1980). Les synchronisations sensori–motrices aux rhythms. In J. Requin (Ed.), *Anticipation et comportement* (pp. 233–257). Paris: Centre National.

Fraisse, P. (1982). Rhythm and tempo. In D. Deutsch (Ed.), *The psychology of music* (pp. 149–180). New York: Academic Press.

Franěk, M., Mates, J., Radil, T., Beck, K. & Pöppel, E. (1991). Sensorimotor synchronization: motor responses to regular auditory patterns. *Perception & Psychophysics, 49*, 509–516.

Franěk, M., Mates, J., Radil, T., Beck, K. & Pöppel, E. (1994). Sensorimotor synchronization: motor responses to pseudoregular auditory patterns. *Perception & Psychophysics, 55*, 204–217.

Franz, E.A., Ivry, R.B. & Helmuth, L.L. (1996). Reduced timing variability in patients with unilateral cerebellar lesions during bimanual movements. *Journal of Cognitive Neuroscience, 8*, 107–118.

Freeman, J.S. (1994). *The role of the basal ganglia in the control of motor timing*. Unpublished doctoral dissertation. University of Manchester, England.

Freeman, J.S., Cody, F.W.J., O'Boyle, D.J., Crauford, D., Neary, D. & Snowden, J.S. (1996). Abnormalities of motor timing in Huntington's disease. *Parkinsonism & Related Disorders, 2*, 81–93.

Freeman, J.S., Cody, F.W.J. & Schady, W. (1993). The influence of external timing cues upon the rhythm of voluntary movements in Parkinson's disease. *Journal of Neurology, Neurosurgery & Psychiatry, 56,* 1078–1084.

Freeman, J.S., O'Boyle, D.J. & Cody, F.W.J. (1994). A longitudinal study of the accuracy of motor timing in a subject with Parkinson's disease, before and during a period of L–dopa medication. *Journal of Physiology (London), 480.P,* 47P.

Georgiou, N., Iansek, R., Bradshaw, J.L., Phillips, J.G., Mattingley, J.B. & Bradshaw, J.A. (1993). An evaluation of the role of internal cues in the pathogenesis of parkinsonian hypokinesia. *Brain, 116,* 1575–1587.

Gibbon, J. (1977). Scalar expectancy theory and Weber's law in animal timing. *Psychological Review, 84,* 279–325.

Gibbon, J. & Church, R.M. (1984). Sources of variance in information processing theories of timing. In H.L. Roitblat, T.G. Bever & H.S. Terrace (Eds.), *Animal cognition* (pp. 465–488). Hillsdale, NJ: Erlbaum

Gottsdanker, R.M. (1954). The continuation of tapping sequences. *Journal of Psychology, 37,* 123–132.

Greene, L.S. & Williams, H.G. (1993). Age–related differences in timing control of repetitive movement: application of the Wing–Kristofferson model. *Research Quarterly for Exercise & Sport, 64,* 32–38.

Halliday, A.M. & Mingay, R. (1964). On the resolution of small time intervals and the effect of conduction delays on the judgement of simultaneity. *Quarterly Journal of Experimental Psychology, 16,* 35–46.

Hardiman, M.J., Ramnani, N. & Yeo, C.H. (1996). Reversible inactivations of the cerebellum with muscimol prevent the acquisition and extinction of conditioned nictitating membrane responses in the rabbit. *Experimental Brain Research, 110,* 235–247.

Hary, D. & Moore, G.P. (1985). Temporal tracking and synchronization strategies. *Human Neurobiology, 4,* 73–77.

Hary, D. & Moore, G.P. (1987a). Synchronizing human movement with an external clock source. *Biological Cybernetics, 56,* 305–311.

Hary, D. & Moore, G.P. (1987b). On the performance and stability of human metronome–synchronization strategies. *British Journal of Mathematical and Statistical Psychology, 40,* 109–124.

Helmuth, L.L. & Ivry, R.B. (1996). When two hands are better than one: reduced timing variability during bimanual movements. *Journal of Experimental Psychology: Human Perception and Performance, 22,* 278–293.

Heuer, H. (1996a). Coordination. In H. Heuer & S.W. Keele (Eds.), *Handbook of perception and action. Volume 3: Motor skills* (pp. 121–180). London: Academic Press.

Heuer, H. (1996b). The timing of human movements. In F. Lacquanti & P. Viviani (Eds.), *Neural bases of motor behaviour.* Dordrecht: Kluwer. In press.

Hirsh, I.J. & Sherrick, C.E. (1961). Perceived order in different sense modalities. *Journal of Experimental Psychology, 62,* 423–432.

Ishida, N. (1995). Molecular biological approach to the circadian clock mechanism. *Neuroscience Research, 23*, 231–240.

Ivry, R.B. (1993). Cerebellar involvement in the explicit representation of temporal information. *Annals New York Academy of Sciences, 682*, 214–230.

Ivry, R.B. & Corcos, D.M. (1993). Slicing the variability pie: component analysis of coordination and motor dysfunction. In K.M. Newell & D.M. Corcos (Eds.), *Variability and motor control* (pp. 416–447). Champaign, Illinois: Human Kinetics.

Ivry, R.B. & Diener, H.C. (1991). Impaired velocity perception in patients with lesions of the cerebellum. *Journal of Cognitive Neuroscience, 3*, 355–366.

Ivry, R.B. & Gopal, H.S. (1992). Speech production and perception in patients with cerebellar lesions. In D.E. Meyer & S. Kornblum (Eds.), *Attention and performance XIV* (pp. 771–802). Cambridge, Massachusetts: Bradford.

Ivry, R.B. & Hazeltine, R.E. (1995). Perception and production of temporal intervals across a range of durations: evidence for a common timing mechanism. *Journal of Experimental Psychology: Human Perception and Performance, 21*, 3–18.

Ivry, R.B. & Keele, S.W. (1989). Timing functions of the cerebellum. *Journal of Cognitive Neuroscience, 1*, 136–152.

Ivry, R.B., Keele, S.W. & Diener, H.C. (1988). Dissociation of the lateral and medial cerebellum in movement timing and movement execution. *Experimental Brain Research, 73*, 167–180.

Jahanshahi, M., Jenkins, I.H., Brown, R.G., Marsden, C.D., Passingham, R.E. & Brooks, D.J. (1995). Self-initiated versus externally triggered movements. I. An investigation using measurement of regional cerebral blood flow with PET and movement–related potentials in normal and Parkinson's disease subjects. *Brain, 118*, 913–933.

Jaśkowski, P., Jaroszyk, F. & Hojan–Jezierska, D. (1990). Temporal–order judgements and reaction time for stimuli of different modalities. *Psychological Research, 52*, 35–38.

Jeffreys, J.G.R., Traub, R.D. & Whittington, M.A. (1996). Neuronal networks for induced '40 Hz' rhythms. *Trends in Neuroscience, 19*, 202–208.

Jueptner, M., Rijntjes, M., Weiller, C., Faiss, J.H., Timmann, D., Mueller, S.P. & Diener, H.C. (1995). Localization of a cerebellar timing process using PET. *Neurology, 45*, 1540–1545.

Keele, S.W. (1968). Movement control in skilled motor performance. *Psychological Bulletin, 70*, 387–403.

Keele, S.W. & Ivry, R.B. (1987). Modular analysis of timing in motor skill. In G. Bower (Ed.), *The psychology of learning and motivation* (pp. 183–228). New York: Academic Press.

Keele, S.W. &, Ivry, R.B. (1990). Does the cerebellum provide a common computation for diverse tasks? *Annals New York Academy of Sciences, 608*, 179–207.

Keele, S.W., Ivry, R.I. & Pokorny, R. (1987). Force control and its relation to timing. *Journal of Motor Behavior, 19*, 96–114.

Keele, S.W., Manchester, D.L. & Rafal, R.D. (1985a). Is the cerebellum involved in motor and perceptual timing? A case study. *University of Oregon Cognitive Science Technical Report 85–5*. Eugene.

Keele, S.W., Pokorny, R.A., Corcos, D.M. & Ivry, R. (1985b). Do perception and motor production share common timing mechanisms: a correlational analysis. *Acta Psychologica, 60*, 173–191.

Kelso, J.A.S., DelColle, J.D. & Schöner, G. (1992). Action–perception as a pattern formation process. *Attention & Performance, 35R*, 139–169.

Kelso, J.A.S., Holt, K.G., Rubin P. & Kugler, P.N. (1981). Patterns of human interlimb coordination emerge from the properties of non–linear, limit cycle oscillatory processes: theory and data. *Journal of Motor Behaviour, 13*, 226–261.

Killeen, P. & Weiss, N. (1987). Optimal timing and the Weber function. *Psychological Review, 94*, 455–468.

Koch, R. (1992). *Sensumotorische Synchronisation: eine Kostenanalyse* [Sensorimotor synchronization: a cost analysis] (internal report #11/92). München: Max–Planck–Institut für Psychologische Forschung (cited by Aschersleben & Prinz (1995, 1996), Vorberg & Wing (1996) and Vos et al. (1995)).

Kolers, P.A. & Brewster, J.M. (1985). Rhythms and responses. *Journal of Experimental Psychology: Human Perception and Performance, 11*, 150–167.

Kosaka, K., Nagasaki, H. & Nakamura, R. (1982). Finger tapping test as a means to differentiate olivo–ponto–cerebellar atrophy among spinocerebellar degenerations. *Tohoku Journal of Experimental Medicine, 136*, 129–134.

Kristofferson, A.B. (1967). Attention and psychological time. *Acta Psychologica, 27*, 93–100.

Kugler, P.N. & Turvey, M.T. (1987). *Information, natural law, and the self–assembly of rhythmic movement*. Hillsdale, NJ: Erlbaum.

Lacruz, F., Artieda, J., Pastor, M.A., & Obeso, J.A. (1991). The anatomical basis of somaesthetic temporal discrimination in humans. *Journal of Neurology, Neurosurgery & Psychiatry, 54*, 1077–1081.

Lashley, K.S. (1951). The problem of serial order in behaviour. In Jeffress, L.A. (Ed.), *Cerebral mechanisms in behavior* (pp. 112–146). New York: Wiley.

Latash, M.L. & Anson, J.G. (1996). What are 'normal movements' in atypical populations? *Behavioral & Brain Sciences, 19*, 55–106.

Lundy–Ekman, L., Ivry, R., Keele, S. & Woollacott, M. (1991). Timing and force control in clumsy children. *Journal of Cognitive Neuroscience, 3*, 367–376.

Lye, R.H., O'Boyle, D.J., Ramsden, R.T. & Schady, W. (1988). Effects of a unilateral cerebellar lesion on the acquisition of eye–blink conditioning in man. *Journal of Physiology (London), 403*, 58P.

MacKay, D.G. (1987). Constraints on theories of sequencing and timing in language perception and production. In A. Allport, D.G. Mackay, W. Prinz & E.

Scheerer (Eds.), *Language perception and production: relationships between listening, speaking, reading and writing* (pp. 407–429). Orlando: Academic Press.

Maquet, P., Lejeune, H., Pouthas, V., Bonnet, M., Casini, L., Macar, F., Timsit-Berthier, M., Vidal, F., Ferrara, A., Degueldre, C., Quaglia, L., Delfiore, G., Luxen, A., Woods, R., Mazziotta, J.C. & Comar, D. (1996). Brain activation induced by estimation of duration: a PET study. *Neuroimage, 3*, 119–126.

Marsden, C.D. (1982). The mysterious motor function of the basal ganglia: The Robert Wartenburg Lecture. *Neurology, 32*, 514–539.

Martin, J.P. (1967). *The basal ganglia and posture*. London: Pitman.

Mates, J. (1991). Extending the model of self-paced periodic responding: comment on Vos and Ellerman (1989). *Journal of Experimental Psychology: Human Perception and Performance, 15*, 877–879.

Mates, J. (1994a). A model of synchronization of motor acts to a stimulus sequence. I. Timing and error corrections. *Biological Cybernetics, 70*, 463–473.

Mates, J. (1994b). A model of synchronization of motor acts to a stimulus sequence. II. Stability analysis, error estimation and simulations. *Biological Cybernetics, 70*, 475–484.

Mates, J., Radil, T., Müller, U. & Pöppel, E. (1994). Temporal integration in sensorimotor synchronization. *Journal of Cognitive Neuroscience, 6*, 332–340.

Mates, J., Radil, T. & Pöppel, E. (1992). Co–operative tapping: time control under different feedback conditions. *Perception & Psychophysics, 52*, 691–704.

McIntosh, R.D. & O'Boyle, D.J. (1995). Effects of perturbation of delay in response-contingent, auditory feedback on the endogenous timing of repetitive movements. *Journal of Physiology (London), 489.P*, 30–31P.

Meck, W.H. (1983). Selective adjustment of the speed of internal clock and memory processes. *Journal of Experimental Psychology: Animal Behavior Processes, 9*, 171–201.

Meck, W.H. (1986). Affinity for the dopamine D2 receptor predicts neuroleptic potency in decreasing the speed of an internal clock. *Pharmacology, Biochemistry & Behaviour, 25*, 1185–1189.

Miall, C. (1996). Models of neural timing. In M.A. Pastor & J. Artieda (Eds.), *Time, internal clocks and movement*. Amsterdam: Elsevier Science.

Michon, J.A. (1967). *Timing in temporal tracking*. Soesterberg, The Netherlands: Institute for Perception RVO–TNO.

Michon, J.A. & van der Valk, N.J.L. (1967). A dynamic model of timing behavior. *Acta Psychologica, 27*, 204–212.

Miller, J.D. (1993). On the nature of the circadian clock in mammals. *American Journal of Physiology, 264*, R821–R832.

Miyake, I. (1902). Researches in rhythmic action. *Studies from the Yale Psychology Laboratory, 10*, 1–48.

Morton, J., Marcus, S.M. & Frankish, C. (1976). Perceptual centres (P–centres).

Psychological Review, 83, 405–408.

Nagasaki, H., Itoh, H., Maruyama, H. & Hashizume, K. (1989). Characteristic difficulty in rhythmic movement with aging and its relation to Parkinson's disease. *Experimental Aging Research, 14*, 171–176.

Nagasaki, H., Nakamura, R. & Taniguchi, R. (1978). Disturbances of rhythm formation in patients with Parkinson's disease: Part II. A forced oscillation model. *Perceptual and Motor Skills, 46*, 79–87.

Nagasaki, H. & Nakamura, R. (1982). Rhythm formation and its disturbances – a study based upon periodic response of a motor output system. *Journal of Human Ergology, 11*, 127–142.

Najeson, T., Ron, S. & Behroozi, K. (1989). Temporal characteristics of tapping responses in healthy subjects and in patients who sustained cerebrovascular accident. *Brain, Behaviour and Evolution, 33*, 175–178.

Nakamura, Y. & Katakura, N. (1995). Generation of masticatory rhythm in the brainstem. *Neuroscience Research, 23*, 1–19.

Nakamura, R., Nagasaki, H. & Narabayashi, H. (1976). Arrhythmokinesia in parkinsonism. In W. Birkmayer & O. Hornykiewicz (Eds.), *Advances in parkinsonism* (pp. 258–268). Basle: Roche.

Nakamura, R., Nagasaki, H. & Narabayashi, H. (1978). Disturbances of rhythm formation in patients with Parkinson's disease: Part I. Characteristics of tapping response to the periodic signals. *Perceptual and Motor Skills, 46*, 63–75.

Nichelli, P. (1993). The neuropsychology of human temporal information processing. In F. Boller & J. Grafman (Eds.), *Handbook of neuropsychology. Vol. 8* (pp. 339–371). Amsterdam: Elsevier Science.

Nichelli, P., Alway, D. & Grafman, J. (1996). Perceptual timing in cerebellar degeneration. *Neuropsychologia, 34*, 863–871.

O'Boyle, D.J. & Clarke, V.L. (1996). On the source of the negative synchronization error during temporal–tracking performance. *Brain Research Association Abstracts, 13*, 40.

O'Boyle, D.J., Freeman, J.S. & Cody, F.W.J. (1995). Sensorimotor synchronization during temporal tracking is impaired in patients with Parkinson's disease (PD). *Journal of Physiology (London), 485.P*, 15P.

O'Boyle, D.J., Freeman, J.S. & Cody, F.W.J. (1996a). The accuracy and precision of timing of self-paced, repetitive movements in subjects with Parkinson's disease. *Brain, 119*, 51–70.

O'Boyle, D.J., Todd, I.L., Lye, R.H., Ramsden, R.T. & Schady, W. (1996b). Effects of a unilateral cerebellar lesion on the acquisition of eye–blink conditioning, and on simple reaction time, interval production and motor timing: a human case–study. Manuscript in preparation.

Paillard, J. (1949). Quelques données psychophysiologiques relatives au déclenchement de la commande motrice. *L'Année Psychologique, 47–48*, 28–47.

Papka, M., Ivry, R.B. & Woodruff-Pak, D.S. (1995). Selective disruption of eyeblink classical conditioning by concurrent tapping. *NeuroReport, 6,* 1493–1497.

Pastor, M.A. & Artieda, J. (1996). Involvement of the basal ganglia in timing perceptual and motor tasks. In M.A. Pastor & J. Artieda (Eds.), *Time, internal clocks and movement* (pp. 235–255). Amsterdam: Elsevier Science.

Pastor, M.A., Artieda, J., Jahanshahi, M. & Obeso J.A. (1992a). Time estimation and reproduction is abnormal in Parkinson's disease. *Brain, 115,* 211–25.

Pastor, M.A., Jahanshahi, M., Artieda, J. & Obeso, J.A. (1992b). Performance of repetitive wrist movements in Parkinson's Disease. *Brain, 115,* 875–91.

Pellionisz, A. & Llinas, R. (1982). Space–time representation in the brain: the cerebellum as a predictive space–time metric tensor. *Neuroscience, 7,* 2949–2970.

Peters, M. (1989). The relationship between variability of intertap intervals and interval duration. *Psychological Research, 51,* 38–42.

Pöppel, E., Schill, K. & von Steinbüchel, N. (1990). Sensory integration within temporally neutral systems states: a hypothesis. *Naturwissenschaften, 77,* 89–91.

Pressing, J. (1995). Testing dynamical and cognitive models of rhythmic pattern production. In D.J. Glencross & J.P. Piek (Eds.), *Motor control and sensory motor integration: issues and directions* (pp. 141–170). Amsterdam: Elsevier Science.

Rothwell, J.C. (1995). The basal ganglia. In F.W.J. Cody (Ed.), *Neural control of skilled human movement* (pp. 13–30). London: Portland Press.

Rutschmann, J. & Link, R. (1964). Perception of temporal order of stimuli differing in sense mode and simple reaction time. *Perceptual and Motor Skills, 18,* 345–352.

Schneider, J.S. & Lidsky, T.I. (Eds.), (1985). *Basal ganglia and behavior: sensory aspects of motor functioning.* Toronto: Hans Huber.

Schugens, M.M., Daum, I., Ackermann, H. & Lutzenberger, W. (1994a). Reaction times and the contingent negative variation (CNV) in cerebellar dysfunction. *Journal of Psychophysiology, 8,* 71–72.

Schugens, M.M., Daum, I., Ackermann, H., Lutzenberger, W. & Birbaumer, N. (1994b). Processing of temporal intervals in patients with cerebellar degeneration. *Journal of Psychophysiology, 8,* 366–367.

Schulze, H–H. (1992). The error correction model for the tracking of a random metronome: statistical properties and an empirical test. In F. Macar, V. Pouthas, & W.J. Friedman (Eds.), *Time, action and cognition. Towards bridging the gap* (pp. 275–286). Dordrecht: Kluwer.

Scripture, E.W. (1899). Observations on rhythmic action. *Studies from the Yale Psychology Laboratory, 8,* 102–108.

Sergent, V., Hellige, J.B. & Cherry, B. (1993). Effects of responding hand and concurrent verbal processing on time–keeping and motor–implementation processes. *Brain & Cognition, 23,* 243–262.

Squire, L., Zola–Morgan, S. & Chen, K. (1988). Human amnesia and animal models of amnesia: performance of amnesic patients on tests designed for the monkey. *Behavioral Neuroscience, 102*, 210–221.

Stelmach, L.B. & Herdman, C.M. (1991). Directed attention and perception of temporal order. *Journal of Experimental Psychology: Human Perception and Performance, 17*, 539–550.

Sternberg, S. & Knoll, R.L. (1973). The perception of temporal order: fundamental issues and a general model. In S. Kornblum (Ed.), *Attention and performance IV* (pp. 629–685). New York: Academic Press.

Stevens, L.T. (1886). On the time–sense. *Mind, 11*, 393–404.

Stone, S. (1926). Prior entry in the auditory–tactual complication. *American Journal of Psychology, 37*, 284–287.

Summers, J.J. (1992). The demise of the motor program. *Behavioral and Brain Sciences, 15*, 800.

Summers, J.J. & Burns, B.D. (1990). Timing in human movement sequences. In R.A. Block (Ed.), *Cognitive models of psychological time* (pp. 181–206). Hillsdale, NJ: Erlbaum.

Thompson, P.D., Berardelli, A., Rothwell, J.C., Day, B.L., Dick, J.P.R., Benecke, R. & Marsden, C.D. (1988). The coexistence of bradykinesia and chorea in Huntington's disease and its implications for theories of basal ganglia control of movement. *Brain, 111*, 223–244.

Thompson, R.F. & Krupa, D.J. (1994). Organization of memory traces in the mammalian brain. *Annual Review of Neuroscience, 17*, 519–549.

Todd, N.P.M. & Brown, G. (1996). Visualization of rhythm, time and metre. *Artificial Intelligence Review, 10*, 253–273.

Ulrich, R. (1987). Threshold models of temporal–order judgements evaluated by a ternary response task. *Perception & Psychophysics, 42*, 224–239.

Voillaume, C. (1971). Modèles pour l'études de la régulation des mouvements cadencés. *L'Année Psychologique, 71*, 347–358.

Vorberg, D. & Hambuch, R. (1978). On the temporal control of rhythmic performance. In J. Requin (Ed.), *Attention and performance VII* (pp. 535–555). Hillsdale, NJ: Erlbaum.

Vorberg, D. & Wing, A.M. (1996). Modelling variability and dependence in timing. In H. Heuer & S.W. Keele (Eds.), *Handbook of perception and action. Volume 3: Motor skills* (pp. 181–262). London: Academic Press.

Vos, P.G. & Ellerman, H.H. (1989). Precision and accuracy in the reproduction of simple tone sequences. *Journal of Experimental Psychology: Human Perception and Performance, 15*, 179–187.

Vos, P.G. & Helsper, H.H. (1992). Tracking simple rhythms: on–beat versus off–beat performance. In F. Macar, V. Pouthas, & W.J. Friedman (Eds.), *Time, action and cognition. Towards bridging the gap* (pp. 287–299). Dordrecht: Kluwer.

Vos, P.G., Mates, J. & van Kruysbergen, N.W. (1995). The perceptual centre of a stimulus as the cue for synchronization to a metronome: evidence from asynchronies. *Quarterly Journal of Experimental Psychology, 48*, 1024–

1040.

Williams, H.G., Woollacott, M.H. & Ivry, R. (1992). Timing and motor control in clumsy children. *Journal of Motor Behavior, 24*, 165–172.

Wing, A.M. (1977a). Effects of type of movement on the temporal precision of response sequences. *British Journal of Mathematical and Statistical Psychology, 30*, 60–72.

Wing, A.M. (1977b). Perturbations of auditory feedback delay and the timing of movement. *Journal of Experimental Psychology: Human Perception and Performance, 3*, 175–86.

Wing, A.M. (1979). A note on the estimation of the autocovariance function in the analysis of timing of repetitive responses. *British Journal of Mathematical & Statistical Psychology 32*, 143–145.

Wing, A.M. (1980). The long and short of timing in response sequences. In G.E. Stelmach & J. Requin, (Eds.), *Tutorials in motor behaviour* (pp. 469–486). Amsterdam: North–Holland.

Wing, A.M. (1982). Timing and co–ordination of repetitive bimanual movements. *Quarterly Journal of Experimental Psychology, 34A*, 339–348.

Wing, A.M., Church, R.M. & Gentner, D.R. (1989). Variability in the timing of responses during repetitive tapping with alternate hands. *Psychological Research, 51*, 28–37.

Wing, A.M., Keele, S. & Margolin, D.I. (1984). Motor disorder and the timing of repetitive movements. *Annals New York Academy of Sciences, 423*, 183–192.

Wing, A.M. & Kristofferson, A.B. (1973a). The timing of interresponse intervals. *Perception & Psychophysics, 13*, 455–460.

Wing, A.M. & Kristofferson, A.B. (1973b). Response delays and the timing of discrete motor responses. *Perception & Psychophysics, 14*, 5–12.

Wing, A.M. & Miller, E. (1984). Basal ganglia lesions and psychological analyses of the control of voluntary movement. In D. Evered & M. O'Connor M. (Eds.), *Functions of the basal ganglia* (pp. 242–257). London (Ciba Foundation Symposium 107): Pitman.

Woodrow, H. (1932). The effect of rate of sequence upon the accuracy of synchronization. *Journal of Experimental Psychology, 15*, 357–379.

Woodruff–Pak, D.S. & Papka, M. (1996). Huntington's disease and eyeblink classical conditioning: normal learning but abnormal timing. *Journal of the International Neuropsychological Society, 2*, 323–334.

CHAPTER 12

5–Hydroxytryptamine and Interval Timing

A.S.A. Al–Ruwaitea, S.S.A. Al–Zahrani, M.–Y. Ho, C.M. Bradshaw & E. Szabadi

This chapter reviews recent evidence for a role of the ascending 5–hydroxtryptaminergic (5HTergic) pathways of the brain in interval timing behaviour. In order to impose some order on our material, we have adopted the taxonomy of timing tasks recommended by Killeen and Fetterman (1988; see also chapter by Killeen et al. in this volume). As will become apparent, this taxonomy has helped us in our efforts to define 5HT's involvement in timing behaviour, although in so doing we have found it necessary to re–interpret, and therefore to re–classify, some traditional timing tasks, at least insofar as they have been performed by our subjects (rats).

The first two sections of the chapter outline the structure of the ascending 5HTergic pathways, and some of their behavioural functions. There follows a summary of the taxonomy of timing schedules and a brief description of the methods employed in our experiments. Then, successive sections consider the possible involvement of the 5HTergic pathways in *prospective*, *immediate* and *retrospective* timing behaviour (cf. Killeen & Fetterman, 1988), as well as acquisition of timing performance and memory for duration. In the final section, we try to draw some general conclusions about 5HT's role in timing behaviour, and discuss some possible directions for future research in this area.

1.   The ascending 5HTergic pathways of the brain

The existence of the indoleamine 5–hydroxytryptamine (5HT, serotonin) in the mammalian brain has been recognized for nearly half a century (see Brody & Shore, 1957). However it was not until the advent of fluorescence histochemistry (Falck et al., 1962) that 5HT's localization in neurones could be firmly established, and a detailed study of the anatomical distribution of 5HT–containing neuronal pathways could be undertaken (Dahlström & Fuxe, 1964). Subsequent biochemical analysis of 5HT's 'life cycle' (synthesis, intraneuronal storage, release from presynaptic terminals, re–uptake and degradation) provided strong supporting evidence for its role as a neurotransmitter (e.g. Ross & Renyi, 1969; Carlsson et al., 1972); physiological/pharmacological studies of its actions on individual neurones (e.g. Roberts & Straughan, 1967; Bloom et al., 1972) finally rendered its status as a central neurotransmitter unassailable.

The 5HTergic pathways, like other monoaminergic pathways of the brain (noradrenergic, dopaminergic), arise from relatively small, circumscribed nuclei in the brainstem; slender unmyelinated fibres project anteriorly via the medial forebrain bundle to innervate most forebrain regions, and posteriorly to innervate other brainstem nuclei and the spinal cord (Ungerstedt, 1971). The origins of the 5HTergic projection are the raphe nuclei of the tegmentum (Figure 1). Fibres projecting to the telencephalon mainly emanate from the dorsal, median and caudal raphe nuclei (areas B7, B8 and B6). Anterograde and retrograde axonal tracing studies (e.g. Geyer et al., 1976; Azmitia & Segal, 1978; Imai et al., 1986) have established that the more anteriorally located neurones of the dorsal raphe nucleus (B7) project mainly to the neostriatum, amygdala and frontal neocortex, whereas the more posteriorly located neurones of the median and caudal raphe nuclei (B8 and B6) are primarily responsible for the 5HTergic innervation of the hippocampus. However, there is evidently considerable overlap between the two branches of the ascending 5HTergic projection, since selective destruction of either the dorsal or median nucleus gives rise to partial depletion of 5HT from many forebrain regions (the caudate–putamen and hippocampus showing the greatest selectivity), whereas destruction of both raphe nuclei typically results in 90% depletion, or more, from most areas (e.g. Fletcher, 1995).

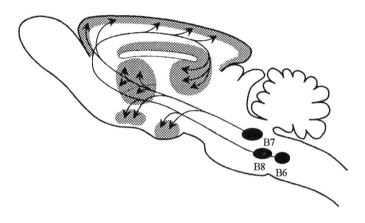

Figure 1. Principal ascending 5HTergic projections of the rat brain.

Most regions of the brain that receive a 5HTergic innervation are also innervated by catecholaminergic fibres. As a general rule, structures whose principal 5HTergic innervation originates from the more rostrally located dorsal raphe nucleus (e.g. frontal neocortex, neostriatum) receive a rich dopaminergic input, whereas those structures whose 5HTergic innervation derives mainly from the median and caudal raphe nuclei are co–innervated principally by noradrenergic fibres (Imai et al., 1986; see also Deakin, 1996). Interestingly, there is reciprocal innervation between the dorsal raphe nucleus and the brainstem dopaminergic nuclei (substantia nigra [A9] and ventral tegmental area [A10]), and between the median/caudal raphe nuclei and the noradrenergic neurones of the locus coeruleus (Imai et al., 1986).

At a functional level, the 5HTergic and catecholaminergic pathways often exert opposing influences on target structures, so that, for example, the effects of facilitating catecholaminergic neurotransmission may resemble the effects of suppressing 5HTergic transmission (e.g. Ögren, 1985; Tanii et al., 1993; Lopez–Rubalcava & Fernandez–Guasti, 1994). However, catecholamine–5HT interactions are far more complex than this generalization might suggest, as each monoamine neurotransmitter operates through a variety of postsynaptic receptor types and subtypes that differ both in their molecular structure and the cellular mechanisms to which they are coupled (see Pazos et al., 1991; Szabadi & Bradshaw, 1991; Strange, 1994). Indeed, direct application of 5HT, noradrenaline or dopamine to individual neurones within a given brain structure may result in either facilitation or suppression of firing, depending upon the balance of excitatory and inhibitory receptors and the 'background' level of activity of the neurone (Bevan et al., 1975, 1977; Szabadi et al., 1977).

The ascending 5HTergic pathways have been implicated in a wide variety of behavioural functions, and a comprehensive review of these functions is beyond the scope of this chapter. Among the more important behavioural functions to which 5HTergic mechanisms are believed to make a significant contribution are locomotor behaviour (for review, see Hillegaart et al., 1988), the acoustic startle reflex (see Davis et al., 1993), the sleep/wakefulness cycle (see Monti et al., 1994), sexual functions (see Mendelson, 1992), aggressive behaviour (see Olivier et al., 1990; Mos et al., 1993; Miczek et al., 1994), food intake (see Clifton, 1994), and the control of behaviour by aversive events (see Deakin & Graeff, 1991; Handley, 1995). A pervasive theme running through the literature on the behavioural role of the 5HTergic pathways is the concept of *inhibitory* regulation of behaviour. Thus, destruction of the 5HTergic pathways has been found to increase the startle response evoked by loud sounds (Davis et al., 1983), suppress sleep (El Kafi et al., 1994), facilitate male sexual performance (Ahlenius & Larsson, 1991), promote offensive aggression (Kantak et al., 1981), increase the intake of certain foodstuffs (Hoebel et al., 1978), and diminish the suppressant effect of punishment on operant behaviour (Tye et al., 1977; Deakin, 1983), suggesting that in the intact organism, activity within the 5HTergic pathways helps to suppress the behaviours in question, possibly via a functional opposition to the catecholaminergic pathways (for discussion, see Crow & Deakin, 1985; Robbins & Everitt, 1995; Deakin, 1996).

2. Summary of timing paradigms

Killeen & Fetterman (1988) have proposed a taxonomy of timing tasks based on the nature of the relation between the organism's behaviour and the flow of time: "A useful way of sorting these [timing tasks] into types is to ask whether the responses that are measured reflect the flow of behavior in real time, are contingent responses based on a time interval that has elapsed, or anticipate a time interval about to occur. We label these types as *immediate*, *retrospective*, and *prospective* timing" (p. 276). *Immediate* timing tasks, with their emphasis on the flow of behaviour during an ongoing interval, are most often represented, in the animal behaviour laboratory, by free–operant schedules such as the interresponse time (IRT) and fixed–interval (FI) schedules. In contrast, *retrospective* timing tasks usually entail training the organism to emit different responses, depending on the duration of some preceding event, in discrete–trials conditional discrimination schedules. The behavioural data generated by such tasks are often suitable for analysis using the techniques of classical psychophysics and signal detection theory. Different again are *prospective* timing tasks, which entail the control of behaviour by events that follow the response by a specified time interval. In most instances the controlling event is reinforcement, and thus tasks of this type frequently take the form of delayed reinforcement schedules.

We will use Killeen & Fetterman's (1988) taxonomy as a framework for our discussion, considering in turn the effects of lesions of the 5HTergic pathways on performance on *prospective*, *immediate* and *retrospective* timing tasks. In concert with the prevailing concerns of interval timing research, most of the evidence for 5HT's involvement in timing behaviour derives from experiments investigating steady–state performance maintained by schedules of these three types. However, we will also consider recent evidence suggesting that the acquisition of such performance may be affected by lesions of the 5HTergic pathways, and that such lesions may alter animals' memory for duration.

3.   General methods

The discussion contained in the following sections includes reference to a number of experiments carried out in our laboratory which attempted to address questions related to 5HT's possible role in timing behaviour. These experiments employed very simple experimental designs: the performance of rats whose ascending 5HTergic pathways had been ablated was compared with that of intact rats. The lesion, which was inflicted prior to any behavioural training, was induced, under halothane anaesthesia, by stereotaxic microinjection of the selective neurotoxin 5,7–dihydroxytryptamine (4 μg of the base dissolved in 2 μl of ascorbate–containing physiological saline solution) into the dorsal and median raphe nuclei; control rats underwent the same surgical and anaesthetic procedures without injection of the neurotoxin (see Wogar et al., 1991, 1992a, for detailed description). After completion of the behavioural experiment, the rats were killed and their brains were immediately dissected for assay of 5HT, its metabolite 5HIAA, and the two catecholamine neurotransmitters noradrenaline and dopamine, in representative brain regions innervated by the 5HTergic pathways – parietal neocortex, hippocampus, amygdala, nucleus accumbens and hypothalamus. The assays were carried out using high–performance liquid chromatography combined with electrochemical detection (see Wogar et al., 1991, 1992a, for details).

4.   5HT and performance of prospective timing tasks

It is well known that the efficacy of a reinforcer diminishes as a function of its temporal separation from the response upon which it is contingent ('delay of reinforcement gradient'). A common ploy in studying the effects of pre–reinforcer delays is to pit a smaller reinforcer, presented after a short delay, against a larger reinforcer, delivered after a longer delay, in a choice schedule. Preference for one reinforcer over the other depends on the sizes and delays of both reinforcers (see Logue, 1988); however, if all variables other than the delay to one reinforcer are held constant, and that delay is systematically varied, the declining efficacy of the reinforcer as a function of pre–reinforcer delay may be charted (see Herrnstein, 1981; Mazur, 1987).

The possibility that 5HTergic function might be involved in the control of behaviour by delayed reinforcers was first proposed by Soubrié (1986). Soubrié reviewed the evidence for a relation between pathological impulsive behaviour in humans and dysfunction of the 5HTergic pathways (see also Linnoila & Virkkünen, 1991), and considered possible 'mediating' behavioural processes that might account for this relation. According to Soubrié (1986; Soubrié & Bizot, 1990), deficient functioning of the 5HTergic system may render the organism 'intolerant of delay of gratification' and therefore liable to prefer small short–term gains to larger delayed gains. Soubrié's proposal is consistent with the operational definition of 'impulsiveness', and its antithesis 'self–control', based on choice between small immediate and large delayed reinforcers (Ainslie, 1974; Rachlin, 1974; Herrnstein, 1981; Mazur, 1987; Logue, 1988). It suggests the eminently testable hypothesis that experimentally induced disruption of 5HTergic function should promote preference for smaller immediate reinforcers over larger delayed ones.

Wogar et al. (1993a) examined the effect of destroying the 5HTergic pathways on rats' choice between reinforcers differing in size and delay, using Mazur's (1987) adjusting–delay schedule. This is a discrete–trials schedule in which the subject chooses between two reinforcers, A and B. The sizes of two reinforcers, q_A and q_B, and the delay to the smaller reinforcer, d_A, are held constant, and the delay to the larger reinforcer, d_B, is adjusted in response to the subject's choices. Repeated choice of the larger delayed reinforcer results in an increase in the value of d_B, whereas repeated choice of the smaller, more immediate reinforcer results in a reduction of d_B. Training is continued until a quasi–stable value of d_B, d_B', is obtained.

Wogar et al. (1993a) found that rats whose 5HTergic pathways had been destroyed showed significantly lower values of d_B' than control rats. This result is consistent with the prediction that 5HT–depleted animals should be prone to selecting small immediate reinforcers in preference to large delayed ones. However it falls short of identifying greater sensitivity to delay as the causal factor, because the use of a single value of d_A, as in Wogar et al.'s experiment, does not allow the effects of delay to be separated from other putative influences on d_B'. The problem can be highlighted by application of the *hyperbolic response–strength* model (see Mazur, 1987; Mazur & Herrnstein, 1988; Bradshaw & Szabadi, 1992; Wogar et al., 1992b; Ho et al., 1997a) to Wogar et al.'s (1993a) finding. According to this model, the 'value' of an immediately delivered reinforcer (the *instantaneous value* of the reinforcer, V_i) is an increasing hyperbolic function of its size, q:

$$V_i = \frac{V_{max} \cdot q}{Q' + q} \tag{1}$$

where V_{max} and Q' are parameters expressing maximum value and the reinforcer magnitude yielding the half–maximal value, respectively. The value of a delayed reinforcer, V_d, is given by

$$V_d = \frac{V_i}{1 + Kd} \tag{2}$$

where d is the length of the delay and K is a time–discounting parameter. Under steady–state conditions in the adjusting–delay schedule, it is assumed that $V_{d(B)} = V_{d(A)}$. The value of d_B' may be derived by substitution into equation 2:

$$d_B' = \frac{1}{K} \cdot \left[\frac{V_{i(B)} - V_{i(A)}}{V_{i(A)}} \right] + d_A \cdot \left[\frac{V_{i(B)}}{V_{i(A)}} \right] \qquad [3]$$

(see Ho et al., 1997a, for derivation). Equation 3 shows that although Wogar et al.'s (1993a) finding of reduced 'tolerance' of delay to reinforcement in 5HT-depleted rats *might* reflect a greater sensitivity to delay of reinforcement (i.e., a higher value of K), other explanations are not ruled out. Both Q' and V_{max} influence the empirical value of d_B', since they are both incorporated into the values of V_i; thus the effect of the lesion on d_B' could reflect an effect on either of these parameters. The possibility that Q' might have been affected by loss of central 5HT deserves serious consideration, because Wogar et al. (1991) previously observed that lesions of the 5HTergic pathways reduced the value of a homologous parameter in Herrnstein's (1970) hyperbolic equation that describes the relation between response rate and reinforcement rate in variable-interval schedules (K_H: see Wogar et al., 1991). One way of identifying which of the above parameters is sensitive to loss of 5HT would be to subject lesioned and control rats to training using a range of delays to the smaller reinforcer. Equation 3 specifies a linear relation between d_B' and d_A. The slope of this relation reflects the ratio of the instantaneous values of the small and large reinforcers ($V_{i(B)}/V_{i(A)}$), and should be impervious to changes in the value of K; however, K may be determined from the formula, $K =$ [slope−1]/intercept. Until such an experiment has been undertaken, it would be premature to assume that the 5HTergic pathways regulate the efficacy of delayed reinforcers via a direct effect on the 'delay of reinforcement gradient' (see Ho et al., 1997b, for further discussion).

5. 5HT and performance of immediate timing tasks

In immediate timing tasks, the datum of interest is the subject's behaviour during an *on-going* time interval. The start of the interval may be marked by the subject's own behaviour (as in the case of

interresponse time schedules), or may be heralded by the onset of a
signal (as in the case of the fixed–interval peak procedure).

5.1.   *Interresponse–time schedules*

One of the best known immediate timing tasks is the delayed response
task specified by the *interresponse– time–greater–than–t* (IRT>t)
schedule. In this schedule, reinforcer delivery follows every response
that is separated from the previous response by an interval of at least
t s (see Zeiler, 1977). Various measures have been used to characterize
the behaviour maintained by IRT>t schedules. The simplest of these
are overall response rate and reinforcement rate, which obviously tend
to be inversely related (e.g. McGuire & Seiden, 1980); some authors
also use the ratio of these measures ('efficiency') (e.g. Fletcher, 1995).
These relatively coarse measures of IRT>t schedule performance are
sufficiently sensitive to detect the effects of acute drug treatment and
some brain lesions (see below); however they neglect the temporal
structure that is the hallmark of behaviour maintained by these
schedules. This temporal structure is revealed by the frequency
distribution of IRTs, which is most commonly bimodal, the lower
mode representing very short ('burst') IRTs, and the upper mode
approximating the criterion IRT, t (Harzem, 1969; Platt, 1979). The
relation between the upper mode, or 'mean IRT', and t is a power
function (power $\approx$ 0.8), and the coefficient of variation ('Weber
fraction') is roughly constant across a wide range of values of t (see
Harzem, 1969; Platt, 1979; Wearden, 1990). The mean IRT and
coefficient of variation provide convenient measures of the central
tendency and variability of temporal differentiation (e.g. Wogar et al.,
1992b, 1993b). More complex analytic methods have also been
recommended, based on deviation of the IRT frequency distribution
from randomness (Richards & Seiden, 1991) and exponential or
Weibull–distribution functions fitted to the cumulative IRT probability
data (Stephens & Voet, 1994); the indices of temporal differentiation
derived from both these methods have proved to be highly sensitive to
pharmacological interventions (Richards & Seiden, 1991; Stephens &

Voet, 1994).

Destruction of the 5HTergic pathways markedly impairs performance on IRT>*t* schedules. Wogar et al. (1992a) found that 5,7–dihydroxytryptamine–induced lesions of the median and dorsal raphe nuclei resulted in retarded acquisition of temporal control (see below, Section 8); even after extended training, lesioned rats showed a greater proportion of very short ('burst') IRTs, a lower mean IRT, and a higher coefficient of variation, than sham–lesioned control rats (see Figure 2). Moreover, in well trained animals, destruction of the 5HTergic pathways resulted in marked deterioration of temporal control (Wogar et al., 1993b). These findings have recently been replicated and extended by Fletcher (1995), who found that the effects of total ablation of the ascending 5HTergic pathways could be reproduced by selective destruction of the median, but not the dorsal raphe, nucleus.

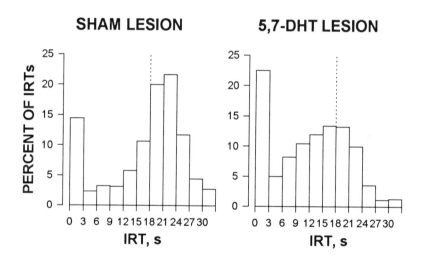

Figure 2. Mean interresponse time distributions obtained from a group of rats whose ascending 5HTergic pathways had been lesioned by injection of 5,7–dihydroxytryptamine (5,7–DHT) into their median and dorsal raphe nuclei and a sham–lesioned control group responding on an IRT>15s schedule. *Ordinates*: percentage of total IRTs emitted; *abscissae*: IRT (s); columns represent 3–s bins, except the extreme right–hand bin, which contains all IRTs longer than 30 s. Data from Wogar et al. (1992a), reproduced by permission.

Consistent with the effects of the lesions, Fletcher (1993, 1994) observed that acute treatment with the $5HT_{1A}$ receptor agonist 8–hydroxy–2(di–n–propylamino)tetralin (8–OH–DPAT), injected directly into the median raphe nucleus, disrupted IRT>t schedule performance, reducing the mean IRT and increasing IRT variability, an effect that was not apparent when the drug was injected into the dorsal raphe nucleus. 8–OH–DPAT is assumed to inhibit 5HTergic function via its action at somatodendritic $5HT_{1A}$ autoreceptors in the raphe nucleus (Hjörth & Magnusson, 1988); thus Fletcher's (1993, 1994) findings with this compound are consistent with the effects of destroying the 5HTergic pathways (Wogar et al., 1992a, 1993b; Fletcher, 1995).

5.2.   Fixed–interval peak procedure

This procedure (Catania, 1970; Roberts, 1981) is a variant of the classical fixed–interval schedule (Ferster & Skinner, 1957). Each trial starts with the onset of a signal (e.g. insertion of a lever into the chamber, illumination of a lamp). In 'standard' trials, reinforcement is scheduled to follow the first response to be emitted after a fixed time has elapsed since the onset of the signal. However, in 'probe' trials, the reinforcer is omitted, and the signal continues for a period three or four times the length of the fixed interval. Behaviour in these probe trials takes the form of a progressively increasing response rate up to the end of the fixed interval, followed by a declining rate of responding (a secondary rise in response rate towards the end of the trial is a common, but not invariable, feature of performance: Church et al., 1991). The indices of timing that may be derived from this performance are the *peak time* (the time from the onset of the signal until response rate reaches its highest point), the *spread time* (conveniently measured as the time from when response rate reaches 70% of its maximum value until it first falls below that level: Church et al., 1991), and the *spread–time/peak–time ratio*; the last of these is regarded as an expression of the Weber fraction, and has been found to assume an approximately constant value over a wide range of peak times (Gibbon, 1977; Church et al., 1991).

Morrissey et al. (1994) examined the effect of destruction of the ascending 5HTergic pathways on rats' performance on a peak fixed interval 40–s schedule. The peak time exhibited by the lesioned group was virtually identical to that of the control group. However, their spread time was greater than that of the control group, and this was reflected in a significant inflation of the Weber fraction (see Figure 3).

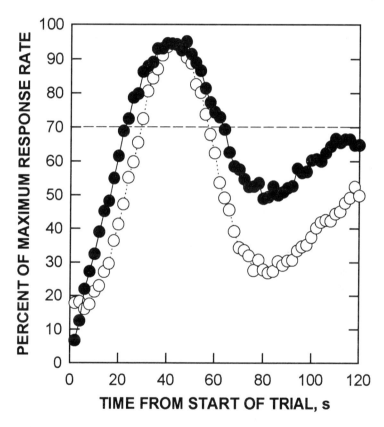

Figure 3. Response rate functions obtained from a group of rats whose ascending 5HTergic pathways had been lesioned by injection of 5,7–dihydroxytryptamine into their median and dorsal raphe nuclei (filled symbols) and a sham–lesioned control group (open symbols) responding on a fixed–interval peak schedule. The fixed interval was 40 s and the probe trials were 120s long. *Ordinate*: percentage of maximum response rate; *abscissa*: time from trial onset. Data from Morrissey et al. (1994), reproduced by permission.

5.3.   Free-operant psychophysical procedure

This schedule (Stubbs, 1976; 1980) consists of a series of trials in which reinforcement is provided, usually under a variable-interval schedule, for responding on one of two continuously available operanda. Reinforcer availability is allocated to operandum A during the first half of each trial, and to operandum B during the second half of the trial. The typical pattern of responding on this schedule consists of an increasing response rate on operandum B, and a concomitantly declining response rate on operandum A, during the course of the trial. This is reflected in an increasing relative rate of responding on operandum B, which passes the 'indifference point' (50% responding on operandum B) approximately midway through the trial, when reinforcer availability is transferred from operandum A to operandum B (Stubbs, 1976, 1980; Bizo & White, 1994a, 1994b). The relative response rate data, which may be described by a logistic function (Bizo & White, 1994a, 1994b), can be used to derive the limen (half the difference between the times corresponding to 25% and 75% responding on operandum B) and the Weber fraction (ratio of the limen to the indifference point).

   One feature of this schedule that is not shared by the other immediate timing schedules discussed above, is the availability of an explicit measure of switching between response alternatives; 'switching' may be defined as a response on operandum B following a response on operandum A, and *vice versa*.

   Al-Zahrani et al. (1996c) examined the effect of destruction of the 5HTergic pathways on rats' performance on the free-operant psychophysical procedure. Some of their results are shown in Figure 4. In agreement with previous results with pigeons (Stubbs, 1968; Bizo & White, 1994a, 1994b), the rats showed a progressive increase in relative response rate on lever B during the course of the trial, the mean value of the 'indifference point' occurring somewhat earlier than the mid-point of the trial. Additionally, the rats showed a remarkable propensity for switching between the two levers, the rate of switching reaching a peak at a time that corresponded approximately to the 'indifference point' of the relative response rate function. The lesion did not alter the indifference point or the Weber fraction derived from the psychophysical function. However, it significantly increased the rate of switching between levers.

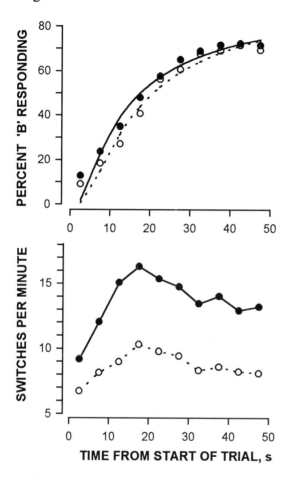

Figure 4. Performance of rats whose ascending 5HTergic pathways had been lesioned by injection of 5,7–dihydroxytryptamine into their median and dorsal raphe nuclei (filled symbols) and a sham–lesioned control group (open symbols) in a free–operant psychophysical task. Reinforcers were allocated according to a variable–interval 25 s schedule to lever A in the first half, and to lever B in the second half, of each 50-s trial. *Ordinates*: Upper panel, responding on lever B, expressed as percent of overall response rate; lower panel, rate of switching between lever A and lever B. *Abscissae*: time from trial onset (s). Data from Al–Zahrani et al. (1996c), reproduced by permission.

5.4.   Comment

Performance on all three immediate timing tasks discussed in this section proved to be sensitive to loss of central 5HT. However, it is not clear whether the apparently disparate effects observed in the three schedules may be interpreted in terms of a unitary effect of the lesion. For example the shortening of the mean IRT in the IRT>t schedule (Wogar et al., 1992b) seems at first sight to be at variance with the lack of effect of the lesion on the peak time in the fixed–interval peak procedure (Morrissey et al., 1994) and the indifference point in the free–operant psychophysical procedure (Al–Zahrani et al., 1996c).

One, admittedly speculative, explanation, which might be able to embrace most of the behavioural phenomena, relies on an extrapolation from Al–Zahrani et al.'s (1996c) finding that the lesion resulted in increased switching between response alternatives in the free–operant psychophysical procedure. Ho et al. (1997b) suggested that destruction of the 5HTergic pathways may facilitate or 'disinhibit' switching between 'behavioural states'. In single–response schedules, like the fixed–interval peak procedure and the IRT>t schedule, the 'behavioural states' are represented by responding and not–responding; it may be expected that the impact of facilitated switching on the performance measures will differ between schedules, according to the way in which behaviour is measured. In the case of the peak procedure, an increased rate of switching into the responding state may reveal itself in both 'tails' of the response rate function, thus symmetrically broadening the function and increasing the size of the Weber fraction. However, in the case of the IRT>t schedule, an increased rate of switching into the responding state must inevitably result in a leftward shift of the IRT frequency distribution (i.e. a reduction of the mean IRT). In the case of the free–operant psychophysical procedure where two response alternatives, corresponding to two sources of reinforcement, are provided, an increased rate of switching between the two responses would not be expected to alter the indifference point.

We will return to the status of 'facilitated switching' as a putative explanation of the effect of central 5HT depletion on performance on immediate timing schedules in the final section of this chapter.

6.   5HT and performance of retrospective timing tasks

In retrospective timing tasks, the datum of interest is the subject's behaviour after a specified interval has elapsed. The usual format of tasks of this type is a conditional discrimination procedure, in which the subject is trained to emit one of two mutually exclusive responses following the offset of a signal, the two responses being differentially reinforced depending upon the duration of the signal.

Retrospective timing tasks have been extensively employed in studies of temporal psychophysics in animals and man (for recent references, see Allan & Gibbon, 1991; Fetterman & Killeen, 1992; Fetterman, 1995; Penton–Voak et al., 1996; also the chapter by Killeen, Fetterman & Bizo in this volume). The *interval bisection task* (Church & Deluty, 1977) has been especially popular, because it yields quantitative temporal discrimination data that are thought to allow rather direct inferences to be drawn about the hypothetical 'pacemaker' which plays a pivotal role in some current models of timing (Gibbon, 1977, 1991; Killeen & Fetterman 1988). In the interval bisection task, the subject is first trained to discriminate two durations ('short' and 'long') in a discrete–trials conditional discrimination schedule. When accurate performance has been attained, 'probe' trials, in which stimuli of intermediate duration are presented, are introduced into each session. In the case of each duration, the percentage of occasions on which the subject responds on the lever appropriate to the 'long' stimulus (%L) is recorded. There is abundant evidence that %L is a sigmoid function (approximately logistic: see chapter by Killeen et al.) of stimulus duration, and that the *bisection point* (the duration corresponding to %L=50) occurs at about the geometric mean of the two standard durations. The sigmoid function may be used to calculate the *difference limen* (half the difference between the durations corresponding to %L=25 and %L=75), and a Weber fraction may be computed from the ratio of the limen to the bisection point (Church & Deluty, 1977). Performance under the interval bisection procedure conforms to rules of scalar timing, in that the Weber fraction remains approximately constant across a wide range of values of the bisection point (e.g. Allan & Gibbon, 1991; Fetterman & Killeen, 1992), and the sigmoid psychophysical functions can be superposed when the abscissa (duration) is re–scaled in fractional units of the bisection point (e.g. Gibbon, 1977; Allan & Gibbon, 1991).

In this section we describe three experiments which examined the effects of central 5HT depletion on interval bisection performance.

Morrissey et al. (1993) trained rats with lesions of their 5HTergic pathways and sham–lesioned control rats on the interval bisection task, using 2–s and 8–s standard stimuli. Their results are shown in Figure 5A. The lesioned rats showed no deficit in discriminative precision as indexed by the Weber fraction; however, their sigmoid psychophysical function was displaced to the left, this being reflected in a significant reduction of the bisection point compared to that of the control group. According to pacemaker/accumulator models of timing, the location of the bisection point is determined by the rate of operation of the pacemaker (see Gibbon, 1991). At first blush, therefore, it appears that 5HT depletion might have resulted in 'speeding up' of the pacemaker. Unfortunately, this explanation is incompatible with current pacemaker/accumulator theories, which assume that, although an acute change in pacemaker speed may reduce the bisection point, a chronic change should be compensated for by gradual adjustment of the accumulator criterion, allowing the bisection point to drift back towards its 'natural' position at the geometric mean of the two standard intervals (Meck, 1983; Killeen & Fetterman, 1988; Bizo & White, 1994a).

Graham et al. (1994) offered an alternative explanation for the lesion–induced shift of the bisection point. They noted that rats performing the interval bisection task often assume a position close to lever A (the lever appropriate to the short stimulus) at the start of trial, and then migrate across the chamber towards lever B (the lever appropriate to the long stimulus) as the period of stimulus presentation progresses (see also Chatlosh & Wasserman, 1987). Graham et al. suggested that destruction of the 5HTergic pathways might have

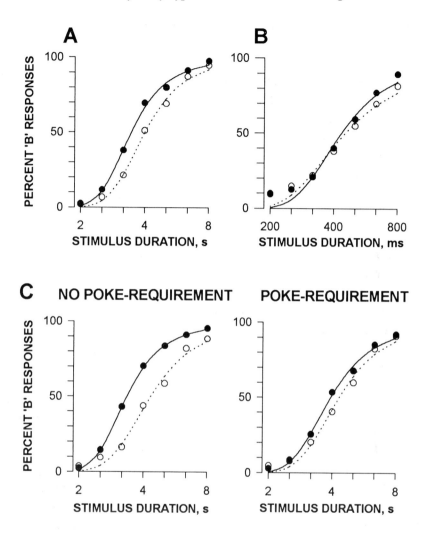

Figure 5. Performance of rats whose ascending 5HTergic pathways had been lesioned by injection of 5,7–dihydroxytryptamine into their median and dorsal raphe nuclei (filled symbols) and a sham–lesioned control group (open symbols) in interval bisection tasks. *Ordinates*: Percent choice of lever B (the lever appropriate to the long standard stimulus. *Abscissae*: stimulus duration (log scale). In **A**, the standard stimuli were 2 s and 8 s (data from Morrissey et al., 1993, reproduced by permission); in **B**, the stimuli were 200 ms and 800 ms (data from Graham et al., 1994, reproduced by permission). **C** shows the effect of introducing a nose–poke requirement between stimulus offset and opportunity to make a response (left–hand panel, no poke requirement; right–hand panel, poke requirement) (data from Ho et al., 1995, reproduced by permission).

facilitated the rats' movement across the chamber from lever A to lever B. This could account for Morrissey et al.'s (1993) finding of a reduction of the bisection point in the lesioned rats, because the termination of any given probe stimulus would be more likely to find a lesioned subject in the vicinity of lever B than an intact subject. Graham et al. (1994) provided circumstantial evidence favouring this interpretation. They trained 5HT–depleted and normal rats to discriminate intervals that were too short to allow movement across the chamber within the period of stimulus presentation (200 ms *vs* 800 ms). In this case, the bisection points of the lesioned and control groups were virtually identical (Figure 5B). Like Morrissey et al. (1993), Graham et al. (1994) found no adverse effect of the lesion on discriminative accuracy; indeed, the lesioned rats showed somewhat *smaller* Weber fractions than the control rats in the case of the millisecond–range stimuli used in Graham et al.'s experiment.

Further evidence that the lesion–induced reduction of the bisection point observed by Morrissey et al. (1993) reflected facilitation of rats' movement from lever A to lever B was obtained by Ho et al. (1995). These authors trained two sets of lesioned and control rats in the interval bisection task using second–range durations (2 s *vs* 8 s). For one set, the task was identical to that used by Morrissey et al. (1993); however, for the other set, a 'nose–poke' response on a panel placed midway between the levers was required after stimulus offset in order to gain access to the levers. This additional requirement presumably reduced the rats' tendency to station themselves close to lever B at the end of the 8–s stimulus presentation. The results obtained using the two versions of the interval bisection task are shown in Figure 5 C; it is apparent that the lesion–induced reduction of the bisection point was effectively prevented by the introduction of the nose–poke requirement. Neither the nose–poke requirement nor the lesion affected discriminative precision, as indexed by the Weber fraction.

These experiments indicate that 5HT depletion had no deleterious effect on temporal discrimination in the interval bisection task. However, it facilitated the rats' tendency to migrate from lever A to lever B during the period of stimulus presentation. This movement across the operant chamber may be thought of as an instance of temporal differentiation, since it was evidently related in an orderly way to the passage of time during an ongoing interval (the 8–s stimulus presentation period). Thus the facilitatory effect of the lesion on this movement is consistent with the evidence reviewed in the previous section, suggesting that the 5HTergic pathways may contribute to behavioural regulation in immediate timing tasks.

The effect of the lesion on the bisection point also has a more general implication for the interpretation of the interval bisection performance. It seems that the task may entail immediate as well as retrospective timing, and that the bisection point does not necessarily reflect the 'point of subjective equality' of temporal discrimination (for further discussion, see Gibbon, 1977; Gibbon & Church, 1990).

7.   5HT and the acquisition of interval timing

7.1.   *Acquisition in immediate timing tasks*

Wogar et al. (1992a) and Fletcher (1993) observed that destruction of the ascending 5HTergic pathways resulted in marked retardation of the acquisition of accurate temporal differentiation under IRT>t schedules. In both studies, the indices of temporal differentiation were overall response rate and overall reinforcement rate; response rate was consistently higher, and reinforcement rate consistently lower, in the lesioned group than in the sham–lesioned control group. Wogar et al.'s data are summarized in Figure 6. As noted above (Section 5.1), the performance of the lesioned rats failed to 'catch up' with that of the control group, even after extended exposure to the schedule. Analysis of performance attained after 50 sessions of training revealed a shorter mean IRT and a larger Weber fraction in the lesioned group than in the control group (cf. Figure 2).

These findings were extended by Fletcher (1995), who found that the deleterious effect of total destruction of the 5HTergic pathways on acquisition could be reproduced by selective destruction of the median raphe nucleus, but not by destruction of the dorsal raphe nucleus.

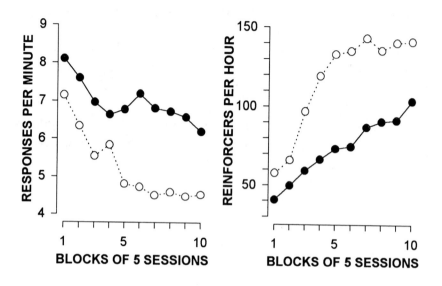

Figure 6. Acquisition of temporal differentiation under an IRT>15s schedule by rats whose ascending 5HTergic pathways had been lesioned by injection of 5,7–dihydroxytryptamine into their median and dorsal raphe nuclei (filled symbols) and a sham–lesioned control group (open symbols). *Left–hand panel:* Overall response rate in successive 5–session blocks. *Right–hand panel:* Obtained reinforcement rate in successive 5–session blocks. (Data from Wogar et al., 1992a, reproduced by permission).

Morrissey et al. (1994) examined the effect of destruction of the 5HTergic pathways on behaviour maintained under a peak fixed–interval 40–s schedule. The effect of the lesion on stable performance under the schedule has been discussed earlier (Section 5.2). However, the transitional behaviour during the 60 sessions of training under the schedule (not presented in Morrissey et al.'s, 1994, paper) provides further evidence for an effect of the lesion on acquisition. The upper panels of Figure 7 show the response rate functions derived from representative 5–session blocks from the early, intermediate and late stages of training. A striking feature of these functions is that the rising phase of the curves developed relatively early in training, whereas the falling phase appeared later. Early in training, the lesioned group's peak response rate was considerably higher than that of the control

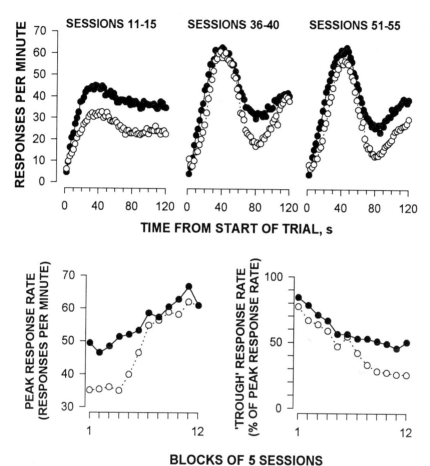

Figure 7. Acquisition of temporal differentiation under a peak fixed–interval 40–s schedule by rats whose ascending 5HTergic pathways had been lesioned by injection of 5,7–dihydroxytryptamine into their median and dorsal raphe nuclei (filled symbols; n=12) and a sham–lesioned control group (open symbols: n=12). *Upper panels:* Response rate functions obtained at early (sessions 11–15), intermediate (sessions 36–40) and late (sessions 51–55) stages of training: *ordinates,* response rate; *abscissae,* time from trial onset (s). *Lower left–hand panel*: Peak response rate in successive 5–session blocks (responses per minute). *Lower right–hand panel*: 'Trough' response rate, expressed as % of peak response rate.

group; however, the difference became less marked as both group's peak response rates rose progressively during the course of training. The gradual development of the 'trough' of the response rate function about two–thirds of the way through the trial (approximately 80 s after trial onset), was clearly attenuated in the lesioned group.

These trends are quantified in the lower panels of Figure 7, which show the mean peak response rates of the two groups, and the 'trough' response rate (expressed as a percentage of the peak rate), during the course of training. Analysis of variance of the latter measure (subject group × session block) revealed significant main effects of both factors (subject group: $F_{1,22}$ = 5.54, $P<0.05$; session block: $F_{11,242}$ = 589.61, $P<0.001$) and no significant interaction ($F<1$).

The processes that determine the gradual development of the characteristic response rate function seen in the peak procedure and the characteristic IRT frequency distribution seen in IRT>t schedules are not fully understood (see, e.g., Wearden, 1990; Kirkpatrick–Steger et al., 1996). However, it is clear that in both cases successful acquisition of precise temporal differentiation entails a progressive reduction of an initially high rate of responding. This 'suppression' of responding appears to be compromised in rats depleted of 5HT.

7.2.   Acquisition in retrospective timing tasks

In contrast to its adverse effects on acquisition in immediate timing tasks, 5HT depletion has not been found to retard the learning of temporal discrimination in retrospective timing tasks. Indeed, under some circumstances, loss of 5HT may even facilitate learning.

Morrissey et al. (1993) found that rats depleted of 5HT acquired accurate discrimination between 2–s and 8–s light stimuli just as rapidly as sham–lesioned control animals (Figure 8A). However, Graham et al. (1994), using much briefer discriminanda (200 ms vs 800 ms), observed that the lesioned group acquired accurate performance significantly more rapidly than the control group (Figure 8B).

A possible interpretation of Graham et al.'s (1994) finding is that destruction of the 5HTergic pathways renders organisms better able to discriminate brief sensory stimuli. Alternatively, the lesion may facilitate the learning of any 'difficult' conditional discrimination

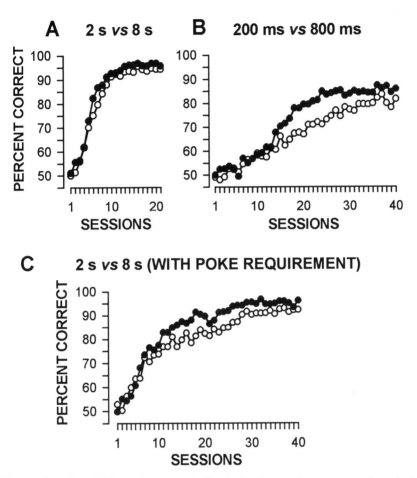

Figure 8. Acquisition of temporal discrimination under retrospective timing (conditional discrimination) tasks by rats whose ascending 5HTergic pathways had been lesioned by injection of 5,7–dihydroxytryptamine into their median and dorsal raphe nuclei (filled symbols) and a sham–lesioned control group (open symbols). *Ordinates:* percent correct responding; *abscissae:* sessions. *Upper left panel:* The discriminanda were 2 s and 8 s (data from Morrissey et al., 1993, reproduced with permission). *Upper right–hand panel:* The discriminanda were 200 ms and 800 ms (data from Graham et al., 1994, reproduced with permission). *Lower panel:* The discriminanda were 2 s and 8 s, and a 'nose–poke' response was required between stimulus offset and presentation of the levers (data from Al–Zahrani et al., 1996a, reproduced by permission).

('difficulty' being operationally definable in terms of the number of trials needed to attain some accuracy criterion). Circumstantial evidence favouring the latter possibility is provided by Al–Zahrani et al.'s (1996a) finding that, in the case of a 2–s *vs* 8–s discrimination, made more 'difficult' by the imposition of a 'nose–poke' requirement after stimulus presentation, lesioned rats again out–performed intact rats (Figure 8C). As discussed earlier (Section 6), the imposition of this additional response requirement has the effect of reducing the rats' tendency to station themselves close to the 'correct' operandum at the moment when the two operanda are offered. The requirement resulted in more protracted acquisition curves for both groups, but the 5HT–depleted rats' learning was significantly faster than that of the intact rats.

7.3. Comment

The data reviewed above reveal a consistent disjunction between the effects of destruction of the 5HTergic pathways on acquisition in immediate and retrospective timing schedules, acquisition being impeded by loss of 5HT in the former case and facilitated in the latter. These observations suggest that different processes are entailed by the two types of timing task, and that the 5HTergic pathways play divergent roles in these different processes.

As mooted earlier, the adverse effects of the lesion upon acquisition in immediate timing schedules seem to reflect disruption of the 'suppression' or 'inhibition' of operant responding which is integral to the acquisition of accurate temporal differentiation.

The basis for the apparently beneficial effect of the lesion on acquisition in retrospective timing schedules is uncertain. There is some evidence that the effect may not be peculiar to temporal discrimination, but may also apply to the acquisition of discriminations along other stimulus dimensions. However, the very wide ranges of tasks and neuropharmacological interventions that have been used to investigate the putative role of 5HTergic function in learning and memory make it difficult to reach a firm conclusion at this time (for discussion and review, see Altman & Normile, 1988; McEntee and Crook, 1991; Sirvio et al., 1994).

8.   5HT and memory for duration

The neural mechanisms of memory are highly complex, and it is generally recognized that no single neurotransmitter should be credited with overall responsibility for memory processes. 5HT is an unlikely candidate for a major role in memory; however, the fact that 5HTergic influences infiltrate many areas of the brain that are believed to subserve some memory processes (e.g. hippocampus, amygdala) provides some *prima facie* evidence for a supporting role of this neurotransmitter in memory (see Altman & Normile, 1988; McEntee & Crook, 1990; Sirviö et al., 1994). Previous studies of the effects of 5HT depletion on memory in animals have yielded mixed results, and it has been noted that the effects of the lesion are highly task dependent. For instance, destruction of the 5HTergic pathways impaired performance on a delayed spatial alternation task (Wenk et al., 1987) and on a passive avoidance task (Harroutunian et al., 1990); however, no delay–dependent effect of 5HT depletion was found in a delayed non–matching–to–position task (Sakurai & Wenk, 1990; Jakälä et al., 1993). A detailed review of this area is beyond the scope of this chapter (see Sirviö et al., 1994, for a recent review); here we will consider only the question of whether 5HT is involved in *memory for duration*.

8.1.   The Choose–Short Effect

Al–Zahrani et al. (1996a) examined the effect of lesions of the ascending 5HTergic pathways on memory for duration, using a conventional delayed conditional discrimination task, with stimuli (light presentations) of 2–s and 8–s durations as the discriminanda. After initial training without delays, which established >90% overall accuracy of discrimination by both lesioned and control groups, post–

stimulus delays ranging from 2 to 32 s were introduced into half the trials. Both groups showed progressively decreasing accuracy as a function of the length of the post–stimulus delay. There was no difference between the two groups in terms of the delay–dependent decline in *overall* accuracy; however, a difference was revealed when the accuracies of responding to the shorter and longer stimuli were considered separately. Both groups showed a delay–dependent bias towards responding on the lever appropriate to the shorter stimulus, which resulted in well–preserved accuracy in 'reporting' that stimulus, and markedly impaired accuracy (extending below chance levels at long delays) in 'reporting' the longer stimulus. This bias was more pronounced in the lesioned group than in the control group.

The delay–dependent bias observed by Al–Zahrani et al. (1996a) is a well known phenomenon in the literature on temporal memory, where it is known as the *choose–short effect* (Spetch & Wilkie, 1983). It has an important status in research on temporal memory because of its purported implications for the way in which the durations of stimuli are 'coded' in memory (for discussion, see chapter by Grant et al. in this volume). However, for the present purposes, it is sufficient to note that the effect suggests that the delay–dependent degradation of stimulus control by stimuli of different durations is not uniform along the duration dimension.

The choose–short bias is usually revealed, as in Al–Zahrani et al.'s (1996a) experiment, in the biased 'reporting' of two temporal discriminanda (in this case, 2– and 8–s durations). However, more detailed information about the biased degradation of stimulus control may be obtained by combining the delayed conditional discrimination task with the interval bisection paradigm. This approach was used by Al–Zahrani et al. (1996b). Rats whose 5HTergic pathways had been ablated and sham–lesioned control rats were trained to discriminate 2–s and 8–s stimuli; then, when accurate performance (>90% correct choices) had been attained, post–stimulus delays were introduced in 50% of the trials, and stimuli of intermediate duration were presented in 10% of both the 'no–delay' and 'delay' trials. Post–stimulus delays of 8 and 12 s were tested in separate phases of the experiment. The results are shown in Figure 9. As expected, both groups showed poorer

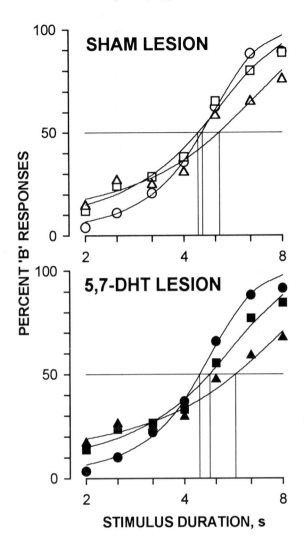

Figure 9. Performance on a delayed interval bisection task by rats whose ascending 5HTergic pathways had been lesioned by injection of 5,7–dihydroxytryptamine into their median and dorsal raphe nuclei (filled symbols) and a sham–lesioned control group (open symbols). The standard stimuli were 2 s and 8 s in duration. *Ordinates:* percent choice lever B (the lever appropriate to the 8–s stimulus); *abscissae:* stimulus duration (log scale). *Circles:* No delay between stimulus offset and presentation of the levers; *squares:* 8–s delay; *triangles:* 12–s delay. Curves are best–fit logistic functions; Vertical lines indicate bisection points for each curve. Data from Al–Zahrani et al. (196b), reproduced with permission.

discrimination when a delay was interposed between the end of the stimulus and the opportunity to make a response; this was reflected in a flattening of the sigmoid psychophysical function and an increase in the size of the Weber fraction, an effect that was especially evident in the case of the 12–s delay. In addition, the bisection point was increased (i.e. the sigmoid function was displaced to the right) in the delay conditions compared to the no–delay condition. The lesion had no effect on discriminative precision, expressed by slope of the curve and the Weber fraction, but significantly enhanced the delay–dependent increase in the bisection point.

The conclusion from these experiments would seem to be that loss of central 5HT has no serious deleterious effect on *overall* accuracy on temporal memory tasks. However, 5HT– depleted rats show a consistent exaggeration of the choose–short bias that is evident to a lesser degree in normal rats. Presumably an explanation for this effect of the lesion may be found in the mechanisms that are responsible for the choose–short effect itself. Unfortunately, the basis of the choose–short effect is complex, and some controversies remain unresolved (for review, see chapter by Grant et al.). In the following section we consider one hypothetical process that is believed to influence the choose–short effect, *proactive interference*, and ask the question, could the enhancement of the choose–short effect that results from loss of central 5HT be mediated by altered susceptibility to proactive interference?

8.2.   Proactive Interference

It is well established that the choose–short effect is reduced by certain procedural manipulations, such as reducing the length of the intertrial interval (Church, 1980; Spetch & Sinha, 1989), which are believed to increase the impact of proactive interference from previous trials on delayed conditional discrimination (Church, 1980; Herremans et al., 1994). Thus it seems that proactive influences may suppress the choose–short effect, and elimination of such influences may enhance the effect (cf. Killeen 1994). Could the enhanced choose–short effect seen in 5HT–depleted rats (Al–Zahrani et al., 1996a, 1996b) have arisen from a reduced sensitivity of these animals to proactive interference? This is an attractive hypothesis, because it offers the prospect of explaining not only the enhanced choose–short effect, but also the faster acquisition of temporal discrimination shown by 5HT–depleted rats compared to their unlesioned counterparts (see above, Section 7.2; also Graham et al., 1994; Al–Zahrani et al., 1996a). It is possible that in normal rats acquisition of accurate conditional discrimination is hampered by proactive interference, choice accuracy on trial N being diminished by the influence of trial $N-1$. If this were the case, then reduced sensitivity to proactive interference could result in faster acquisition.

In this section we describe two hitherto unpublished experiments that explored the possible effect of 5HT depletion on sensitivity to proactive interference in a temporal memory task. These experiments exploited the *intratrial proactive interference method* developed by Grant, Spetch and their colleagues (Grant, 1982; Spetch & Sinha, 1989). This method entails presenting a 'prestimulus' shortly before the stimulus on some trials. Proactive interference is signified by a reduction in the accuracy of discriminative responding in those trials that contain a prestimulus, compared to the accuracy observed in the standard trials without prestimuli. Two types of proactive interference may occur, *symmetric* and *asymmetric*. In symmetric interference, the prior occurrence of a prestimulus of either type (in our case, short or long) reduces the accuracy of discriminative responding to the stimulus only when the prestimulus and stimulus are *discordant* (e.g. a 2–s prestimulus followed by an 8–s stimulus). This type of interference may be thought of as 'confusion' between the stimulus and the prestimulus (see White, 1991). In contrast, in asymmetric interference, the occurrence of a prestimulus impedes accurate discrimination of the subsequent stimulus, even when the prestimulus and stimulus are concordant (e.g. an 8–s prestimulus followed by a 8–s stimulus); however, the inaccuracy induced by the prestimulus differs between the two stimuli. Spetch and Sinha (1989) reported this type of proactive interference in the case of duration discrimination: prestimuli selectively reduced the accuracy of responding to the shorter of two stimuli, greater inaccuracy being induced by longer prestimuli. The occurrence of this type of interference, which has been dubbed the *temporal summation* effect, has been taken as evidence for analogical coding of durations in working memory (Spetch & Rusak, 1992; see chapter by Grant et al.).

Experiment 1

This experiment examined whether *symmetric* and *asymmetric* proactive interference can be identified in rats' performance on a delayed temporal discrimination task, how the interpolation of a post–stimulus delay affects the two types of interference, and whether destruction of the ascending 5HTergic pathways influences these phenomena.

Method

Subjects. Sixteen rats that had undergone 5,7–dihydroxytryptamine–induced lesions of their 5HTergic pathways and 13 sham–lesioned control rats were used (for details of the surgical and anaesthetic procedures, and the protocol for intracerebral injection of the neurotoxin, see Section 3, above).

Behavioural training. The rats were trained in a conventional conditional discrimination task to press lever A following a 2–s presentation of a light stimulus and lever B following an 8–s presentation of the same stimulus (see Al–Zahrani et al., 1996a, for further details). Once accurate performance (>90% correct choices) had been attained, prestimuli and post–stimulus delays were introduced into some trials, as shown in Table 1. The interstimulus interval, timed from the end of the prestimulus until the start of the stimulus, was 2 s.

Table 1: Numbers of trials of each type in Experiment 1*

Post–stimulus delay (s)	Prestimulus		
	none	2–s	8–s
0	66	2	2
2	2	2	2
4	2	2	2
8	2	2	2
16	2	2	2
32	2	2	2

* For each type of trial, the stimulus duration was 2 s in 50% of the trials and 8 s in the other 50%

Biochemical assay. After the completion of the behavioural experiment, the concentrations of 5HT, 5HIAA, noradrenaline and dopamine in the parietal neocortex, hippocampus, amygdala, nucleus accumbens and hypothalamus of each rat were determined by high–performance liquid chromatography combined with electrochemical detection (for details, see Wogar et al., 1992a).

Results
The analysis was based on data obtained during 50 successive sessions.

The left–hand panel of Figure 10 shows the data obtained in the stimulus–alone trials (i.e., the trials in which no prestimulus was presented). Accuracy of performance is expressed as *log d*, a bias–free index of discriminative accuracy ('discriminability') derived from signal detection theory (Davison & McCarthy, 1988; White, 1991; Herremans et al. 1994):

$$log\, d = 0.5\, log\, [(P_{HIT} \times P_{CORRECT\ REJECTION})/(P_{MISS} \times P_{FALSE\ ALARM})$$

(For the purposes of the analysis, the 2–s stimulus was chosen to be the 'signal'; thus P_{HIT} and P_{MISS} were the probabilities of a response on lever A and lever B, respectively, following a 2–s stimulus, and $P_{CORRECT\ REJECTION}$ and $P_{FALSE\ ALARM}$ were the probabilities of a response on lever B and lever A, respectively, following an 8–s stimulus.) In both groups, accuracy declined as a function of the length of the post–stimulus delay, approaching chance levels ($log\, d = 0$) at 16– and 32–s delays. Analysis of variance of these data (subject group × delay) showed a significant main effect of delay ($F_{5,135} = 133.52$, $P < 0.001$), but no significant main effect of subject group ($F_{1,27} = 1.25$, $P > 0.05$) and no significant group × delay interaction ($F_{5,135} = 2.03$, $P > 0.05$).

For each rat, an exponential 'forgetting function' was derived for the relationship between $log\, d$ and the length of the post–stimulus delay, t (White, 1991; Al–Zahrani et al., 1997):

$$log\, d = log\, d_0 \cdot exp\, (-m.t).$$

The parameters of this function, $log\, d_0$ and m, express the estimated discriminability in the absence of a delay and the rate of decay of discriminability, respectively. The latter parameter, m, may be regarded as a 'forgetting constant', although it should be emphasized that the 'decay' of discriminability refers simply to the form of the forgetting curve, and does not imply any particular mechanism of forgetting (e.g. trace decay *vs* interference). The right hand panels of Figure 10 shows the mean (± s.e.m.) values of $log\, d_0$ and m derived for the two groups. There was no significant difference between the two groups in the case of either parameter ($log\, d_0$: $t_{27} = 1.76$, $P > 0.05$; m: $t < 1$). The proportion of the data variance accounted for by the exponential decay function was $0.81 ± 0.06$ (sham lesioned group) and $0.89 ± 0.02$ (lesioned group) ($t < 1$).

Figure 11 and 12 show the results of the proactive interference testing. Figure 11 shows the analysis of *symmetric* proactive interference. The circles show the relation between percent accuracy and the length of the post–stimulus delay under 'control' conditions (i.e.

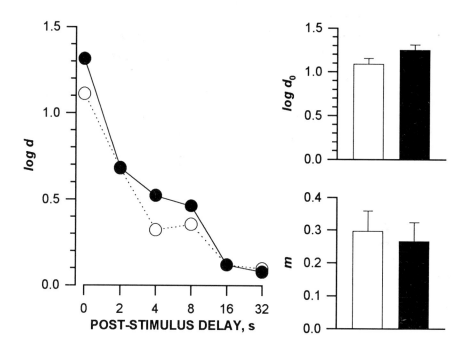

Figure 10. Experiment 1. Data from 'control' trials, in which no prestimulus was presented. *Left–hand panel*. *Ordinate*: 'discriminability', *log d*; *abscissa*: length of post–stimulus delay (s). Points are group mean data (*open symbols*, sham–lesioned control group; *filled symbols*, lesioned group). *Right hand panels*. Group mean values of the parameters of the 'exponential decay function' fitted to the data from individual subjects: 'initial discriminability' (*log d_0*) and the 'forgetting constant' (*m*). Columns show means + s.e.mean (*open column*, sham–lesioned group; *filled column*, lesioned group.

in those trials in which no prestimulus was presented), the triangles show the relation in 'concordant' trials (i.e. when the prestimulus and the stimulus were identical), and the squares show the relation in 'discordant' trials (i.e. when the prestimulus and stimulus were discrepant). It is clear that the occurrence of a prestimulus had a deleterious effect upon discrimination even in the 'concordant' trials. However, there was a tendency for the disruption to be greater in 'discordant' trials than in 'concordant' trials. There was no obvious difference between the performances of the lesioned and control groups. These trends are suported by the analysis of variance of the data from all the trials containing prestimuli (subject group $\times$ trial type $\times$ delay), which revealed significant main effects of trial type ($F_{1,27}$ = 29.26, $P < 0.001$) and delay ($F_{5,135}$ = 35.12, $P < 0.001$), and a significant trial type $\times$ delay interaction ($F_{5,135}$ = 7.11, $P < 0.001$); however, there was no significant main effect of subject group ($F < 1$), and none of the interaction terms involving the subject–group factor achieved statistical significance (subject group $\times$ trial type: $F_{1,27}$ = 3.82, $P > 0.05$; subject group $\times$ delay: $F < 1$; subject group $\times$ trial type $\times$ delay: $F < 1$).

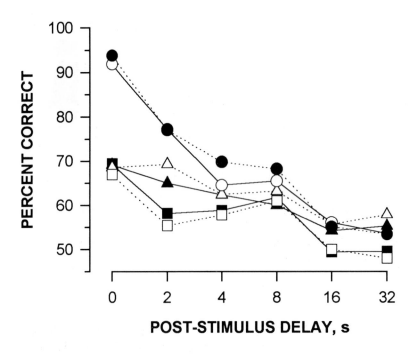

Figure 11. Experiment 1: Analysis of *symmetric* proactive interference effects. *Ordinate*: percent correct choices; *abscissa*: length of post–stimulus delay (s). Points are group mean data (*open symbols*, sham–lesioned control group; *filled symbols*, lesioned group): *circles*, data from trials in which no prestimulus was presented; *triangles*, data from trials in which the stimulus was preceded by a prestimulus of the same duration ('concordant' trials); *squares*, data from trials in which the stimulus was preceded by a prestimulus of different duration ('discordant' trials).

Figure 12 shows the analysis of *asymmetric* proactive interference. Percentage accuracy of discrimination has been plotted separately for the trials containing 2–s and 8–s stimuli (left– and right–hand panels). In each case, the circles represent performance in the 'control' condition (i.e. in the trials without prestimuli), the squares represent performance in trials in which the stimulus was preceded by a 2–s prestimulus, and the triangles performance in trials in which the stimulus was preceded by an 8–s prestimulus. It is apparent that the effect of the prestimulus depended jointly on its duration and on the duration of the subsequent stimulus. Indeed, prestimuli had little effect on the accuracy of responding to the longer (8–s) stimulus, but markedly reduced the accuracy of responding to the shorter (2–s) stimulus. Moreover, the degree of disruption imposed by the 8–s

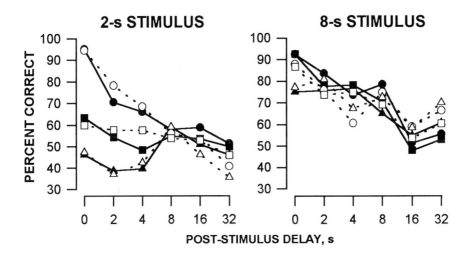

Figure 12. Experiment 1: Analysis of *asymmetric* proactive interference effects. *Ordinates*: percent correct choices; *abscissae*: length of post–stimulus delay (s). *Left–hand graph*: data from the trials in which the 2–s discriminative stimulus was presented; *right–hand graph*: data from the trials in which the 8–s stimulus was presented. Points are group mean data (*open symbols*, sham–lesioned group; *filled symbols*, lesioned group): *circles*, data from trials in which no prestimulus was presented; *squares*, data from trials in which the stimulus was preceded by a 2–s prestimulus; *triangles*, data from trials in which the stimulus was preceded by an 8–s prestimulus.

prestimulus was much greater than that imposed by the 2–s prestimulus. The disruptive effects of both types of prestimulus were apparent only when the post–stimulus delay was relatively brief (< 8 s); with longer delays, accuracy approached chance levels (50%) in all types of trial. These effects seem to be equivalent in the lesioned and control groups. Analysis of variance (subject group × stimulus type × prestimulus type × delay) supports these impressions. There were significant main effects of stimulus type ($F_{1,27}$ = 6.76, P < 0.02), prestimulus type ($F_{2,54}$ = 108.06, P < 0.001) and delay ($F_{5,135}$ = 94.24, P < 0.001). The stimulus × prestimulus, stimulus × delay, prestimulus × delay, and stimulus × prestimulus × delay interactions were all statistically significant ($F_{2,54}$ = 30.41, $F_{5,135}$ = 3.49, $F_{10,270}$ = 26.11 and $F_{10,270}$ = 14.07, respectively; all P's < 0.01). There was no significant main effect of subject group (F < 1), and none of the interaction terms involving group achieved statistical significance (all P's >0.05).

The biochemical data summarized in Table 2 confirm the

Table 2: Experiment 1: Concentrations of 5HT, 5HIAA, noradrenaline and dopamine (ng g^{-1} wet weight of tissue) in brain regions (mean ± s.e.m.)

Region	control	lesion	% control	control	lesion	% control
		5HT			*5HIAA*	
Parietal cortex	253±16	26±4	10*	136±7	4±2	3*
Hippocampus	390±16	26±4	7*	240±11	12±2	5*
Amygdala	833±21	39±8	5*	350±15	16±5	5*
N. accumbens	863±34	31±19	4*	411±17	12±8	3*
Hypothalamus	935±41	154±23	16*	407±21	55±10	10*
		noradrenaline			*dopamine*	
Parietal cortex	395±18	357±20	90	62±18	81±16	129
Hippocampus	509±20	446±24	88	18±2	16±2	91
Amygdala	743±21	705±29	95	391±34	423±31	108
N. accumbens	402±47	358±52	89	5606±254	5802±284	103
Hypothalamus	2124±73	2088±55	98	368±19	361±18	98

Difference between lesioned and sham–lesioned groups (t–test): * P < 0.001

effectiveness and selectivity of the lesion. The levels of 5HT and its metabolite 5HIAA in all brain regions examined were greatly reduced in the lesioned group compared to the control group, whereas the concentrations of the catecholamines were not significantly altered.

Experiment 2

This experiment extended the analysis of asymmetric proactive interference by varying the length of the prestimulus–stimulus interval. Again, we examined whether the interference phenomenon could be affected by destruction of the 5HTergic pathways.

Method

Subjects. Thirteen rats that had undergone the same lesion as described above and 14 sham–lesioned control rats were used.

Behavioural training. The rats were trained in the same conditional discrimination task as in Experiment 1. Once accurate performance (>90% correct choices) had been attained, prestimuli (2 s or 8 s) were introduced into some of the trials, as indicated in Table 3. The interstimulus interval (time between the end of the prestimulus and the start of the stimulus) was varied between 1 s and 8 s.

Table 3:   Numbers of trials of each type in Experiment 2*

Interstimulus interval (s)	Prestimulus		
	none	2–s	8–s
–	68	–	–
1	–	4	4
2	–	4	4
4	–	4	4
8	–	4	4

* For each type of trial, the stimulus duration was 2 s in 50% of the trials and 8 s in the other 50%

   Biochemical assay. The same procedure was used as in Experiment 1.

Results

   The analysis was based on the data obtained during 25 consecutive sessions.

   Figure 13 summarizes the data. The disconnected points on the right–hand side of the figure show the percentage of correct responses to each of the two stimuli (2 s and 8 s) under the 'control' condition (i.e. the standard trials in which no prestimulus was presented). The remaining points show the performance in the trials that contained prestimuli, subdivided according to the four prestimulus/stimulus combinations. It is apparent that accuracy of responding to the 8–s stimulus was not appreciably affected by the prior occurrence of either a 2–s or an 8–s prestimulus (triangles, squares). In contrast, accuracy of responding to the 2–s stimulus was somewhat reduced by the prior presentation of a 2–s prestimulus (circles) and, to a greater extent, by the prior presentation of an 8–s prestimulus (inverted triangles). There was an clear inverse relation between the degree of disruption and the length of the interstimulus interval. The effects of the prestimuli appeared to be equivalent in the lesioned and sham–lesioned groups. Analysis of variance of the data obtained in the trials containing prestimuli (group × stimulus type × prestimulus type × interstimulus interval) confirmed these trends. There were significant main effects of stimulus type ($F_{1,25} = 30.38$, $P < 0.001$), prestimulus type ($F_{1,25} = 10.54$, $P < 0.01$) and interstimulus interval ($F_{3,75} = 20.32$, $P < 0.001$). The stimulus type × prestimulus type interaction was statistically significant ($F_{1,25} = 11.93$, $P < 0.01$), as were the interactions between interstimulus interval and prestimulus type ($F_{3,73} = 6.69$, $P < 0.001$) and between interstimulus interval and stimulus type ($F_{3,75} = 18.54$, $P < 0.001$). The three way interaction between interstimulus interval, prestimulus type and stimulus type was statistically significant ($F_{3,75} = 3.77$, $P < 0.05$). However, neither the main effect of subject group nor any of the interactions involving the subject group factor was statistically significant ($P > 0.1$ in each case).

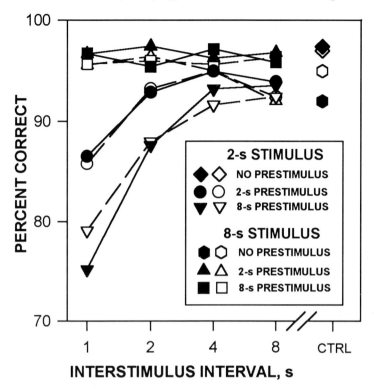

Figure13. Experiment 2: Effect of the interval between prestimulus and stimulus presentation on the *asymmetric* proactive interference effect exerted by the prestimulus. *Ordinate*: percent correct choices; *abscissae*: length of the interstimulus interval (s). Points are group mean data (*open symbols*, sham–lesioned control group; *filled symbols*, lesioned group): see inset.

The biochemical data summarized in Table 4 confirm the effectiveness and selectivity of the lesion. The extent of the depletion of 5HT and 5HIAA was similar to that seen in Experiment 1.

Discussion

The data obtained under the 'control' condition in Experiment 1 (i.e. in those trials that did not contain a prestimulus) showed an orderly decline in discriminative accuracy as a function of the length of the post–stimulus delay. Exponential 'forgetting functions' (cf.

Table 4:   Experiment 2: Concentrations of 5HT, 5HIAA, noradrenaline and dopamine (ng g^{-1} wet weight of tissue) in brain regions (mean ± s.e.m.)

Region	control	lesion	% control	control	lesion	% control
		5HT			*5HIAA*	
Parietal cortex	362±15	42±7	12*	190±7	5±3	3*
Hippocampus	508±19	36±5	7*	320±14	5±3	2*
Amygdala	1049±39	58±8	6*	475±21	13±5	3*
N. accumbens	1171±65	63±17	5*	577±37	43±12	7*
Hypothalamus	1260±51	163±28	13*	611±26	55±16	9*
		noradrenaline			*dopamine*	
Parietal cortex	365±11	335±10	92	164±27	178±46	109
Hippocampus	431±13	410±14	95	20±3	18±5	90
Amygdala	624±34	614±18	98	606±63	567±42	93
N. accumbens	400±59	368±68	92	6452±267	6247±177	97
Hypothalamus	1900±87	2052±116	108	423±20	434±22	103

Difference between lesioned and sham–lesioned groups (*t*–test): * $P < 0.001$

White, 1991; Al–Zahrani et al., 1997) fitted to the data obtained from the lesioned and control rats revealed no significant effect of the lesion on the value of either of the two parameters of the function, *log d$_0$* ('initial discriminability') and *m* ('decay constant').

Experiment 1 revealed the occurrence of both *symmetric* and *asymmetric* proactive interference in rats' performance on the delayed temporal discrimination task. At a qualitative level, it seems that asymmetric interference exerted the greater disruptive effect on choice accuracy, although quantitative comparison of the two effects is not feasible using the present data.

The pattern of asymmetric interference seen in both experiments is consistent with the phenomenon of *temporal summation* reported by Grant, Spetch and colleagues (e.g. Spetch & Sinha, 1989). Noteworthy features of the interference include (i) the selective disruption of the accuracy of responding to the shorter of the two stimuli, (ii) the greater effect produced by the longer of the two prestimuli, (iii) the declining magnitude of the interference effect as a function of the length of the post–stimulus delay, and (iv) the inverse relation between the magnitude of the effect and the length of the interstimulus interval. None of these phenomena was affected by ablation of the ascending 5HTergic pathways.

In summary, these experiments have failed to identify any effect of 5HT depletion on the phenomena of proactive interference.

8.3.   Comment

The evidence reviewed above suggests that the ascending 5HTergic pathways do not play a major role in temporal memory. Indeed, traditional measures of 'working memory' in delayed conditional discrimination tasks (e.g. delay–dependent decline in overall accuracy or discriminability [*log d*]) appear to be impervious to destruction of the 5HTergic projection.

Nevertheless, the enhanced choose–short effect observed in 5HT–depleted rats by Al–Zahrani et al. (1996a, 1996b) suggests that the 5HTergic pathways may play a more subtle role in memory for duration. It may be premature to speculate on the exact nature of this role; however, the two new experiments presented in this section would seem to cast doubt on one potential interpretation. Based on the known properties of the choose–short effect, we anticipated that the enhancement of this effect in 5HT–deficient rats would be associated with reduced sensitivity to proactive interference. Using the intratrial proactive interference method (Grant, 1982; Spetch & Sinha, 1989), we were able to identify two types of proactive interference, *symmetric* and *asymmetric* (cf. White, 1991), the latter corresponding to the temporal summation effect described by Spetch, Grant and colleagues (see chapter by Grant et al. in this volume). The theoretical importance of the temporal summation effect lies in its presumed revelation of 'analogical coding' of durations in memory (Grant, 1982; Spetch & Sinha, 1989). In our experiments, neither type of proactive interference

was affected by destruction of the 5HTergic pathways. The implication
of this negative finding, we suggest, is that the basis of the enhanced
choose–short effect shown by 5HT–depleted rats probably does not
reside in an effect of the lesion on the memorial 'coding' of durations.

9   Conclusions and caveats

Rats whose 5HTergic pathways have been lesioned are not rendered
totally incapable of performing any of the timing tasks with which they
have been confronted thus far. Since the lesion is 90–95% effective in
depleting the telencephalon of 5HT and its metabolite 5HIAA, one has
to conclude that 5HTergic mechanisms play only a supporting or
modulatory role, rather than an imperative role, in interval timing
behaviour.

In a single study (Wogar et al., 1993a), destruction of the
5HTergic pathways has been found to affect prospective timing
performance, promoting preference for a smaller immediate reinforcer
over a larger delayed one. This is a provocative finding, because this
type of prospective timing task constitutes a cornerstone of behavioural
models of 'self–control', preference for small immediate reinforcers in
such tasks being designated 'impulsive'. The notion that 5HT depletion
may render rats abnormally 'impulsive' is enticing, in view of the
copious evidence for deficient 5HTergic function in humans with
clinical 'impulse control disorders' (see Linnoila & Virkkünen, 1991;
Soubrié, 1986). However, the homology between human pathological
'impulsiveness' and 'impulsiveness' defined on the basis of performance
in delayed reinforcement paradigms remains to be established (see Ho
et al., 1997b). Moreover, as discussed above, concurrent delayed
reinforcement paradigms entail highly complex contingencies, and, on
the basis of Wogar et al.'s (1993a) experiment, it remains uncertain
whether 5HT depletion specifically alters sensitivity to delay of
reinforcement.

Destruction of the 5HTergic pathways results in reliable
disruption of steady–state performance on immediate timing schedules,
and deficient acquisition of such performance. However, the effects of

the lesion on the traditional indices of timing are not uniform across schedules. This diversity of effect suggests that 5HT is probably involved in processes other than 'pure timing', processes which are nonetheless important for effective performance on the various immediate timing tasks. The diversity of effect also highlights the need for cautious interpretation of the traditional indices of timing derived from immediate timing tasks (cf. Platt, 1979). We suggested, tentatively, that an increase in the propensity of rats to 'switch' between behavioural states might provide a unifying principle to account for the effects of 5HT depletion on immediate timing performance (see also Ho et al., 1997b). However, it should be noted that the concept of 'switching' lacks a generally accepted definition. Formal models have been proposed to account for switching between response alternatives in conventional concurrent schedules (changeover behaviour) (e.g. Myerson & Miezen, 1980; Davison & McCarthy, 1988; Gibbon, 1995), but the identity of the principal controlling variables governing changeover behaviour remains controversial (e.g. Williams & Bell, 1996), and it remains to be demonstrated that such models can account for 'switching' in situations other than the concurrent interval schedules for which they were originally proposed.

Performance on the interval bisection task was affected in a paradoxical manner by destruction of the 5HTergic pathways. Although the task is considered to be a retrospective timing task *par excellence*, it seems that it can have some features of immediate timing tasks, in that overt behaviour may change systematically during the period of stimulus presentation, and this may affect the discriminative response emitted after the offset of the stimulus. When steps were taken to minimize the influence of behaviour occuring during stimulus presentation upon post–stimulus discriminative responding, the effect of the lesion was virtually abolished. Once again, these results counsel caution in interpreting indices of timing derived from this and similar interval timing tasks.

Retrospective timing tasks were used to investigate the effects of 5HT depletion on memory for duration. It appears that although the conventional indices of 'working memory' are not affected by loss of 5HT, the lesion can alter response bias in a manner that suggests an effect on mnemonic processes. To date our attempts at identifying the nature of this effect have not met with success, save in showing that a deficit in proactive interference is unlikely to be involved.

In conclusion, the multifarious effects of manipulating 5HTergic function on performance on a range of timing tasks attest to the very complex and varied nature of interval timing behaviour. The ascending 5HTergic pathways probably contribute to the regulation of timing behaviour via several mechanisms, which may operate either in concert or in opposition, depending on the particular schedule under consideration. Clarification of 5HT's role in timing behaviour is likely to ensue from a systematic behaviour analysis aimed at teasing apart controlling variables which are, unfortunately, intertwined in many currently available timing schedules.

The authors' work reviewed in this chapter was supported by the Medical Research Council. We are grateful to Miss M.S. Fisher and Mr R.W. Langley for technical help with the experiments described in Section 8, and to Miss Cheryl Davis and Miss Naomi Hardwick for help with the data analysis.

References

Ahlenius, S. & Larsson, K. (1991). Physiological and pharmacological implications of specific effects by $5-HT_{1A}$ agonists on rat sexual behavior. In Rodgers, R.J. & Cooper, S.J. (eds) *$5HT_{1A}$ agonists, $5-HT_3$ antagonists and benzodiazepines: their comparative behavioural pharmacology.* Wiley, Chichester.

Ainslie, G.W. (1974) Self-control in pigeons. *Journal of the Experimental Analysis of Behavior, 21,* 485-489.

Allan, L.G. & Gibbon, J. (1991). Human bisection at the geometric mean. *Learning and Motivation, 22,* 39–58.

Al-Zahrani, S.S.A., Ho, M.-Y., Velazquez Martinez, D.N., Lopez Cabrera, M., Bradshaw, C.M. & Szabadi, E. (1996a). Effect of destruction of the 5-hydroxytryptaminergic pathways on the acquisition of temporal discrimination and temporal memory in a delayed conditional discrimination task. *Psychopharmacology, 123,* 103-110.

Al-Zahrani, S.S.A., Ho, M.-Y., Al-Ruwaitea, A.S.A., Bradshaw, C.M. & Szabadi, E. (1996b). Effect of destruction of the 5-hydroxytryptaminergic pathways on temporal memory: quantitative analysis with a delayed interval bisection task. *Psychopharmacology*, in press.

Al-Zahrani, S.S.A., Ho, M.-Y., Velazquez Martinez, D.N., Lopez Cabrera, M., Bradshaw, C.M. & Szabadi, E. (1996c). Effect of destruction of the 5-hydroxytryptaminergic pathways on behavioural timing and 'switching' in a free-operant psychophysical procedure. *Psychopharmacology, 127*, 346-352.

Al-Zahrani, S.S.A., Al-Ruwaitea, A.S.A., Ho, M.-Y., Bradshaw, C.M. & Szabadi, E. (1997). Destruction of central noradrenergic neurones with DSP4 impairs the acquisition of temporal discrimination but does not affect memory for duration in a delayed conditional discrimination task. *Psychopharmacology*, in press.

Altman, H.G. & Normile, H.J. (1988). What is the nature of the role of the sertonergic nervous system in learning and memory: prospects for development of an effective treatment strategy for senile dementia. *Neurobiology of Aging, 9*, 627-638.

Azmitia, E.C. & Segal, M. (1978). An autoradiographic analysis of the ascending projections of the dorsal and median raphe nuclei in the rat. *Journal of Comparative Neurology, 179*, 641-668.

Bevan, P., Bradshaw, C.M. & Szabadi, E. (1975). Effects of desipramine on neuronal responses to dopamine, noradrenaline, 5–hydroxytryptamine and acetylcholine in the caudate nucleus of the rat. *British Journal of Pharmacology, 54*, 285-293.

Bevan, P., Bradshaw, C.M. & Szabadi, E. (1977). The pharmacology of adrenergic neuronal respones in the cerebral cortex: evidence for excitatory α– and inhibitory β–receptors. *British Journal of Pharmacology, 59*, 635-641.

Bizo, L.A. & White, K.G. (1994a). Pacemaker rate and the behavioral theory of timing. *Journal of Experimental Psychology [Animal Behavior Processes], 20*, 308-32.

Bizo, L.A. & White, K.G. (1994b). The behavioral theory of timing: reinforcer rate determines pacemaker rate. *Journal of the Experimental Analysis of Behavior, 61*, 19-33.

Bloom, F.E., Hoffer, B.J., Siggins, G.R., Barker, J.L. & Nicoll, R.A. (1972). Effects of serotonin on central neurones: microiontophoretic administration. *Federation Proceedings, 31*, 97-106.

Bradshaw, C.M. & Szabadi, E. (1992). Choice between delayed reinforcers in a discrete-trials schedule: the effect of deprivation level. *Quarterly Journal of Experimental Psychology, 44B*, 1-16.

Brody, B.B. & Shore, P.A. (1957). A concept for a role of serotonin and norepinephrine as chemical mediators in the brain. *Annals of the New York Academy of Sciences, 66*, 631-642.

Carlsson, A., Kehr, W., Lindqvist, M., Magnusson, T. & Atack, C.V. (1972). Regulation of monoamine metabolism in the central nervous system. *Pharmacological Reviews, 24*, 371-384.

Catania, A.C. (1970). Reinforcement schedules and psychophysical judgements: a study of some temporal properties of behavior. In Schoenfeld, W.N. (ed) *The theory of reinforcement schedules*. New York, Appleton–Century–Crofts.

Chatlosh, D.L. & Wasserman, E.A. (1987). Delayed temporal discrimination in pigeons: a comparison of two procedures. *Journal of the Experimental Analysis of Behavior, 47*, 299–309.

Church, R.M. (1980). Short–term memory for time intervals. *Learning and Motivation, 11*, 208–219.

Church, R.M. & Deluty, M.Z. (1977). Bisection of temporal intervals. *Journal of Experimental Psychology [Animal Behavior Processes], 3*, 216–228.

Church, R.M., Miller, K.D., Meck, W.H., Gibbon, J. (1991). Symmetrical and asymmetrical sources of variance in temporal generalization. *Animal Learning and Behavior, 19*, 207–214.

Clifton, P.G. (1994). The neuropharmacology of meal patterning. In Cooper, S.J. & Hendrie, C.A. (eds) *Ethology and psychopharmacology*, Wiley, Chichester.

Crow, T.J. & Deakin, J.F.W. (1985). Neurohumoral transmission, behaviour and mental disorder. In Shepherd M (ed) *Handbook of psychiatry, vol. 5: The scientific foundations of psychiatry*, Cambridge University Press, Cambridge.

Dahlström, A. & Fuxe, K. (1964). Evidence for the existence of monoamine–containing neurons in the central nervous system. 1. Demonstration of monoamines in the cell bodies of brain stem neurons. *Acta Physiologica Scandanavica, 62*, Suppl. 232, 1–55.

Davis, M., Falls, W.A., Campeau, S. & Kim, M. (1993). Fear–potentiated startle: a neural and pharmacological analysis. *Behavioural Brain Research, 58*, 175–198.

Davison, M. & McCarthy, D. (1988). *The matching law.* Hillsdale, Erlbaum

Deakin, J.F.W. (1983). Roles of serotonergic systems in escape, avoidance and other behaviours. In: Cooper, S.J. (ed) *Theory in psychopharmacology, vol. 2.* Academic Press, New York.

Deakin, J.F.W. (1996). 5–HT, antidepressant drugs and the psychosocial origins of depression. *Journal of Psychopharmacology, 10*, 31–38.

Deakin, J.F.W. & Graeff, F.G. (1991). 5–HT and mechanisms of defence. *Journal of Psychopharmacology, 5*, 305–315.

El–Kafi, B., Cespuglio, R., Leger, L., Marinesco, S. & Jouvet, M. (1994). Is the nucleus dorsalis a target for the peptides possessing hypnogenic properties? *Brain Research, 637*, 211–221.

Falck, B., Hillarp, N.–A., Thieme, G. & Thorp, A. (1962). Fluorescence of catecholamines and related compounds condensed with formaldehyde. *Journal of Histochemistry and Cytochemistry, 10*, 348–354.

Ferster, C.B. & Skinner, B.F. (1957). *Schedules of reinforcement.* New York, Appleton–Centrury–Crofts

Fetterman, J.G. (1995). The psychophysics of remembered duration. *Animal*

Learning and Behavior, 23, 49–62.

Fetterman, J.G. & Killeen, P.R. (1992). Time discrimination in Columba livia and Homo sapiens. *Journal of Experimental Psychology [Animal Behavior Processes], 18,* 80–94.

Fletcher, P.J. (1993). A comparison of the effects of dorsal or median raphe injections of 8–OH–DPAT in three operant tasks measuring response inhibition. *Behavioral Brain Research, 54,* 187–197.

Fletcher, P.J. (1994). Effects of 8–OH–DPAT, 5–CT and muscimol on behaviour maintained by a DRL 20s schedule of reinforcement following microinjection into the dorsal or median raphe nuclei. *Behavioural Pharmacology, 5,* 326–336.

Fletcher, P.J. (1995). Effects of combined and separate 5,7–dihydroxytryptamine lesions of the dorsal and median raphe nuclei on responding maintained by a DRL 20s schedule of food reinforcement. *Brain Research, 675,* 45–54.

Geyer, M.A., Puerto, A., Dawsey, W.J., Knapp, S., Bullard, W.P. & Mandell, A.J. (1976). Histologic and enzymatic studies of the mesolimbic and mesostriatal serotonergic pathways. *Brain Research, 106,* 241–256.

Gibbon, J. (1977). Scalar expectancy theory and Weber's law in animal timing. *Psychological Review, 84,* 278–325.

Gibbon, J. (1991). Origins of scalar timing. *Learning and Motivation, 22,* 3–38.

Gibbon, J. (1995). Dynamics of time matching: arousal makes better seem worse. *Psychonomic Bulletin & Review, 2,* 208–215.

Gibbon, J. & Church, R.M. (1990). Representations of time. *Cognition, 37,* 23–54.

Graham, S., Ho, M.-Y., Bradshaw, C.M. & Szabadi, E. (1994). Facilitated acquisition of a temporal discrimination following destruction of the ascending 5–hydroxytryptaminergic pathways. *Psychopharmacology, 116,* 373–378.

Grant, D.S. (1982). Prospective versus retrospective coding of samples of stimuli, responses, and reinforcers in delayed matching with pigeons. *Learning and Motivation, 13,* 265–280.

Handley, S.L. (1995). 5–Hydroxytryptamine pathways in anxiety and its treatment. *Pharmacology and Therapeutics, 66,* 103–148.

Haroutunian, V., Santucci, A.C., Davis, K. (1990). Implications of multiple transmitter system lesions for cholinomimetic therapy in Alzheimer's disease. In Aquilonius, S.M. & Gillberg, B.G. (eds) *Cholinergic neurotransmission: functional and clinical aspects.* Elsevier, New York.

Harzem, P. (1969). Temporal discrimination. In Gilbert, R.M. & Sutherland, N.S. (eds) *Animal discrimination learning.* London, Academic Press

Herremans, A.H.J., Hijzen, T.H. & Slangen, J.L. (1994). Validity of a delayed conditional discrimination as a model for working memory in the rat. *Physiology and Behavior, 56,* 869–875.

Herrnstein, R.J. (1970). On the law of effect. *Journal of the Experimental Analysis of Behavior, 13,* 243–266.

Herrnstein, R.J. (1981). Self–control as response strength. In Bradshaw, C.M.,

Szabadi, E. & Lowe, C.F. (eds) *Quantification of steady-state operant behaviour*. Amsterdam: Elsevier/North Holland.

Hillegaart, V., Ahlenius, S. & Larsson, K. (1988). Different roles of the median and dorsal raphe nuclei in rat serotonergic motor functions. *Neurochemistry International, 13 (suppl. 1),* 126.

Hjörth, S. & Magnusson, T. (1988). The 5-HT$_{1A}$ receptor agonist 8-OH-DPAT preferentially activates cell body 5HTautoreceptors in the rat brain in vivo. *Naunyn-Schmiedeberg's Archives of Pharmacology, 338,* 463-471.

Ho, M.-Y., Al-Zahrani, S.S.A., Velazquez Martinez, D.N., Lopez Cabrera, M., Bradshaw, C.M. & Szabadi, E. (1995). The role of the ascending 5-hydroxytryptaminergic pathways in timing behaviour: further observations with the interval bisection task. *Psychopharmacology, 120,* 213-219.

Ho, M.-Y., Wogar, M.A., Bradshaw, C.M. & Szabadi, E. (1997a). Choice between delayed reinforcers: interaction between delay and deprivation level. *Quarterly Journal of Experimental Psychology,* in press

Ho, M.-Y., Al-Zahrani, S.S.A., Al-Ruwaitea, A.S.A., Bradshaw, C.M. & Szabadi, E. (1997b). 5-Hydroxytryptamine and impulse control: prospects for a behaviour analysis. *Journal of Psychopharmacology,* in press.

Hoebel, B.G., Zemlan, F.P., Trulson, M.E., MacKenzie, R.G., DuCret, R.P. & Norelli, C. (1978). Different effects of p-chlorophenylalanine and 5,7-dihydroxytryptamine on feeding in rats. *Annals of the New York Academy of Sciences, 305,* 590-594.

Imai, H., Steindler, D.A. & Kitai, S.T. (1986). The organization of divergent axonal projections of the midbrain raphe nuclei in the rat. *Journal of Comparative Neurology, 243,* 363-380.

Jakälä, P., Sirviö, J., Riekkinen, P. Jr & Riekkinen, P. Sr (1993). Effects of p-chlorophenylalanine-induced serotonin synthesis inhibition and muscarinic blockade on the performance of rats in a 5-choice serial reaction time task. *Pharmacology, Biochemistry & Behavior, 44,* 411-418.

Kantak, K.M., Hegstrand, L.R. & Eichelman, B. (1981). Facilitation of shock-induced fighting following intraventricular 5,7-dihydroxytryptamine and 6-hydroxydopamine. *Psychopharmacology, 74,* 157-160.

Killeen, P.R. (1994). Mathematical principles of reinforcement. *Behavioral and Brain Sciences, 17,* 105-134.

Killeen, P.R. & Fetterman, J.G. (1988). A behavioral theory of timing. *Psychological Review, 95,* 274-295.

Kirkpatrick-Steger, K., Miller, S.S., Betti, C.A. & Wasserman, E.A. (1996). Cyclic responding by pigeons on the peak timing procedure. *Journal of Experimental Psychology [Animal Behavior Prosesses], 22,* 447-460.

Linnoila, M. & Virkkünen, M. (1991). Monoamines, glucose metabolism and impulse control. In Sandler, M., Coppen, A. & Harnett, S. (eds) *5-Hydroxytryptamine in psychiatry.* New York, Oxford University Press.

Logue, A.W. (1988). Research on self-control: an integrated framework. *Behavioral and Brain Sciences, 11,* 665-709.

Lopez–Rubalcava, C. & Fernandez–Guasti, A. (1994). Noradrenaline–serotonin interactions in the anxiolytic effects of $5HT_{1A}$ agonists. *Behavioural Pharmacology, 5*, 42–51.

Mazur, J.E. (1987). An adjusting procedure for studying delayed reinforcement. In Commons, M.L., Mazur, J.E., Nevin, J.A. & Rachlin, H. (eds) *Quantitative analyses of behavior, Vol V: The effect of delay and intervening events.* Hillsdale: Erlbaum.

Mazur, J.E. & Herrnstein, R.J. (1988). On the functions relating delay, reinforcer value and behavior. *Behavioral and Brain Sciences 11,* 690–691.

McEntee, W.J. & Crook, T.H. (1990). Serotonin, memory and aging brain. *Psychopharmacology, 103,* 143–149.

McGuire, P.S. & Seiden, L.S. (1980). The effects of tricyclic antidepressants on performance under a differential–reinforcement–of–low–rates schedule in rats. *Journal of Pharmacology and Experimental Therapeutics, 214,* 635–641.

Meck, W.H. (1983). Selective adjustment of the speed of internal clock and memory process. *Journal of Experimental Psychology [Animal Behavior Processes],* 9, 171–201.

Mendelson, S.D. (1992). A review and reevaluation of the role of serotonin in the modulation of lordosis behavior in the female rat. *Neuroscience and Biobehavioral Reviews, 16,* 309–350.

Miczek, K.A., Weerts, E., Haney, M. & Tidey, J. (1994). Neurobiological mechanisms controlling aggression: preclinical developments for pharmacotherapeutic interventions. *Neuroscience and Biobehavioral Reviews, 18,* 94–110.

Monti, J.M., Jantos, H., Silveira, R., Reyes–Parada, M., Scorza, C. & Prunell, G. (1994). Depletion of brain serotonin by 5,7–DHT: effects on the 8–OH–DPAT–induced changes of sleep and waking in the rat. *Psychopharmacology, 115,* 273–277.

Morrissey, G., Ho, M.-Y., Wogar, M.A., Bradshaw & C.M., Szabadi, E. (1994). Effect of lesions of the ascending 5–hydroxytryptaminergic pathways on timing behaviour investigated with the fixed–interval peak procedure. *Psychopharmacology, 114,* 463–468.

Morrissey, G., Wogar, M.A., Bradshaw, C.M. & Szabadi, E. (1993). Effect of lesions of the ascending 5–hydroxytryptaminergic pathways on timing behaviour investigated with the interval bisection task. *Psychopharmacology, 112,* 80–85.

Mos, J., Olivier, B., Poth, M., van Oorschot, R. & van Aken, H. (1993). The effects of dorsal raphe administration of eltoprazine, TFMPP and 8–OH–DPAT on resident intruder aggression in the rat. *European Journal of Pharmacology, 238,* 411–415.

Myerson, J. & Miezen, F.M. (1980). The kinetics of choice: an operant systems analysis. *Psychological Review, 87,* 160–174.

Ögren, S.-O. (1985). Central serotonin neurones in avoidance learning: interaction with noradrenaline and dopamine neurones. *Pharmacology, Biochemistry and Behavior, 23,* 107–123.

Olivier, B., Mos, J. & Rasmussen, D.L. (1990). Behavioural pharmacology of the serenic, eltoprazine. *Drug Metabolism and Drug Interaction, 8,* 31–85.

Pazos, A., Gonzalez, A.M., Waeber, C. & Palacios, J.M. (1991). Multiple serotonin receptors in the human brain. In: *Receptors in the human nervous system,* pp. 71–101. Eds. Mendelsohn, F.A.O. & Paxinos, G. Academic Press, Sydney.

Penton–Voak, I.S., Edwards, H., Percival, A. & Wearden, J.H. (1996). Speeding up an internal clock in humans? Effects of click trains on subjective duration. *Journal of Experimental Psychology [Animal Behavior Processes], 22,* 307–322.

Platt, J.R. (1979). Temporal differentiation and the psychophysics of time. In Zeiler, M.D. & Harzem, P. (eds) *Advances in the analysis of behaviour, Vol 1: Reinforcement and the organization of behaviour.* Wiley, Chichester.

Rachlin, H.C. (1974). Self–control. *Behaviorism, 2,* 94–107.

Richards, J.B. & Seiden, L.S. (1991). A quantitative interresponse–time analysis of DRL performance differentiates similar effects of the antidepressant desipramine from the novel anxiolytic gepirone. *Journal of the Experimental Analysis of Behavior, 56,* 173–192.

Robbins, T.W. & Everitt, B.J. (1995). Arousal systems and attention. In Gazzaniga, M.S. (ed.) *The cognitive neurosciences.* MIT Press, Cambridge MA.

Roberts, M.H.T. & Straughan, D.W. (1967). Excitation and depression of cortical neurones by 5–hydroxytryptamine. *Journal of Physiology, 193,* 269–294.

Ross, S.B. & Renyi, A.L. (1969).Inhibition of the uptake of 5–hydroxytryptamine in brain tissue. *European Journal of Pharmacology, 7,* 270–277.

Roberts, S. (1981). Isolation of an internal clock. *Journal of Experimental Psychology [Animal Behavior Processes], 7,* 242–268.

Sakurai, Y. & Wenk , G.L. (1990). The interaction of acetylcholinergic and serotonergic neuronal systems on performance in a continuous non-matching to sample task. *Brain Research, 519,* 118–121.

Sirviö, J., Riekkinen, P., Jakälä, P. & Riekkinen, P.J. (1994). Experimental studies on the role of serotonin in cognition. *Progress in Neurobiology, 43,* 363–379.

Soubrié, P. (1986). Reconciling the role of central serotonin neurons in human and animal behavior. *Behavioral and Brain Sciences, 9,* 319–364.

Soubrié, P. & Bizot, J.C. (1990). Monoaminergic control of waiting capacity (impulsivity) in animals. In van Praag HM, Plutchik R, Apter A (eds) *Violence and suicidality.* New York, Brunner/Mazel.

Spetch, M.L. & Rusak, B. (1992). Time present and time past. In Honig, W.K. & Fetterman, J.G. (eds) *Cognitive aspects of stimulus control.* Erlbaum, Hillsdale N.J.

Spetch, M.L. & Sinha, S.S. (1989). Proactive effects in pigeons' memory for event duration: evidence for analogical retention. *Journal of Experimental*

Psychology [Animal Behavior Processes], 15, 347–357.

Spetch, M.L. & Wilkie, D.M. (1983). Subjective shortening: a model of pigeons' memory for event durations. *Journal of Experimental Psychology [Animal Behavior Processes], 9,* 14–30.

Stephens, D.N. & Voet, B. (1994) Differential effects of anxiolytic and non-anxiolytic benzodiazepine receptor ligands on performance of a differential reinforcement of low rate (DRL) schedule. *Behavioural Pharmacology, 5,* 4–14.

Strange, P. (1994). Multiple dopamine receptors: relevance for neurodegenerative disorders. *Biochemical Society Transactions, 22.* 155–159.

Stubbs, D. A. (1968) The discrimination of stimulus duration by pigeons. *Journal of the Experimental Analysis of Behavior, 11,* 223–238.

Stubbs, D.A. (1976). Scaling of stimulus duration by pigeons. *Journal of the Experimental Analysis of Behavior, 26,* 15–25.

Stubbs, D.A. (1980). Temporal discrimination and a free–operant psychophysical procedure. *Journal of the Experimental Analysis of Behavior, 33,* 167–185.

Szabadi, E. & Bradshaw, C.M. (1991) (Ed.) *Adrenoceptors: Structure, mechanisms and function.* Birkhäuser Verlag, Basel.

Szabadi, E., Bradshaw, C.M. & Bevan, P. (1977). Excitatory and depressant responses to noradrenaline and mescaline: the role of the baseline firing rate. *Brain Research, 126,* 580–583.

Tanii, H., Huang., J. & Hashimoto, K. (1993). Involvement of noradrenergic and 5–hydroxytryptaminergic systems in allylnitrile–induced head twitching. *Brain Research, 626,* 265–271.

Tye, N.C., Everitt, B.J. & Iversen, S.D. (1977). 5–Hydroxytryptamine and punishment. *Nature, 268,* 741–743.

Ungerstedt, U. (1971). Stereotaxic mapping of the monoamine pathways in the rat brain. *Acta Physiologica Scandinavica, suppl. 367,* 1–48.

Wearden, J.H. (1990). Maximizing reinforcement rate on spaced responding schedules under conditions of temporal uncertainty. *Behavioural Processes, 22,* 47–60.

Wenk, G., Hughey, D., Boundy, V., Kim, A., Walker, L. & Olton, D. (1987). Neurotransmitters and memory: role of cholinergic, serotoninergic, and noradrenergic systems. *Behavioral Neuroscience, 101,* 325–332.

White, K.G. (1991). Psychophysics of direct remembering. In Commons, M.L., Nevin, J.A. & Davison, M.C. (eds) *Signal detection: mechanisms, models and applications.* Erlbaum, Hillsdale N.J.

Williams, B.A. & Bell, M.C. (1996). Changeover behavior and preference in concurrent schedules. *Journal of the Experimental Analysis of Behavior, 65,* 513–526.

Wogar, M.A., Bradshaw, C.M. & Szabadi, E. (1991). Evidence for an involvement of 5–hydroxytryptaminergic neurones in the maintenance of operant behaviour by positive reinforcement. *Psychopharmacology, 105,* 119–124.

Wogar, M.A., Bradshaw, C.M. & Szabadi, E. (1992a). Impaired acquisition of temporal differentiation performance following lesions of the ascending 5–hydroxytryptaminergic pathways. *Psychopharmacology, 107,* 373–378.

Wogar, M.A., Bradshaw, C.M. & Szabadi, E. (1992b). Choice between delayed reinforcers in an adjusting–delay schedule: the effects of absolute reinforcer size and deprivation level. *Quarterly Journal of Experimental Psychology, 45B,* 1–13.

Wogar, M.A., Bradshaw, C.M. & Szabadi, E. (1993a). Effects of lesions of the ascending 5–hydroxytryptaminergic pathways on choice between delayed reinforcers. *Psychopharmacology, 113,* 239–243.

Wogar, M.A., Bradshaw, C.M. & Szabadi, E. (1993b). Does the effect of central 5–hydroxytryptamine depletion depend on motivational change? *Psychopharmacology, 112,* 86–92.

Zeiler, M.D. (1977). Schedul es of reinforcement. In Honig W.K. & Staddon J.E.R. (eds) *Handbook of operant behavior.* Englewood Cliffs, Prentice–Hall.

Index

DATE DUE